Forest Ecohydrology: From Theory to Practice

Forest Ecohydrology: From Theory to Practice

Editors

Yanhui Wang
Karl-Heinz Feger
Lulu Zhang

MDPI

Basel • Beijing • Wuhan • Barcelona • Belgrade • Novi Sad • Cluj • Manchester

Editors
Yanhui Wang
Chinese Academy of Forestry
Beijing
China

Karl-Heinz Feger
Technische Universität
Dresden
Tharandt
Germany

Lulu Zhang
United Nations University Institute
for Integrated Management of
Material Fluxes and of Resources
Dresden
Germany

Editorial Office
MDPI
St. Alban-Anlage 66
4052 Basel, Switzerland

This is a reprint of articles from the Special Issue published online in the open access journal *Forests* (ISSN 1999-4907) (available at: https://www.mdpi.com/journal/forests/special_issues/forest_ecohydrology).

For citation purposes, cite each article independently as indicated on the article page online and as indicated below:

Lastname, A.A.; Lastname, B.B. Article Title. *Journal Name* **Year**, *Volume Number*, Page Range.

ISBN 978-3-0365-9636-5 (Hbk)
ISBN 978-3-0365-9637-2 (PDF)
doi.org/10.3390/books978-3-0365-9637-2

Contents

About the Editors

Yanhui Wang

Prof. Dr. Yanhui Wang is a leading researcher at the Ecology and Nature Conservation Institute of Chinese Academy of Forestry. He is interested in the research of forest ecohydrology, forest soil, and multifunctional forest management, especially in the dryland regions. His current research areas mainly focus on the effects of site and stand conditions on forest growth and related hydrological and ecosystem services, forest–water coordination, and the multifunctional management of forests at different scales. The main goals of his current research are to improve forest management for a balanced supply of multiple desired forest ecosystem services based on enhanced forest stability against environmental stresses.

Karl-Heinz Feger

Prof. Dr. Karl-Heinz Feger is a leading scientist at the Institute of Soil Science and Site Ecology, Technische Universität Dresden, Germany. He is interested in the interacting fields of hydrology, soil science, and forest and landscape ecology. His current research areas mainly focus on soil–water–forest interactions, site effects on water flow after forestation, ecosystem services, and the carbon sequestration of forests. The main goals of his current research are to improve the understanding of site–water–forest–landscape interactions, ecosystem services of forests, and integrated watershed management.

Lulu Zhang

Dr. Lulu Zhang at the United Nations University Institute for Integrated Management of Material Fluxes and of Resources (UNU-FLORES) specializes in the impact assessment of land-based interventions, anthropogenic activities, and climate change on soil–water–vegetation–atmosphere interactions. At UNU, she further grows her expertise into the integrated resource management of agriculture and forest systems under global change conditions by applying a Resource Nexus approach at various scales, and across disciplines and sectors. The aim is to unlock the potential of terrestrial and aquatic ecosystems to meet future demands without further depleting finite natural resources.

Preface

The aim of this collection of papers is to present new research achievements in the field of forest ecohydrology and the forest–water interaction to related researchers and practitioners. It consists of 12 papers, referring to drought tolerance mechanisms, the hydrological properties of soil and litter layers, variation in solar radiation within stands, snowpack isotopic composition in open and forested areas, canopy transpiration and response to the main influencing factors, and water balance components of typical forests at different regions with various climates. We hope that this book can contribute to the theoretical development of forest ecohydrology and the innovative practice of integrated forest–water management to enhance the balanced supply of hydrological and other ecosystem services from forests.

Yanhui Wang, Karl-Heinz Feger, and Lulu Zhang
Editors

forests

Article

Integration of the Physiology, Transcriptome and Proteome Reveals the Molecular Mechanism of Drought Tolerance in *Cupressus gigantea*

Pei Lei [1,2,†], Zhi Liu [1,3,†], Jianxin Li [1], Guangze Jin [1], Liping Xu [1], Ximei Ji [1], Xiyang Zhao [2], Lei Tao [1] and Fanjuan Meng [1,2,*]

1 College of Life Science, Northeast Forestry University, Harbin 150040, China; lppaper@163.com (P.L.); 13313626779@163.com (Z.L.); 15776685360@163.com (J.L.); jiangyaxuan1206@163.com (G.J.); 18846831735@163.com (L.X.); 18800465846@163.com (X.J.); 18800465845@163.com (L.T.)
2 State Key Laboratory of Tree Genetics and Breeding, Northeast Forestry University, Harbin 150040, China; zhaoxyphd@163.com
3 Zhongshan School of Medicine, Sun Yat-sen University, Guangzhou 510000, China
* Correspondence: mfjtougao@163.com; Tel.: +86-4-518-210-2170
† These authors contributed equally to this work.

Abstract: Drought stress can dramatically impair woody plant growth and restrict the geographical distribution of many tree species. To better understand the dynamics between the response and mechanism of *Cupressus gigantea* to drought and post-drought recovery, a comparative analysis was performed, relying on physiological measurements, RNA sequencing (RNA-Seq) and two-dimensional gel electrophoresis (2-DE) proteins. In this study, the analyses revealed that photosynthesis was seriously inhibited, while osmolyte contents, antioxidant enzyme activity and non-enzymatic antioxidant contents were all increased under drought stress in seedlings. Re-watering led to a recovery in most of the parameters analyzed, mainly the photosynthetic parameters and osmolyte contents. Transcriptomic and proteomic profiling suggested that most of the differentially expressed genes (DEGs) and differentially expressed proteins (DEPs) were specifically altered, and a few were consistently altered. Drought induced a common reduction in the level of DEGs and DEPs associated with photosynthesis. Notably, DEGs and DEPs involved in reactive oxygen species (ROS) scavenging, such as ascorbate oxidase and superoxide dismutase (SOD), showed an inverse pattern under desiccation. This study may improve our understanding of the underlying molecular mechanisms of drought resistance in *C. gigantea* and paves the way for more detailed molecular analysis of the candidate genes.

Keywords: *Cupressus gigantea*; oxidative stress; antioxidant enzymes; gene expression; drought responses

Citation: Lei, P.; Liu, Z.; Li, J.; Jin, G.; Xu, L.; Ji, X.; Zhao, X.; Tao, L.; Meng, F. Integration of the Physiology, Transcriptome and Proteome Reveals the Molecular Mechanism of Drought Tolerance in *Cupressus gigantea*. *Forests* **2022**, *13*, 401. https://doi.org/10.3390/f13030401

Academic Editor: Romà Ogaya

Received: 28 January 2022
Accepted: 23 February 2022
Published: 1 March 2022

1. Introduction

Environmental stresses are known to adversely affect plants' growth and distribution [1]. Among these many adverse factors, drought is a serious detrimental environmental factor constraining seed germination, plant growth and the economic value of crops [2]. However, being sessile organisms, plants are unable to escape environmental stresses and thus have evolved sophisticated strategies to acclimate to dehydration, including morphological, anatomical, physiological and molecular adaptive strategies [3]. Accordingly, the initial response of plants is closely related to the reduction in water evaporation, and consequent reductions in photosynthesis, transpiration, stomatal closure and the accumulation of osmolytes [4]. Of these, the decline in photosynthetic process under drought stress is mainly attributed to stomatal closure, reductions in CO_2 fixation and disturbances in photosynthetic electron transport (PET) [5]. PET is composed of multi-subunit protein complexes, such as Photosystems I and II (PSI, PSII), the cytochrome b6/f complex (Cytb6/f) and ATPase synthase, which absorb light energy and transduce solar energy into chemical energy [6]. Among these, PSII functions as a water-plastoquinone oxidoreductase embedded within the thylakoid membrane, which contains a series of peripheral light-harvesting complexes (LHC), and is vital to the initiation of photosynthesis and electron transport [7]. The Cyt b6/f complex, a rate-limiting step in photosynthesis, is involved the linear electron transport from PSII to PSI, giving rise to the production of ATP and NADPH [8]. Accumulation of some osmolytes such as proline and soluble sugar can play a prominent roles in preventing membrane disintegration and enzyme inactivation under water deficit [9].

When a reduction in water evaporation is insufficient to mitigate the stress stimulus, plants mainly respond to dehydration by activating a defense system of enzymatic and non-enzymatic antioxidants to cope with the overaccumulation of reactive oxygen species (ROS) [10]. Superoxide dismutase (SOD) is the first line of plant ROS defense, catalyzing the conversion of oxygen ions ($O_2^{\cdot -}$) to oxygen (O_2) and hydrogen peroxide (H_2O_2), which is then eliminated by the coordinated action of peroxidase (POD), catalase (CAT), ascorbate peroxidase (APX) and the AsA-GSH cycle [11]. Additionally, some enzymes, including APX, glutathione reductase (GR), dehydroascorbate reductase (DHAR) and monodehydroascorbate reductase (MDHAR), maintain the balance of AsA-GSH [12]. In addition, numerous drought-related transcripts and proteins are induced, which involve signaling transduction, activation/regulation of transcription, antioxidant capacity and ROS scavengers [13]. Thus, joint transcriptomic and proteomic profiling is considered to be effective and acceptable for disentangling the sophisticated processes of plants' responses to water deficit at the molecular levels [14,15].

More recently, to understand such complex regulatory mechanisms, integrated multi-omics approaches (e.g., transcriptomics, proteomics and metabonomics) have been applied because multi-omics techniques offer powerful ways to reveal the relationships between genotype and phenotype. For example, comparative proteomic and transcriptomic analyses have provided insight into the formation mechanisms of seed size in castor bean [16]. Similarly, comparative transcriptomic and proteomic analyses were performed to determine the effect of pigment content on drought resistance in a wheat mutant [17]. In maize, an integrated transcriptomic, proteomic and metabolomic analysis found that its UV-B stress response involves signal transduction and signal molecules [18]. Dry-farm plants are promising candidates for studies on drought-related genes, proteins and metabolites [19]. While candidate genes and proteins have been discovered for wheat, rice, and soybean under dehydration conditions, no such reports are yet available for *C. gigantea* [20–22].

Being a rare tree species that is remarkably drought-tolerant, *Cupressus gigantea* W.C. Chen et L.K. Fu has high ecological and medicinal values due to its extensive use in afforestation, traditional Tibetan medicine and the construction industry [23]. It was added to the Red List of Threatened Species of the International Union for Conservation of Nature (IUCN), given the slow natural regeneration of this species, the anthropogenic disturbances it faces and its geographic isolation [24]. To date, studies on *C. gigantea* have mainly focused on its genetic diversity and population structure [25], total protein extraction [26], photosynthetic capacity [27], determination of the complete chloroplast genome [28] and comprehensive transcriptome characterization [29]. Nevertheless, to our best knowledge, no study has comprehensively investigated the molecular mechanisms enabling *C. gigantea* to tolerate drought. Overall, the mining of genes and proteins related to drought is an indispensable step towards deciphering the adaptive mechanisms of *C. gigantea.*

In this study, a cross-disciplinary approach combining classical physiological measurements, RNA sequencing (RNA-Seq) and two-dimensional gel electrophoresis (2-DE) was carried out to gain insight into the responses of *C. gigantea* to drought stress. This study identified differentially expressed genes (DEGs) and differentially expressed proteins (DEPs) linked to the tree's drought stress responses, which are of great and timely importance for understanding the mechanisms by which *C. gigantea* ameliorates the effects of water deficit at the physiological and molecular levels.

2. Materials and Methods

2.1. Plant Material and Drought Treatments

In this study, we selected *C. gigantea* from the Tibet Agriculture and Animal Husbandry College (Linzhi, Tibet, China). Experiments were conducted at the Northeast Forestry University of (Harbin, Heilongjiang, China). Three-year-old seedlings were sown in plastic pots (25 cm in diameter and 25 cm high) containing a 1:1 mix of perlite and soil. The soil was humus soil, with a pH in H_2O of 4.5–5.5, an organic matter content of 12.5 g/kg, and available nitrogen (N), phosphorous (P) and potassium (K) contents of 100, 40 and 90 mg/kg, respectively. All seedlings were grown in a greenhouse at a temperature of 17 °C during the night and 22 °C during the daytime with a relative humidity of 60% and a 12 h photoperiod. Three-year-old seedlings were exposed to the drought treatment (DTs), in which irrigation of the plants was suspended for 21 consecutive days (hereafter referred to as DT0, DT7, DT14 and DT21). At the end of the stress treatment, the seedlings were re-watered over a period of 7 days (hereafter referred to as RW). Well-watered control plant seedlings (CKs) irrigated every 7 days served as the control group (hereafter referred to as CK0, CK7, CK14, CK21 and CK28). Leaves were collected from DT and CK plants. Three biological replicates were used for the total RNA and protein extractions, with another 5 biological replicates measured for their photosynthesis and physiology parameters.

2.2. Physiological Measurements

To evaluate the physiological changes in *C. gigantea* under drought stress, the net photosynthetic rate (*Pn*), stomatal conductance (*Gs*), intercellular CO_2 concentration (Ci) and transpiration rate (*Tr*) of the leaves were measured using a Li-6400 portable photosynthesis system (LI-COR Inc, Lincoln, NE, USA) at 9:00–11:00 a.m. The CO_2 concentration, stimulation light intensity, and gas flow rate were set as 450 $\mu mol \cdot mol^{-1}$, 1000 $\mu mol \cdot m^{-2} \cdot s^{-1}$ and 500 $\mu mol \cdot s^{-1}$, respectively. Five plants were chosen from the control and drought-stressed groups, and each plant was measured 5 times.

2.3. Biochemical Analysis

The leaves were infiltrated with 0.1 mg/mL 3, 3′-diaminobenzidin (DAB) (Sigma, USA) in 50 mM Tris-acetate buffer (pH 5.0) for H_2O_2 staining or with 0.1 mg/mL nitro blue tetrazolium (NBT) (Sigma, USA) in 25 mM K-HEPES (pH 7.6) for superoxide staining, according to the method described by Chen [30].

Proline from the leaves was extracted by the method described by Zhang [31]. The leaves (0.3 g) were extracted in 4 mL of 1% sulfosalicylic acid. After centrifugation at 6000× *g* for 5 min, 1 mL of the supernatant was added to 1.5 mL of glacial acetic acid and 2 mL of a ninhydrin solution. The mixture was incubated at 100 °C for 30 min. After cooling, the contents were separated with 2 mL toluene, and the optical density was measured at 520 nm.

The soluble sugars of the leaves were analyzed by the anthrone method as described by Zhao [32]. Tissue oven-dried at 65 °C for 24 h (100 mg) was added to 10 mL of 80% ethanol and incubated in a thermostat water bath at 90 °C for 10 min, then the supernatant was collected. The pellet was extracted again as described above, and the supernatant was obtained and combined with the previous aliquot. After adding 3.5 mL of anthrone reagent to 0.1 mL of the supernatant, the mixture was heated in a thermostat water bath at 90 °C for 10 min. After the mixture had cooled to room temperature, the absorbance was measured at 620 nm.

The SOD (EC 1.15.1.1) activity of the leaves was detected with NBT [33]. To measure enzyme activity, leaves (0.3 g) were ground to a fine powder in liquid nitrogen and dissolved in 4 mL of potassium phosphate buffer (PBS) (50 mM, pH = 7.8). The assay mixture contained 50 mM PBS (pH = 7.8), 195 mM methionine, 0.3 mM ethylene diamine tetra-acetic acid, 1.125 mM NBT, 70 μL of the extracting solution and 60 μM riboflavin. Enzyme activity was detected at 560 nm by a spectrophotometer.

POD (EC 1.11.1.7) and CAT (EC 1.11.1.6) activity was measured in leaves according to the method of Kosar et al. [34]. The activity of POD was assayed in a mix containing 50 μL of the extracting solution, 50 mM PBS, 14 μL guaiacol and 19 μL H_2O_2 (30%, *v/v*). Enzyme activity was measured at 470 nm. The activity of CAT was assayed in a mix containing 0.2 mL of the extracting solution, 50 mM PBS (50 mM, pH = 7.8), 500 μL water and 10 mM H_2O_2, and the decrease in absorbance at 240 nm was monitored for 3 min.

The contents of ascorbate and glutathione were estimated in leaves according to the protocol of Sun et al. [35]. The activity of ASA was assayed in a mix containing 0.1 mL of the extracting solution, 0.7 mL ddH_2O, 10% TCA, 44% H_3PO_4, 4% 2,2′-dipyridyl and 3% $FeCl_3$, then the mixture was heated in a thermostat water bath at 37 °C for 60 min. After the mixture had cooled to room temperature, the absorbance was measured at 525 nm. The activity of ASA + DHA was assayed in a mix containing 0.1 mL of the extracting solution, 0.5 mL ddH_2O, 10 mM DTT, 10% TCA, 5% N-ethylmaleimide, 44% H_3PO_4, 4% 2,2′-dipyridyl and 3% $FeCl_3$, then the mixture was heated in a thermostat water bath at 37 °C for 60 min. After the mixture had cooled to room temperature, the absorbance was measured at 525 nm. The activity of GSH was assayed in a mix containing 0.2 mL of the extracting solution, 150 mM (pH = 7.4) NaH_2PO_4 and 0.6 mM DTNB, then the mixture was heated in a thermostat water bath at 30 °C for 5 min. After the mixture had cooled to room temperature, the absorbance was measured at 412 nm. The activity of GSH + GSSG was assayed in a mix containing 0.2 mL of the extracting solution, 50 mM (pH = 7.4) PBS, 0.6 mM DTNB, 2 mM NADPH and 2 U GR, then the mixture was heated in a thermostat water bath at 25 °C for 10 min. After the mixture had cooled to room temperature, the absorbance was measured at 412 nm.

The activity of APX (EC 1.11.1.11), DHAR, GR and MDHAR was determined in leaves as described previously [36]. To measure the enzyme activity, leaves (0.3 g) were ground to a fine powder in liquid nitrogen and dissolved in 4 mL of a potassium phosphate buffer (PBS) (50 mM, pH = 7.8). To measure APX, the extraction solution was added to a mix containing 50 mM PBS (50 mM, pH = 7.8), 5 mM ASA and 20 mM H_2O_2, and the decrease in absorbance at 290 nm was monitored for 3 min. To measure DHAR, the extraction solution was added to a mix containing 50 mM PBS (50 mM, pH = 7.8), 0.5 mM DHA and 5 mM GSH, and the decrease in absorbance at 265 nm was monitored for 3 min. To measure GR, the extraction solution was added to a mix containing 50 mM PBS (50 mM, pH = 7.8), 2 mM NaDPH and 10 mM GSSG, and the decrease in absorbance at 340 nm was monitored for 3 min. To measure MDHAR, the extraction solution was added to a mix containing 50 mM PBS (pH = 7.8), 0.1 mM ASA and 0.55 U AAO, and the decrease in absorbance at 340 nm was monitored for 3 min.

2.4. Protein Preparation and Quantitative Proteome Analysis

In each sample, protein was extracted from the leaves using the TCA/acetone method [37]. Plant tissue (2 g) frozen by liquid nitrogen was ground into powder, then suspended in 4 mL of a cold extraction buffer (0.5 M Tris-HCl (SRL, India) pH = 7.5, 0.7 M sucrose (Himedia, India)) for protein solubilization. Samples were incubated overnight at 4 °C, after which an equal volume of phenol saturated with Tris-HCl (pH = 7.5) was added to each sample. The homogenate was centrifuged (5000× *g* at 4 °C for 30 min) and the clear white pellet was washed three times with ice-cold acetone (Himedia, Mumbai, India) at −20 °C, followed by centrifugation at 5000× *g* for 5 min. The pellet was then freeze-dried and stored at −80 °C. The protein powder was dissolved in a rehydration buffer (8 M urea (Sigma, St. Louis, MO, USA), 2% (*w*/*v*) CHAPS (Sigma, St. Louis, MO, USA), 50 mM DTT (Sigma, St. Louis, MO, USA), and 0.2% (*v*/*v*) Biolyte (Bio-Rad, Hercules, CA, USA)) for 1 h at 37 °C. After centrifugation, the protein concentration of the supernatant was determined by the Bradford assay (Bio-Rad, Hercules, CA, USA).

The 2-DE was carried out according to the methodology of Wang et al. [38]. First, the immobilized pH gradient (IPG) strips [pH 4–7, 13 cm, (GE Healthcare, Chicago, IL, USA)] were separated. Next, 1.5 mg of the protein sample was applied to each IPG strip and this was covered with 800 μL of mineral oil. The voltages used were as follows: 30 V for 13 h, 100 V for 1 h, 500 V for 1 h, 1000 V for 1 h and 3000 V for 1 h, and then 8000 V for 7 h. After focusing, the strips were equilibrated in Equilibration Buffer I (0.1 g DTT and 10 mL of a SDS balancing buffer) and in Equilibration Buffer II (0.15 g iodoacetamide (IAM) (Sigma, St. Louis, MO, USA) and 10 mL of a SDS balancing buffer). After 15 min, the strips were washed with a SDS balancing buffer. The equilibrated strips were analyzed by 12.5% sodium dodecyl sulfate–polyacrylamide gel electrophoresis. The electrophoresis was run on Electrophoreses Power Supply EPS 601 (GE Healthcare, Chicago, IL, USA) at a constant current of 2 W for 30 min and then modulated to 8 W. The gels were visualized by staining them overnight with colloidal Coomassie brilliant blue R-250. Stained gels were scanned on an Image Scanner II (GE Healthcare, Chicago, IL, USA), and the captured images were analyzed in Melanie 7.0 software (GeneBio, Geneva, Switzerland). Three independent replicated gels were analyzed and characterized. Protein spot detection was based on a fold-change of ≥ 2 or ≤ 0.5, for which a threshold of $p \leq 0.05$ was used to distinguish the differentially expressed protein spots.

The protein spots were carefully excised from the gels. The gel spots were washed twice for 20 min with deionized water, then incubated and dehydrated with acetonitrile (ACN). Proteins were digested for 18 h at 37 °C in 10 μL of a trypsin solution (15 ng μL^{-1}). Next, the supernatants were collected, and the gel spots were extracted twice with a 50 μL extraction buffer (50% ACN (Sigma, St. Louis, MO, USA) and 5% TFA) for 1 h at 37 °C. The extractions and the trypsin supernatant of the gel spots were combined and then vacuum-dried. The peptides' mass spectra were detected via matrix-assisted laser desorption ionization time-of-flight/time-of-flight (MALDI-TOF/TOF) (ABI 4700, AB Systems, CA, USA) and the proteins were identified using the UniProt database (http://www.uniprot.org, accessed on 22 February 2020). The gene ontology of identified proteins was determined using the Blast2Go v2.3.6, with the Target P program used for their functional classification.

2.5. RNA-Seq Library Construction and Transcriptome Analysis

The RNA-Seq libraries' preparation and their sequencing were carried out in leaves by Novogene equipment (Beijing, China). Briefly, RNA degradation and contamination were monitored on 1% agarose gels, RNA purity was checked using a NanoPhotometer® spectrophotometer (Implen, Westlake Village, CA, USA) and RNA integrity was assessed using the RNA Nano 6000 Assay Kit of the Bioanalyzer 2100 system (Agilent Technologies, Palo Alto, CA, USA), with 1 μg RNA per sample used as the input material for RNA sample preparation. The sequencing libraries were generated using the NEBNext® UltraTM RNA Library Prep Kit for Illumina® (NEB, Ipswich, MA, USA), following the manufacturer's recommendations, with unique index codes added to attribute the sequences to each sample. The clustering of the index-coded samples was performed on a cBot Cluster Generation System, using TruSeq PE Cluster Kit v3-cBot-HS (Illumina, San Diego, CA, USA), according to the manufacturer's instructions. After cluster generation, the libraries of the preparations were sequenced on an Illumina NovaSeq platform, from which 150-bp paired-end reads were generated. Both the reference genome and the gene model annotation files were downloaded from the genome website directly. Through use of Hisat2 v2.0.5 (Johns Hopkins University, Baltimore, MD, USA), the index of the reference genome was built, and paired-end clean reads were aligned to it. The clean data were deposited in NCBI Sequence Read Archive (SRA; https://www.ncbi.nlm.nih.gov/sra, accessed on 18 July 2021) under the accession number SRR15183944-SRR15183948.

Gene ontology (GO) enrichment analysis of the DEGs was implemented by the R package 'clusterProfiler', which corrected any gene length bias. GO terms with a corrected p-value of <0.05 were considered to be significantly enriched by the DEGs. The Kyoto Encyclopedia of Genes and Genomes (KEGG) is a database resource for understanding the high-level functions and utilities of biological systems (http://www.genome.jp/kegg, accessed on 22 March 2020). The clusterProfiler package in R was used to statistically test for the enrichment of DEGs in the KEGG pathways.

2.6. Quantitative Real-Time PCR (RT-PCR)

Total RNA was extracted by using the OmniPlant RNA Kit (DNase I) (Cowin Biosciences, Beijin, China). A summary of the procedure is as follows: the samples were individually milled in a mortar with liquid nitrogen and then incubated with 500 μL of RLS Buffer. The samples were centrifuged at 12,000 rpm for 2 min at 4 °C, and a half volume of absolute ethanol was added to the supernatant after centrifuging, then 52 μL RNase-Free Water, 8 μL 10 × Reaction Buffer and 20 μL DNase I was then added to the spin column after incubating at room temperature for 15 min. The samples were centrifuged at 12,000 rpm for 2 min at 4 °C, then the precipitates were washed in RNA Wash Buffer II and dried. RNA was dissolved with RNase-Free Water, then 0.5 μg of total RNA was used to synthesize the first-strand cDNA by using the ReverTra Ace® qPCR RT Master Mix with the gDNA Remover (Toyobo, Osaka, Japan). Amplifications were performed on a Roche 480 PCR System (Roche, Rotkreuz, Switzerland) with a THUNDERBIRD® qPCR Mix (Toyobo, Osaka, Japan). The primer sequences used for the real-time qPCR are listed in Table S7.

2.7. Statistical Analysis

The data were subject to analysis of variance (ANOVA) ($p < 0.05$). Multiple comparisons of the treatment effects were analyzed using the least significant difference (LSD). All the tests were performed using SPSS 22.0 software (Windows, USA). All figures were developed using GraphPad Prism 8.2 (GraphPad Software, CA, USA) and Adobe Illustrator CC2018 (Adobe, CA, USA).

3. Results

3.1. Drought Stress Induced Growth and Physiological Changes in C. gigantea

To investigate the morphological and physiological response mechanisms of *C. gigantea* under drought stress, 3-year-old seedlings were planted in the same environment and then subjected to a drought treatment for 3 weeks, then re-watered for 1 week (Figure 1a). At the onset of stress, no significant differences were observed in plant growth. After 21 days, seedlings in the drought treatment had leaves and stems that were badly curled and wilted when compared with those of the control group. Interestingly, the drought-stressed seedlings were able to largely restore their viability during the re-watering period. Upon closer examination, the seedlings' *Pn*, *Gs*, and *Tr* significantly decreased by 38.8%, 72.3% and 57.9%, respectively, whereas Ci gradually increased by 20.9% after completing the drought treatment compared with the control ($p < 0.05$). Surprisingly, RW induced a rapid recovery in *C. gigantea*, reaching parameters similar to those recorded in the control group (Figure 1b–e), indicating that photosynthesis was inhibited by drought.

Physiological indexes such as histochemical staining, soluble sugar content, proline content and antioxidant enzymes are commonly used to evaluate the stress resistance capacity of plants in response to drought stress. Here, DAB and NBT histochemical staining analyses were conducted to measure the H_2O_2 and O_2^- contents. Under normal conditions, there was no apparent difference between the treatment and control; however, during prolonged drought stress, the DAB and NBT staining intensities of the treatment plants' leaves were stronger than those of the control leaves (Figure 2a,b). Importantly, we found that the proline and soluble sugar contents were higher in the seedlings under the drought stress treatment than in the control (Figure 2c,d). Furthermore, sharp increases of 17.9%, 68.6% and 82.6%, respectively, were observed in SOD, POD and CAT activity ($p < 0.05$) after 21 days of the drought treatment (Figure 2e–g), indicating that *C. gigantea* promoted its ROS scavenging by modulating the activity of key antioxidant enzymes, such as SOD, POD and CAT. Similarly, non-enzymatic antioxidant contents, namely those of ascorbic acid (ASA), dehydroascorbic acid (DHA), reduced glutathione (GSH) and oxidized glutathione (GSSG), were starkly increased under the drought stress treatment relative to the control seedlings ($p < 0.05$, Figure 2h–k). The antioxidant enzyme activities of APX, DHAR, GR and MDHAR were further analyzed; this showed that they increased in content after the drought treatment ($p < 0.05$), and their accumulation was highest at 21 days of the drought treatment ($p < 0.05$, Figure 2i–o). Subsequent to rehydration of the seedlings', the parameters of those in the treatment returned to similar levels to those of the control group. Overall, these results revealed that seedlings of *C. gigantea* enhanced their drought resistance by increasing their osmolyte content, activating the antioxidant system and maintaining ROS homeostasis.

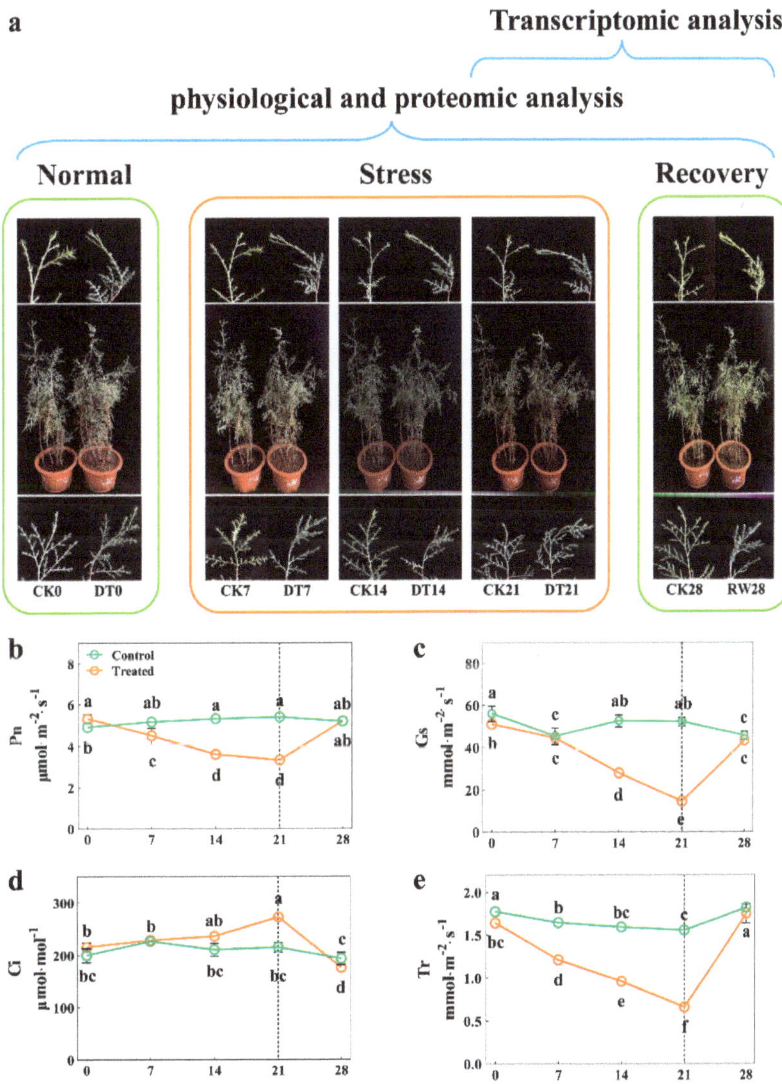

Figure 1. The effect of drought stress on the morphological and photosynthesis traits of *C. gigantea*. (**a**) Photographs showing the phenotypes of the seedlings over time. (**b**–**e**) Line charts showing the physiological changes corresponding to *Pn*, *Gs*, *Tr*, Ci, respectively. Values are the means ± SD ($n = 5$). Columns headed by different letters denote means differing significantly from one another ($p < 0.05$).

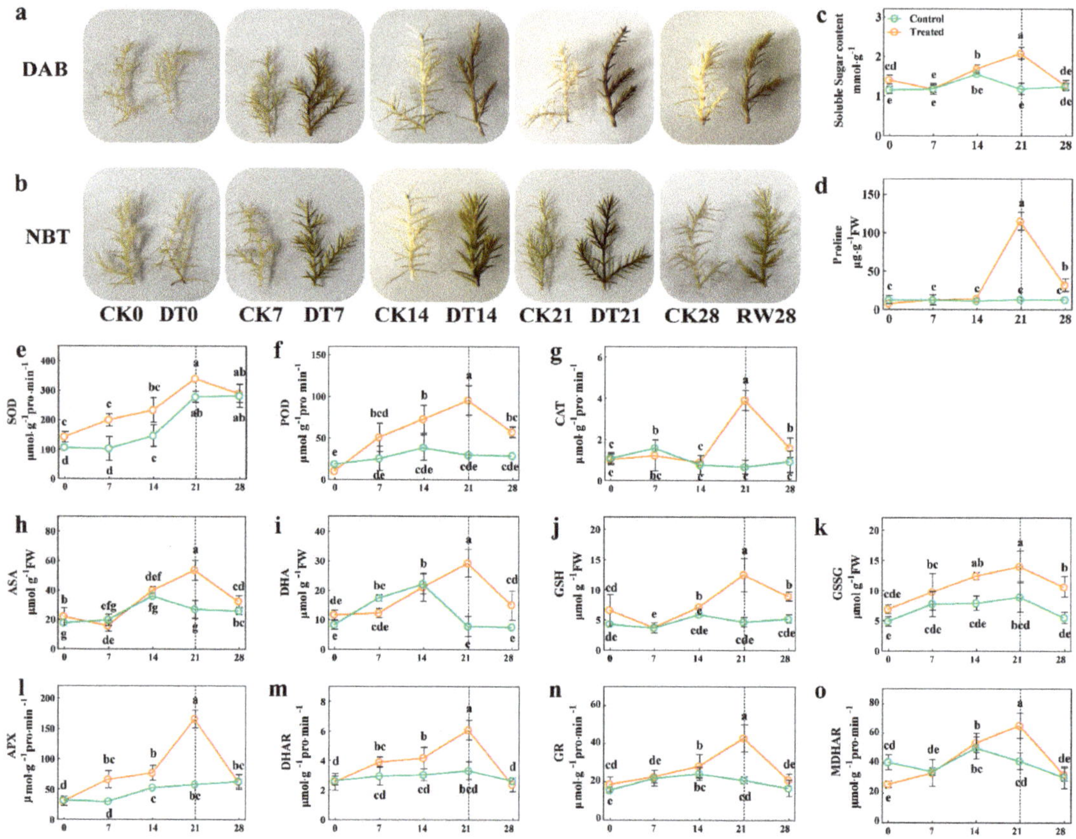

Figure 2. The effect of drought stress on the physiological indexes of *C. gigantea*. (**a**,**b**) Photographs showing DAB and NBT histochemical staining of the seedlings over time. (**c**–**o**) Line charts showing the physiological changes corresponding to soluble sugar, proline, superoxide dismutase (SOD), peroxidase (POD), catalase (CAT), ascorbic acid (ASA), dehydroascorbic acid (DHA), reduced glutathione (GSH), oxidized glutathione (GSSG), ascorbate peroxidase (APX), dehydroascorbate reductase (DHAR), glutathione reductase (GR) and monodehydroascorbate reductase (MDHAR), respectively. Values are the means ± SD ($n = 3$). Columns headed by different letters denote means differing significantly from one another ($p < 0.05$).

3.2. Characterization and Functional Distribution of the *C. gigantea* Leaf Proteome

To identify the drought-induced proteins of *C. gigantea*, we performed a 2-DE analysis coupled with MALDI-TOF/TOF at different time points (7, 14, 21, and 28 days) in seedlings under drought stress. Total protein extracts were separated in all samples, with a pI range of 4–7 and a molecular mass range of 14.4–116 kDa. All these protein spots were localized and detected on Coomassie brilliant blue-stained gels. Accordingly, 66 protein spots changed significantly in abundance between the drought-stressed and control seedlings of *C. gigantea* (Figure S1). Temporally, of these, 26 (7 days), 16 (14 days), 21 (21 days) and 3 (28 days) protein spots were found to be altered by drought stress.

In total, 66 DEPs were identified in the control vs. treatment comparison: 61 (i.e., 23 upregulated and 39 downregulated) in the 7DT vs. 7CK comparison, 62 (21 upregulated and 41 downregulated) in the 14DT vs. 14CK comparison, 66 (36 upregulated and 30 downregulated) in the 21DT vs. 21CK comparison and 64 (29 upregulated and 35 downregulated) in the 28RW vs. 28CK comparison (Figure 3a,b). In addition, there were four proteins

(Spots 6, 10, 11 and 17) whose expression was significantly upregulated by drought, but was upregulated/unchanged during the recovery period. Conversely, the expression of two proteins (Spots 36 and 39) was upregulated/unchanged by drought yet downregulated during the recovery phase of the seedlings (Figure S2).

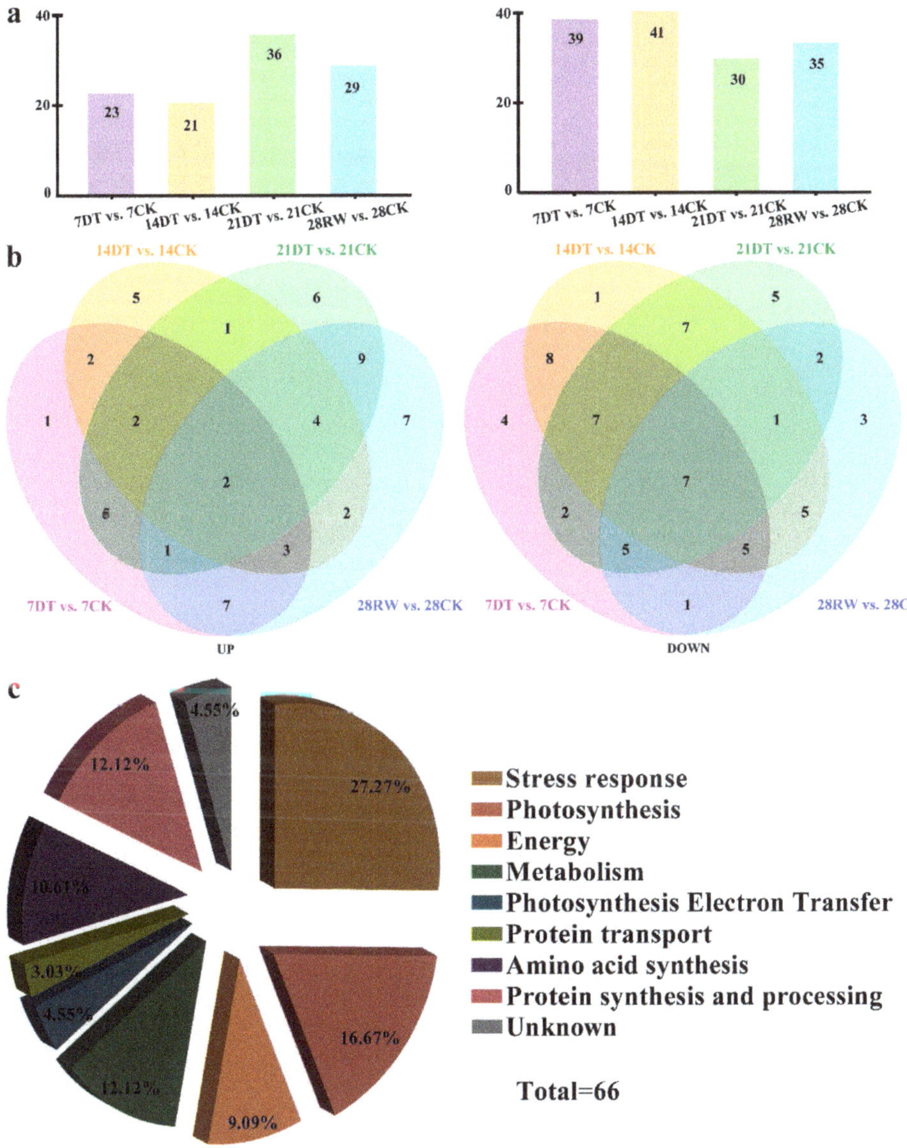

Figure 3. Identification and functional classification analysis of the proteomics data from *C. gigantea* seedlings. (**a**) Bar graph showing the upregulated and downregulated differentially expressed proteins (DEPs) for each pairwise comparison. (**b**) Venn diagrams depicting the overlaps of upregulated and downregulated differentially expressed proteins (DEPs) across four comparisons. The total number of differentially expressed proteins (DEPs) is provided in parentheses. (**c**) Functional categorization of the differentially expressed proteins (DEPs) identified in *C. gigantea* in response to drought.

The experimental MW and pI values of the protein spots were then estimated and compared with the theoretical counterparts of the corresponding proteins. As Table S1 shows, most of the experimental values matched up well with the theoretical ones, indicating unambiguous identifications. According to their biological function, these identified proteins were classified into nine major groups: stress response (27.27%), photosynthesis (16.67%), energy (9.09%), metabolism (12.12%), photosynthesis electron transfer (4.55%), protein transport (3.03%), amino acid synthesis (10.61%), protein synthesis and processing (12.12%), and unknown (4.55%) under the drought stress treatment (Figure 3c).

3.3. *Characterization and Functional Distribution of C. gigantea's Leaf Transcriptome*

To further investigate the regulatory network of *C. gigantea*'s response to drought stress, cDNA libraries were built using leaves from seedlings under the two treatments: severe stress (21 days) and re-watering (28 days), and these were sequenced on an Illumina NovaSeq platform (Illumina Inc, San Diego, CA, USA). High-quality reads of the treatment and control samples were acquired after implementing data-cleaning procedures (Table S2). The length distribution of the assembled transcripts of *C. gigantea* was obtained (Figure S3a). To validate the accuracy and reproducibility of our RNA-Seq data, we selected 15 DEGs for a follow-up qRT-PCR; their gene expression levels as inferred by RNA-Seq were strongly correlated with those from the qRT-PCR (R^2 = 0.92; Figure S3c).

Many DEGs were identified via two pairwise comparisons between the treatment and control (i.e., 21DT vs. 21CK, and 28RW vs. 28CK). Under severe stress, 215 genes were expressed differentially in 21DT vs. 21CK, including 36 upregulated genes and 84 downregulated genes, while 442 genes were expressed differentially under re-watering in 28R vs. 28CK, including 250 upregulated and 97 downregulated genes ($p < 0.05$, fold-change ≥ 2). Another 30 upregulated and 65 downregulated genes overlapped in the pairwise comparisons between 21DT and 21CK, and between 28RW and 28CK (Table S3, Figure S3b).

To illustrate the biological functions and pathways of the DEGs, GO annotation (at $p < 0.05$) was performed to analyze the drought-responsive genes in *C. gigantea* (Table S4), and 24 enriched GO terms were subsequently selected (Table S4, Figure 4). According to the GO classification graphic, "GO:0050896: response to stimulus", "GO:0016020: membrane" and "GO:0003824: catalytic activity" were highly enriched during both severe drought stress and re-watering of the seedlings, suggesting that the DEGs identified are likely involved in the modulation of ROS scavenging. All DEGs were further examined for KEGG pathway enrichment (Table S5, Figure 5). The top 30 enriched KEGG pathways of DEGs between the 21DT and 21CK, and between the 28RW and 28CK groups are shown in Figure 5a,b. The pathway of phototransduction, starch and sucrose metabolism, as well as plant hormone signal transduction, were also significantly enriched for DEGs between 21DT and 21CK, and between 28RW and 28CK. Accordingly, we performed a careful manual annotation of our list of DEGs to identify candidate genes that were directly involved in ROS scavenging and photosynthesis. Based on a combination of the GO and KEGG pathway results, we identified 12 such candidate DEGs (Table S6). According to our manual annotation, eight DEGs could be assigned generically to ROS scavenging, with another four DEGs assigned to photosynthesis; this suggested that these 12 candidate DEGs are probably key regulatory components of the drought stress response in *C. gigantea*.

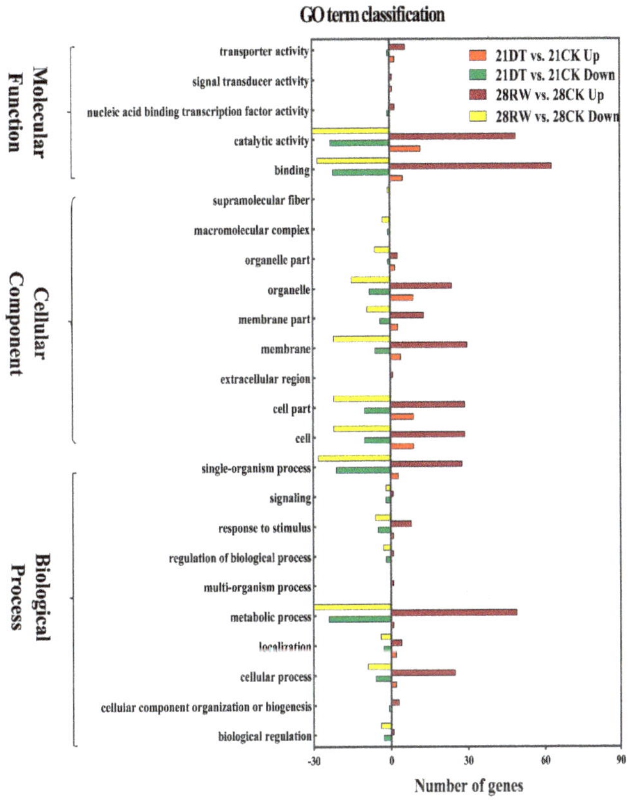

Figure 4. Differentially expressed genes (DEGs) identified by Gene Ontology (GO) classification in *C. gigantea*.

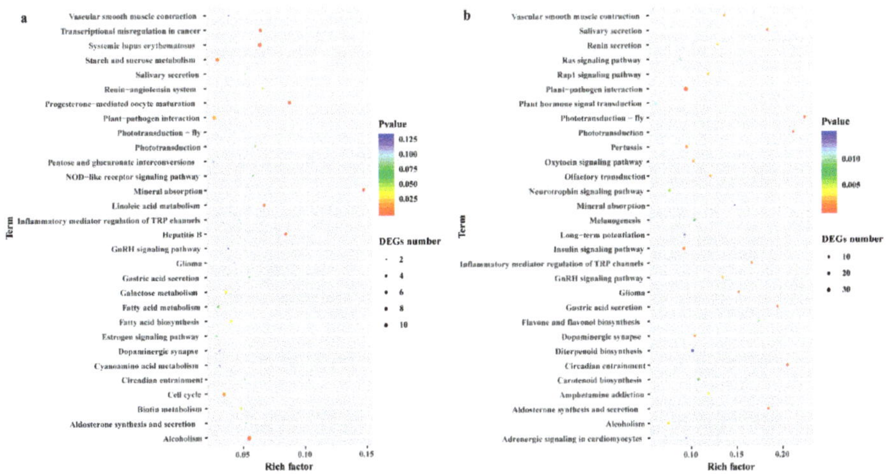

Figure 5. Differentially expressed genes (DEGs) identified by Kyoto Encyclopedia of Genes and Genomes (KEGG) pathway analysis in *C. gigantea*. (**a**) Top 30 KEGG enrichment results in 21DT vs. 21CK. (**b**) Top 30 KEGG enrichment results in 28RW vs. 28CK.

4. Discussion

Much accumulated empirical evidence has indicated that drought can alter cell membrane fluidity, disrupt protein complexes and inhibit photosynthesis. Therefore, a comprehensive study that combined physiological measurements, RNA-Seq and 2-DE analysis was conducted here to screen out some important candidate drought-induced changes in key physiological parameters, DEGs and DEPs. Our results revealed that the transcriptional regulation of proteins related to photosynthesis and the post-transcriptional regulation of genes involved in ROS scavenging might be the reasons for drought resistance in *C. gigantea*.

As it is an important secondary messenger, the rapid accumulation of ROS leads to rapid stomatal closure to limit photosynthesis under conditions of water deficit [39]. A recent study found that the *Pn*, *Gs* and *Tr* of rubber trees decreased significantly under dehydration conditions [40]. In our study, similar results were detected in the drought-stressed seedlings (Figure 1b–e). Additionally, how a plant responds to stress not only entails changed photosynthesis but also the accumulation of osmolytes, such as soluble sugars and proline, which reflect the degree of injury [41]. We found that the content of soluble sugars and proline both enlarged as the drought stress continued (Figure 2c,d). Similar findings were reported recently for *Scutellaria baicalensis* Georgi [42].

In the present study, histochemical staining by both DAB and NBT together revealed a pronounced drought-induced accumulation of H_2O_2 and $O_2^{\cdot -}$ in *C. gigantea* leaves (Figure 2a,b). This result is consistent with earlier findings that drought promotes the generation of ROS [43]. Moreover, we demonstrated that the drought-stressed seedlings showed higher SOD, POD and CAT activity compared with the control group under drought stress (Figure 2e–g). This is consistent with another analysis of antioxidant enzymes in plants under drought conditions [44]. Similarly, drought stress significantly enhanced the contents of AsA, DHA, GSH and GSSG during drought, and these antioxidant molecules peaked under severe drought conditions (Figure 2h–o), which also is in line with prior research [45]. Compared with the control group, the characteristic withered and curved growth phenotype of the treatment seedlings suggested that *C. gigantea* likely relies on limited photosynthesis, the accumulation of osmolytes and ROS scavenging strategy to adapt to drought conditions.

Underlying many of these changes is the differential expression of genes in response to drought stress. Thus, both RNA-Seq and 2-DE were used to further explore the critical genes and pathways to obtain a better understanding of the molecular mechanisms underlying the response to drought. Here, we successfully quantified and compared 551 DEGs and 66 DEPs in the leaves of *C. gigantea* seedlings under drought stress. However, only a few genes were commonly regulated at the transcriptomic and proteomic levels, although this is consistent with a previous study [46], whose data indicated that post-transcriptional regulation plays a crucial role in the drought response of cassava plants. In the case of *C. gigantea*, the categories of overlapping DEGs and DEPs were most related to photosynthesis and ROS scavenging, suggesting their involvement in the drought response of this tree species.

Photosynthesis the most fundamental and intricate physiological process in plants, and it is highly susceptible to drought [47]. Consistent with previous studies of photosynthesis-related genes' responses to drought, we found that genes involved in the Cyt b6/f complex and ATPase synthase were all downregulated after seedlings experienced the drought treatment [48]. Similarly, we found that many DEPs associated with PSII, LHC and ATP synthase were also downregulated, as inferred by the proteomic analysis (Figure S2). Similar responses in maize plants were reported recently [49]. Taken together, our results all implied that modulating PET is probably crucial to increasing the tolerance of *C. gigantea* to drought.

The ROS scavenging system, an important defense mechanism against drought, maintains the cellular homeostasis of ROS. Overproduction of ROS during drought not only affects the stress-related proteins but also modulates the transcription level of genes [50]. In the current work, four of these enzymes, including GST, SOD, APX and POD, were identified. Among these genes, the *SOD* gene was downregulated in *C. gigantea* seedlings under drought stress. A significant reduction in the expression of the *APX* gene was obtained under drought stress conditions. Notably, our proteomics data showed that the abundance of APX, SOD [Cu-Zn], ASR and GST was also enhanced in seedlings of *C. gigantea* to cope with the imposed drought stress. These results are consistent with the observed activities of antioxidant enzymes. These proteins could also work together to maintain the dynamic balance of ROS levels and thereby avoid the oxidative damage incurred by plants. This type of coordination occurs in alkaligrass [51], rapeseed [52] and cassava [53], perhaps due to the fact that enzyme activity is regulated by post-translational modifications [54]. Finally, POD can also contribute to how plants respond to drought stress [5]. Nevertheless, in this study, drought stress led to downregulation of POD at both the protein and gene levels, but, interestingly, the activity of POD was significantly increased by drought stress. Evidently, to explain the changes in enzyme activity and the expression of genes and proteins of *C. gigantea* under drought stress more precisely, further in-depth research investigations and analyses are still needed.

5. Conclusions

The main aim of this study was to expand our knowledge about the mechanism underlying the impact of drought stress in seedlings of *C. gigantea*. An attempt was made to identify the critical genes and pathways by comparing the findings from transcriptomic and proteomic analytical methods. On the basis of results, here, we propose a schematic diagram for the regulation of drought resistance in *C. gigantea* (Figure 6). When plants are subjected to drought, ROS induces oxidative stress and activates the ROS scavenging system. Simultaneously, oxidative stress impairs photosynthesis and inhibits PET, ultimately inhibiting plant growth. Taken together, the omics data suggest that post-transcriptional regulation of PET and ROS scavenging plays a crucial role in the drought response of *C. gigantea* seedlings.

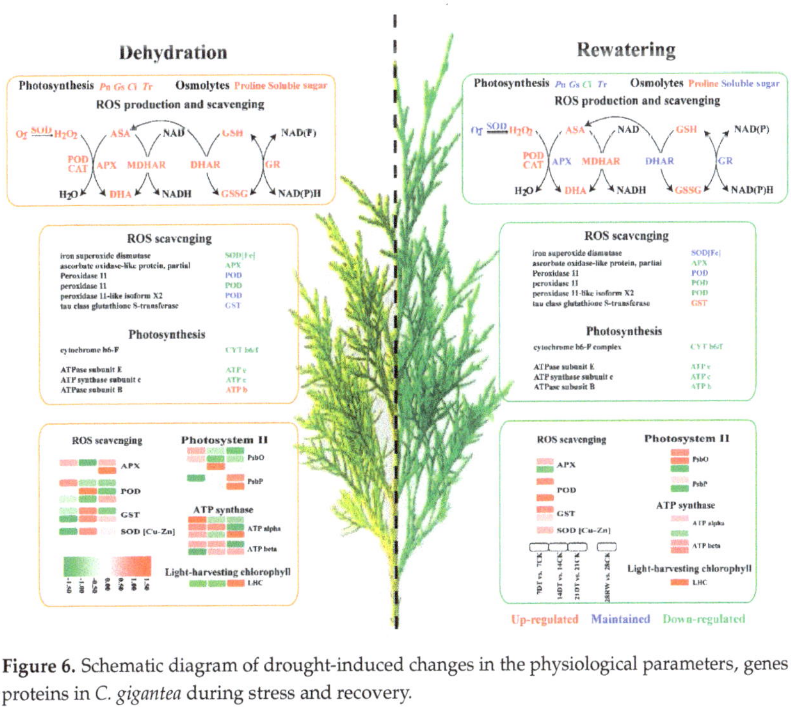

Figure 6. Schematic diagram of drought-induced changes in the physiological parameters, genes and proteins in *C. gigantea* during stress and recovery.

Supplementary Materials: The following are available online at https://www.mdpi.com/article/10.3390/f13030401/s1. Figure S1: 2-DE gel images of the control and treatment seedlings under drought extracted at different time points (7 d, 14 d, 21 d and recovery). The squares represent the drought-responsive proteins identified by MS. Figure S2: Expression of DEPs in the control and drought treatment at different time points (7 d, 14 d, 21 d and recovery). Figure S3: Characteristics of *C. gigantea* unigenes. (**a**) Length frequency distribution of all unigenes. (**b**) Venn diagram showing the number of DEGs. (**c**) Correlation analysis between the transcriptome results and the qRT-PCR results, and validation of the RNA-Seq results by qRT-PCR. Table S1: The corresponding induction factor (percent volume of the spot under stress conditions/percent volume of the spots under control conditions) of drought-responsive proteins of seedlings. Table S2: Summary of unigenes identified in *C. gigantea*. Table S3: Summary of DEGs identified in *C. gigantea*. FDR $\leq$ 0.001 log2 fold change ≤ -2 or $\geq$2. Table S4: List of GO enrichment terms of DEGs in *C. gigantea* during drought stress. Table S5: KEGG pathway enrichment analysis of DEGs in *C. gigantea* during drought stress. Table S6: DEGs and DEPs involved in ROS scavenging and photosynthesis in *C. giganteas* during drought stress. Table S7: Primers of the genes used for qRT-PCR and the expression correlation between the RNA-seq and qRT-PCR analyses.

Author Contributions: P.L., Z.L., L.T. and F.M., conceived and designed the experiments; P.L., Z.L. and J.L. performed the experiments; P.L., Z.L. and L.X. analyzed or interpreted the data for the work; P.L., X.Z., L.T. and F.M. wrote the manuscript; P.L., X.J., L.X., G.J., X.Z. and F.M. ensure that questions related to the accuracy or integrity of any part of the work. All authors have read and agreed to the published version of the manuscript.

Funding: This project was supported by the Fundamental Research Funds for the Central Universities (2572021AW15).

Informed Consent Statement: Informed consent was obtained from all subjects involved in the study.

Data Availability Statement: Sequencing data have been deposited at the Short Read Archive (SRA) database (https://www.ncbi.nlm.nih.gov/sra, accessed on 18 July 2021) under accession numbers SAMN20236055 to SAMN20236058 (Bio-Project PRJNA746687). The proteome data can be found at: https://github.com/lppaper/datebase, accessed on 18 February 2022.

Acknowledgments: We thank the reviewers and editors for their work.

Conflicts of Interest: The authors declare no conflict of interest.

References

1. Zhu, J.K. Abiotic Stress Signaling and Responses in Plants. *Cell* **2016**, *167*, 313–324. [CrossRef]
2. Kaya, C.; Senbayram, M.; Akram, N.A.; Ashraf, M.; Alyemeni, M.N.; Ahmad, P. Sulfur-enriched leonardite and humic acid soil amendments enhance tolerance to drought and phosphorus deficiency stress in maize (*Zea mays* L.). *Sci. Rep.* **2020**, *10*, 6432. [CrossRef] [PubMed]
3. Li, J.; Li, Y.; Yin, Z.; Jiang, J.; Zhang, M.; Guo, X.; Ye, Z.; Zhao, Y.; Xiong, H.; Zhang, Z.; et al. OsASR5 enhances drought tolerance through a stomatal closure pathway associated with ABA and H_2O_2 signalling in rice. *Plant Biotechnol. J.* **2017**, *15*, 183–196. [CrossRef] [PubMed]
4. Li, S.; Zhang, J.; Liu, H.; Liu, N.; Shen, G.; Zhuang, H.; Wu, J. Dodder-transmitted mobile signals prime host plants for enhanced salt tolerance. *J. Exp. Bot.* **2020**, *71*, 1171–1184. [CrossRef] [PubMed]
5. Talbi, S.; Antonio Rojas, J.; Sahrawy, M.; Rodriguez-Serrano, M.; Cardenas, K.E.; Debouba, M.; Maria Sandalio, L. Effect of drought on growth, photosynthesis and total antioxidant capacity of the saharan plant *Oudeneya africana*. *Environ. Exp. Bot.* **2020**, *176*, 104099. [CrossRef]
6. Jin, H.L.; Fu, M.; Duan, Z.K.; Duan, S.J.; Li, M.S.; Dong, X.X.; Liu, B.; Feng, D.R.; Wang, J.F.; Peng, L.W.; et al. LOW PHOTOSYNTHETIC EFFICIENCY 1 is required for light-regulated photosystem II biogenesis in *Arabidopsis*. *Proc. Natl. Acad. Sci. USA* **2018**, *115*, E6075–E6084. [CrossRef]
7. Wei, X.P.; Su, X.D.; Cao, P.; Liu, X.Y.; Chang, W.R.; Li, M.; Zhang, X.Z.; Liu, Z.F. Structure of spinach photosystem II-LHCII supercomplex at 3.2 angstrom resolution. *Nature* **2016**, *534*, 69–74. [CrossRef]
8. Xiao, J.; Li, J.; Ouyang, M.; Yun, T.; He, B.; Ji, D.; Ma, J.; Chi, W.; Lu, C.; Zhang, L. DAC Is Involved in the Accumulation of the Cytochrome b(6)/f Complex in *Arabidopsis*. *Plant Physiol.* **2012**, *160*, 1911–1922. [CrossRef]
9. Estravis-Barcala, M.; Gabriela Mattera, M.; Soliani, C.; Bellora, N.; Opgenoorth, L.; Heer, K.; Veronica Arana, M. Molecular bases of responses to abiotic stress in trees. *J. Exp. Bot.* **2020**, *71*, 3765–3779. [CrossRef]
10. Cui, X.-Y.; Gao, Y.; Guo, J.; Yu, T.-F.; Zheng, W.-J.; Liu, Y.-W.; Chen, J.; Xu, Z.-S.; Ma, Y.-Z. BES/BZR Transcription Factor TaBZR2 Positively Regulates Drought Responses by Activation of TaGST1. *Plant Physiol.* **2019**, *180*, 605–620. [CrossRef]

11. Liu, Y.; Yang, T.; Lin, Z.; Gu, B.; Xing, C.; Zhao, L.; Dong, H.; Gao, J.; Xie, Z.; Zhang, S.; et al. A WRKY transcription factor PbrWRKY53 from *Pyrus betulaefolia* is involved in drought tolerance and AsA accumulation. *Plant Biotechnol. J.* **2019**, *17*, 1770–1787. [CrossRef] [PubMed]
12. Mao, C.; Zhu, Y.; Cheng, H.; Yan, H.; Zhao, L.; Tang, J.; Ma, X.; Mao, P. Nitric Oxide Regulates Seedling Growth and Mitochondrial Responses in Aged Oat Seeds. *Int. J. Mol. Sci.* **2018**, *19*, 1052. [CrossRef] [PubMed]
13. Kaya, C.; Ashraf, M.; Wijaya, L.; Ahmad, P. The putative role of endogenous nitric oxide in brassinosteroid-induced antioxidant defence system in pepper (*Capsicum annuum* L.) plants under water stress. *Plant Physiol. Biochem.* **2019**, *143*, 119–128. [CrossRef] [PubMed]
14. Ricroch, A.E.; Berge, J.B.; Kuntz, M. Evaluation of Genetically Engineered Crops Using Transcriptomic, Proteomic, and Metabolomic Profiling Techniques. *Plant Physiol.* **2011**, *155*, 1752–1761. [CrossRef] [PubMed]
15. Osorio, S.; Alba, R.; Damasceno, C.M.; Lopez-Casado, G.; Lohse, M.; Zanor, M.I.; Tohge, T.; Usadel, B.; Rose, J.K.; Fei, Z.; et al. Systems biology of tomato fruit development: Combined transcript, protein, and metabolite analysis of tomato transcription factor (nor, rin) and ethylene receptor (Nr) mutants reveals novel regulatory interactions. *Plant Physiol.* **2011**, *157*, 405–425. [CrossRef] [PubMed]
16. Yu, A.; Li, F.; Liu, A. Comparative proteomic and transcriptomic analyses provide new insight into the formation of seed size in castor bean. *BMC Plant Biol.* **2020**, *20*, 48. [CrossRef]
17. Peremarti, A.; Mare, C.; Aprile, A.; Roncaglia, E.; Cattivelli, L.; Villegas, D.; Royo, C. Transcriptomic and proteomic analyses of a pale-green durum wheat mutant shows variations in photosystem components and metabolic deficiencies under drought stress. *BMC Genom.* **2014**, *15*, 125. [CrossRef]
18. Casati, P.; Campi, M.; Morrow, D.J.; Fernandes, J.F.; Walbot, V. Transcriptomic, proteomic and metabolomic analysis of UV-B signaling in maize. *BMC Genom.* **2011**, *12*, 321. [CrossRef]
19. Wang, N.; Zhao, J.; He, X.; Sun, H.; Zhang, G.; Wu, F. Comparative proteomic analysis of drought tolerance in the two contrasting Tibetan wild genotypes and cultivated genotype. *BMC Genom.* **2015**, *16*, 432. [CrossRef]
20. Koobaz, P.; Ghaffari, M.R.; Heidari, M.; Mirzaei, M.; Ghanati, F.; Amirkhani, A.; Mortazavi, S.E.; Moradi, F.; Hajirezaei, M.R.; Salekdeh, G.H. Proteomic and metabolomic analysis of desiccation tolerance in wheat young seedlings. *Plant Physiol. Biochem.* **2020**, *146*, 349–362. [CrossRef]
21. Anupama, A.; Bhugra, S.; Lall, B.; Chaudhury, S.; Chugh, A. Morphological, transcriptomic and proteomic responses of contrasting rice genotypes towards drought stress. *Environ. Exp. Bot.* **2019**, *166*, 103795. [CrossRef]
22. Tripathi, P.; Rabara, R.C.; Reese, R.N.; Miller, M.A.; Rohila, J.S.; Subramanian, S.; Shen, Q.J.; Morandi, D.; Buecking, H.; Shulaev, V.; et al. A toolbox of genes, proteins, metabolites and promoters for improving drought tolerance in soybean includes the metabolite coumestrol and stomatal development genes. *BMC Genom.* **2016**, *17*, 102. [CrossRef] [PubMed]
23. Xu, H.; Shi, D.; Wang, J.; Xu, T.; Wu, Y. Isolation and characterization of polymorphic microsatellite markers in *Cupressus chengiana* S. Y. Hu (Cupressaceae). *Conserv. Genet.* **2008**, *9*, 1023–1026. [CrossRef]
24. Li, S.; Qian, Z.; Fu, Y.; Zheng, W.; Li, H. Isolation and characterization of polymorphic microsatellites in the Tibetan cypress *Cupressus gigantea* using paired-end Illumina shotgun sequencing. *Conserv. Genet. Resour.* **2014**, *6*, 795–797. [CrossRef]
25. Lu, X.; Xu, H.; Li, Z.; Shang, H.; Adams, R.P.; Mao, K. Genetic Diversity and Conservation Implications of Four *Cupressus* Species in China as Revealed by Microsatellite Markers. *Biochem. Genet.* **2014**, *52*, 181–202. [CrossRef]
26. Yin, Z.; Wang, Y.; Wang, Y.; Sang, L.; Luo, Q.; Meng, F. Improvement on the method of the extraction of total proteins on *Cupressus gigantea* leaves for two-dimensional gel electrophoresis. *J. Cent. South Univ. For. Technol.* **2019**, *39*, 143–146.
27. Xin, F.; Wang, Y.; Li, S.; Danzengluobu; Pubuciren. Effects of different temperatures on photosynthesis and rooting of *Cupressus gigantea* seedlings. *J. Zhejiang Univ. Agric. Life Sci.* **2019**, *45*, 102–108.
28. Li, H.; Guo, Q.; Zheng, W. The complete chloroplast genome of *Cupressus gigantea*, an endemic conifer species to Qinghai-Tibetan Plateau. *Mitochondrial DNA Part A* **2016**, *27*, 3743–3744. [CrossRef]
29. Zhou, S.-S.; Xing, Z.; Liu, H.; Hu, X.-G.; Gao, Q.; Xu, J.; Jiao, S.-Q.; Jia, K.-H.; Jin, Y.Q.; Zhao, W.; et al. In-depth transcriptome characterization uncovers distinct gene family expansions for *Cupressus gigantea* important to this long-lived species' adaptability to environmental cues. *BMC Genom.* **2019**, *20*, 213. [CrossRef]
30. Chen, K.Q.; Song, M.R.; Guo, Y.N.; Liu, L.F.; Xue, H.; Dai, H.Y.; Zhang, Z.H. MdMYB46 could enhance salt and osmotic stress tolerance in apple by directly activating stress-responsive signals. *Plant Biotechnol. J.* **2019**, *17*, 2341–2355. [CrossRef]
31. Zhang, Z.H.; Zhu, L.; Song, A.P.; Wang, H.B.; Chen, S.M.; Jiang, J.F.; Chen, F.D. Chrysanthemum (*Chrysanthemum morifolium*) CmICE2 conferred freezing tolerance in Arabidopsis. *Plant Physiol. Biochem.* **2020**, *146*, 31–41. [CrossRef] [PubMed]
32. Zhao, Y.; Xu, F.; Liu, J.; Guan, F.; Quan, H.; Meng, F. The adaptation strategies of *Herpetospermum pedunculosum* (Ser.) Baill at altitude gradient of the Tibetan plateau by physiological and metabolomic methods. *BMC Genom.* **2019**, *20*, 451. [CrossRef] [PubMed]
33. Giannopolitis, C.N.; Ries, S.K. Superoxide dismutases: I. Occurrence in higher plants. *Plant Physiol.* **1977**, *59*, 309–314. [CrossRef]
34. Kosar, F.; Akram, N.A.; Ashraf, M.; Ahmad, A.; Alyemeni, M.N.; Ahmad, P. Impact of exogenously applied trehalose on leaf biochemistry, achene yield and oil composition of sunflower under drought stress. *Physiol. Plant* **2020**, *172*, 317–333. [CrossRef]
35. Sun, C.; Liu, L.; Yu, Y.; Liu, W.; Lu, L.; Jin, C.; Lin, X. Nitric oxide alleviates aluminum-induced oxidative damage through regulating the ascorbate-glutathione cycle in roots of wheat. *J. Integr. Plant Biol.* **2015**, *57*, 550–561. [CrossRef]

36. Raja, V.; Qadir, S.U.; Alyemeni, M.N.; Ahmad, P. Impact of drought and heat stress individually and in combination on physio-biochemical parameters, antioxidant responses, and gene expression in *Solanum lycopersicum*. *3 Biotech* **2020**, *10*, 208. [CrossRef]
37. Cao, Y.; Luo, Q.; Tian, Y.; Meng, F. Physiological and proteomic analyses of the drought stress response in *Amygdalus Mira* (Koehne) Yu et Lu roots. *BMC Plant Biol.* **2017**, *17*, 53. [CrossRef] [PubMed]
38. Wang, Z.M.; Wang, M.Y.; Liu, L.; Meng, F.J. Physiological and Proteomic Responses of Diploid and Tetraploid Black Locust (*Robinia pseudoacacia* L.) Subjected to Salt Stress. *Int. J. Mol. Sci.* **2013**, *14*, 20299–20325. [CrossRef]
39. Jan, S.; Abbas, N.; Ashraf, M.; Ahmad, P. Roles of potential plant hormones and transcription factors in controlling leaf senescence and drought tolerance. *Protoplasma* **2019**, *256*, 313–329. [CrossRef]
40. Santos, J.O.d.; Oliveira, L.E.M.d.; Souza, T.d.; Lopes, G.M.; Coelho, V.T.; Gomes, M.P. Physiological mechanisms responsible for tolerance to, and recuperation from, drought conditions in four different rubber clones. *Ind. Crop. Prod.* **2019**, *141*, 111714. [CrossRef]
41. Mohasseli, V.; Sadeghi, S. Exogenously applied sodium nitroprusside improves physiological attributes and essential oil yield of two drought susceptible and resistant specie of *Thymus* under reduced irrigation. *Ind. Crop. Prod.* **2019**, *130*, 130–136. [CrossRef]
42. Cheng, L.; Han, M.; Yang, L.M.; Yang, L.; Sun, Z.; Zhang, T. Changes in the physiological characteristics and baicalin biosynthesis metabolism of *Scutellaria baicalensis* Georgi under drought stress. *Ind. Crop. Prod.* **2018**, *122*, 473–482. [CrossRef]
43. Hasanuzzaman, M.; Nahar, K.; Rahman, A.; Inafuku, M.; Oku, H.; Fujita, M. Exogenous nitric oxide donor and arginine provide protection against short-term drought stress in wheat seedlings. *Physiol. Mol. Biol. Plants* **2018**, *24*, 993–1004. [CrossRef]
44. Li, T.; Wang, R.; Zhao, D.; Tao, J. Effects of drought stress on physiological responses and gene expression changes in herbaceous peony (*Paeonia lactiflora* Pall.). *Plant Signal. Behav.* **2020**, *15*, 1746034. [CrossRef] [PubMed]
45. Shan, C.; Zhou, Y.; Liu, M. Nitric oxide participates in the regulation of the ascorbate-glutathione cycle by exogenous jasmonic acid in the leaves of wheat seedlings under drought stress. *Protoplasma* **2015**, *252*, 1397–1405. [CrossRef] [PubMed]
46. Lan, P.; Li, W.; Schmidt, W. Complementary proteome and transcriptome profiling in phosphate-deficient Arabidopsis roots reveals multiple levels of gene regulation. *Mol. Cell. Proteom.* **2012**, *11*, 1156–1166. [CrossRef] [PubMed]
47. Begum, N.; Ahanger, M.A.; Su, Y.; Lei, Y.; Mustafa, N.S.A.; Ahmad, P.; Zhang, L. Improved Drought Tolerance by AMF Inoculation in Maize (*Zea mays*) Involves Physiological and Biochemical Implications. *Plants* **2019**, *8*, 579. [CrossRef]
48. Meyer, E.; Aspinwall, M.J.; Lowry, D.B.; Palacio-Mejia, J.D.; Logan, T.L.; Fay, P.A.; Juenger, T.E. Integrating transcriptional, metabolomic, and physiological responses to drought stress and recovery in switchgrass (*Panicum virgatum* L.). *BMC Genom.* **2014**, *15*, 527. [CrossRef]
49. Wang, B.; Liu, C.; Zhang, D.; He, C.; Zhang, J.; Li, Z. Effects of maize organ-specific drought stress response on yields from transcriptome analysis. *BMC Plant Biol.* **2019**, *19*, 335. [CrossRef]
50. Li, B.; Feng, Y.; Zong, Y.; Zhang, D.; Hao, X.; Li, P. Elevated CO_2-induced changes in photosynthesis, antioxidant enzymes and signal transduction enzyme of soybean under drought stress. *Plant Physiol. Biochem. PPB* **2020**, *154*, 105–114. [CrossRef]
51. Nemati, M.; Piro, A.; Norouzi, M.; Vahed, M.M.; Nistico, D.M.; Mazzuca, S. Comparative physiological and leaf proteomic analyses revealed the tolerant and sensitive traits to drought stress in two wheat parental lines and their F6 progenies. *Environ. Exp. Bot.* **2019**, *158*, 223–237. [CrossRef]
52. Mohammadi, P.P.; Moieni, A.; Komatsu, S. Comparative proteome analysis of drought-sensitive and drought-tolerant rapeseed roots and their hybrid F1 line under drought stress. *Amino Acids* **2012**, *43*, 2137–2152. [CrossRef] [PubMed]
53. Ding, Z.; Fu, L.; Tie, W.; Yan, Y.; Wu, C.; Hu, W.; Zhang, J. Extensive Post-Transcriptional Regulation Revealed by Transcriptomic and Proteomic Integrative Analysis in Cassava under Drought. *J. Agric. Food Chem.* **2019**, *67*, 3521–3534. [CrossRef] [PubMed]
54. Meng, X.; Zhao, Q.; Jin, Y.; Yu, J.; Yin, Z.; Chen, S.; Dai, S. Chilling-responsive mechanisms in halophyte *Puccinellia tenuiflora* seedlings revealed from proteomics analysis. *J. Proteom.* **2016**, *143*, 365–381. [CrossRef] [PubMed]

forests

MDPI

Article

Water Retention Capacity of Leaf Litter According to Field Lysimetry

Taehyun Kim [1], Jungyoon Kim [1], Jeman Lee [1], Hyun Seok Kim [1,2,3], Juhan Park [4] and Sangjun Im [1,2,*]

[1] Department of Agriculture, Forestry and Bioresources, Seoul National University, Seoul 08826, Republic of Korea
[2] Research Institute of Agriculture and Life Sciences, Seoul National University, Seoul 08826, Republic of Korea
[3] Interdisciplinary Program in Agricultural and Forest Meteorology, Seoul National University, Seoul 08826, Republic of Korea
[4] National Center for AgroMeteorology, Seoul National University, Seoul 08826, Republic of Korea
* Correspondence: junie@snu.ac.kr

Abstract: The water retention capacity of forest leaf litter was estimated through lysimeter measurements under field conditions. Six lysimeters were placed in *Pinus koraiensis* and *Quercus acutissima* forests and filled with the surrounding leaf litter to represent the effects of litter type on the water retention capacity. Two years of measurements for rainfall and litter weight have been conducted in all lysimeters at 30 min intervals. Field measurements showed that *P. koraiensis* litter stored more water during rainfall periods than did *Q. acutissima* litter. As a result, immediately after the cessation of rainfall, 1.82 mm and 3.00 mm of water were retained per unit mass of *Q. acutissima* and *P. koraiensis* litter, respectively. Following rainfall, after the gravitational flow had entirely drained, the remaining water adhered to the litter was estimated to be 1.66 ± 1.72 mm and 2.72 ± 2.82 mm per unit mass per rainfall event for *Q. acutissima* and *P. koraiensis* litter, respectively. During the study period, approximately 83.7% of incident rainfall drained into the uppermost soil layer below the *Q. acutissima* litter, whereas 84.5% of rainfall percolated through the *P. koraiensis* litter. The moisture depletion curves indicated that 50% of the water retained in the *Q. acutissima* and *P. koraiensis* litter was lost via evaporation within 27 h and 90 h after the cessation of rainfall, respectively. This study demonstrated the water retention storage of leaf litter and its contribution to the water balance over floor litter according to litter and rainfall characteristics. The results also proved that lysimetry is a reliable method to quantify the variation of litter moisture under natural conditions.

Keywords: litter lysimeter; water retention capacity; litter drainage; *Pinus koraiensis* litter; *Quercus acutissima* litter

Citation: Kim, T.; Kim, J.; Lee, J.; Kim, H.S.; Park, J.; Im, S. Water Retention Capacity of Leaf Litter According to Field Lysimetry. *Forests* **2023**, *14*, 478. https://doi.org/10.3390/f14030478

Academic Editors: Yanhui Wang, Karl-Heinz Feger and Lulu Zhang

Received: 8 February 2023
Revised: 21 February 2023
Accepted: 24 February 2023
Published: 27 February 2023

1. Introduction

Forests occupy approximately 63% of the land area of the Republic of Korea. The national forest restoration project led to an increase in the volume of forests from 7.3 m^3/ha in 1955 to 165.2 m^3/ha in 2020 [1], thereby contributing to the development of a thick layer of leaf litter on the forest floors. Fallen leaves resting on the underlying mineral soil are the source of most plant nutrients and provide habitats for a great diversity of organisms. Hydrologically, leaf litter on the forest floor acts as a porous interface between mineral soil and ambient air and influences the processes of subsurface runoff, overland flow, and soil erosion [2,3]. Leaf litter also influences the physical properties of soils, thereby increasing the ability of water to infiltrate into the soil [2,4].

Rainfall partitioning occurring on the forest floor is a minor but fundamental component of the water balance in forest watersheds [4]. A small amount of rainwater is temporarily captured by the surface of leaf litter, and the retained water is lost via evaporation within a few hours or days after the rainfall. When the amount of intercepted water exceeds the threshold, known as the water retention capacity, water percolates into the soil along litter funnels under the influence of gravity [2,5]. These hydrological processes repeatedly occur during successive rainfall events.

Leaf litter intercepts rainfall much in the same way as canopy interception. The extent of rainfall retained in the litter layer is highly related to the litter's ability of water absorption. Where water retention storage is small, less water is available for evaporation, but more water reaches the upper soil layer for runoff. Water retention capacity also plays a major role in predicting the ignition and spread rate of forest fires [2–4].

The water retention capacity of litter differs among leaf types [6–8]. Owing to the large size and curved shape of the leaves, broadleaf litter has an advantage in storing water on the leaf surface during rainfall [9]. Conversely, the low porosity of the needle shape of coniferous litter contributes to high water retention within the litter as a consequence of the greater resistance of the packed litter to the percolation of water [10]. Considerably more water is required to completely saturate thicker needle litter [11] before it moves into the soil underneath.

Numerous attempts have been made to continuously measure the water retention storage of floor litter over the past couple of decades. According to Helvey and Patric [11], these methods may categorize into laboratory and field experiments. A detailed review of measuring techniques can be found in Gerrits and Savenije [12].

Several laboratory experiments have been conducted to analyze the water dynamics of leaf litter as it relates to retention and drainage capacity [9,10,13,14]. These experiments have provided a reliable estimation; however, accuracy is limited by several factors. First, the hydrologic response of leaf litter under simulated rainfall in a lab likely differs from that under real incidents of rainfall. Further, inconsistencies in data from natural rainfall events can be attributed to spatial and temporal variations, which would not occur in a lab setting. In addition, the relatively short duration and high intensity of artificial rainfall in laboratory studies cannot completely saturate the leaf accumulation, thereby leading to low measurements of water retention capacity [10], which may be inaccurate.

To overcome some of the drawbacks of laboratory experiments, field lysimeters have recently been employed in experimental forest hydrology. This device was designed to measure the weight of the litter within a certain time interval. Schaap and Bouten [15] developed a concrete lysimeter equipped with a load cell sensor. The present study is based on a field experiment conducted by Gerrits et al. [16], similar to that of Schaap and Bouten [15], to measure the weight of the litter layer. Lysimetric techniques are scalable and suitable for directly quantifying the water budget of a forest floor through changes in litter weight over the shortest intervals, but the challenges of using this tool include high manufacturing and operational costs [17].

Forests can retain a significant amount of rainfall via foliage, branches, stems, and leaf litter. Although the litter layer can store only a few millimeters of water, water absorption and depletion processes of litter can result in a considerable reduction of rainfall reaching to soil surface. However, the role of litter on rainfall interception and retention remains unclear and may even be disregarded in many hydrological studies because it is commonly considered a minor process and has practical difficulties in making accurate measurements [8,13]. Therefore, an accurate understanding of how rainfall is partitioned over the floor litter can provide a better perspective on rainfall interception and evaporation processes.

The objectives of this study were to measure the water stored in the litter layer under natural conditions and to examine the effects of litter type and rainfall amount on the water balance of the litter layer. In this study, six lysimeters were placed in deciduous and coniferous forests, and litter weight was measured during the experimental period. The reliable estimation of water retention capacity can contribute to understanding the hydrologic processes of the forest floor litter layer and can be utilized to predict fire ignition potential based on fuel moisture variation.

2. Materials and Methods

2.1. Study Area

This study was conducted between November 2015 and November 2017 in the Mt. Taehwa University Forest (TUF) of Seoul National University, located approximately 30 km southeast of the Seoul capital area (N37°18′–20′, E127°17′–20′) (Figure 1). The TUF encompasses a 796-ha mountain forest at a latitude ranging from 150 m to 644 m above sea level. The climate of the TUF is characterized as temperate, with hot wet summers and cold dry winters. The mean monthly temperature ranges from −3.3 °C in January to 25.2 °C in August, according to the most recent 30-year data (1991–2020). The mean annual precipitation is 1308 mm, most of which falls in the summer season between June and September [18].

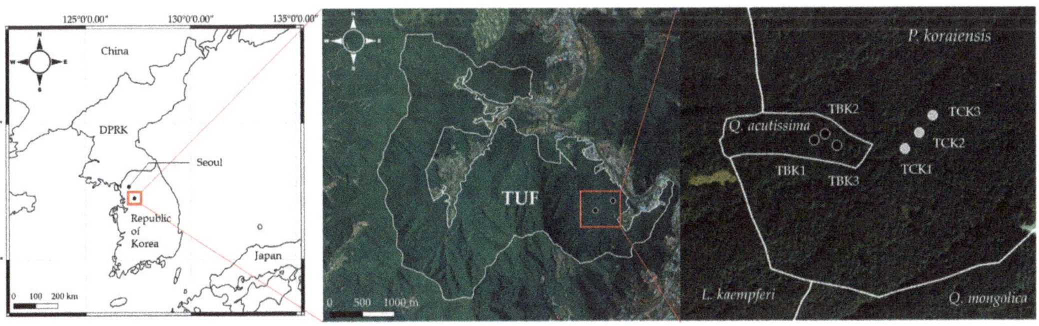

Figure 1. Locations of lysimeters in the Mt. Taehwa University Forest (TUF), Korea.

The TUF comprises 496 ha of natural deciduous forest and 300 ha of the coniferous plantation, with the remaining being composed of mixed forest. The dominant tree species in the deciduous forest include *Quercus acutissima*, *Quercus variabilis*, and *Quercus mongolica*, and the plantation forest comprises *Pinus koraiensis* and *Larix kaempferi*. The bedrock underneath the TUF is composed largely of weathered granite, and the soil texture is primarily sandy loam according to USDA classification. There has been no record of fire in the TUF over the last few decades. A detailed description of the TUF is provided by Im et al. [19].

2.2. Data Collection

Field lysimeters were used to measure the amount of water retained in the litter layer within each forest (Figure 2). The lysimeter comprises an aluminum container, a steel frame, a load cell, and a data storage device. The square litter container has a surface area of 1 m^2 with a permeable wire mesh bottom that allows excess water to drain easily into the underlying soil. Four load cells (BCL-5L, CAS, Korea) were embedded between the container and the steel frame to measure the container weight at 1 min intervals. The container weight, recorded in grams, was averaged over 30 measurements to calculate the variation in litter weight every 30 min.

Each lysimeter was placed in a gap area where other species or understory vegetation were completely absent and filled with the surrounding leaf litter. Three lysimeters, TBK1, TBK2, and TBK3, were located in *Q. acutissima* forest, while three, TCK1, TCK2, and TCK3, were placed in *P. koraiensis* plantation forest that was planted in 1964 (Figure 1). A description of each forest is presented in Table 1. The broadleaf litter lysimeters were set up in a 34-year-old *Q. acutissima* forest with an average tree diameter (n = 12) at breast height (DBH) of 24.1 ± 8.5 cm (mean ± std. dev) and an average tree height of 13.6 ± 3.3 m. The needle litter lysimeters were set up in a *P. koraiensis* plantation with an average DBH of 33.2 ± 4.5 cm and an average tree height of 18.4 ± 1.5 m.

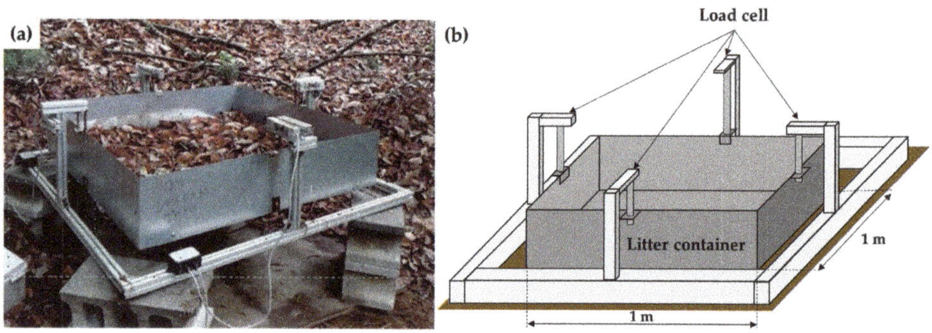

Figure 2. Litter lysimeter (**a**) and schematic diagram of the lysimeter (**b**).

Table 1. Tree and leaf litter characteristics.

Forest	Tree Sample (n = 12)			Leaf Sample			
	Tree Species	DBH (cm)	Height (m)	n	Length (mm)	Width (mm)	Dry Weight (kg/m^2)
Deciduous	*Q. acutissima*	24.1 ± 8.5	13.6 ± 3.3	67	155.1 ± 22.0	41.2 ± 7.5	0.09 ± 0.02
Coniferous	*P. koraiensis*	33.2 ± 1.5	18.4 ± 1.5	600	84.9 ± 16.1	0.8 ± 0.1	0.18 ± 0.02

The physical characteristics of the leaf litter used in the lysimeter experiment are also presented in Table 1. The uneven distribution of leaf size and shape can influence the water retention capacity of the leaf litter [9,10]. Therefore, samples of litter in 10 locations near each lysimeter were collected in a zipper storage bag and immediately moved to the laboratory for analysis. Surface area parameters of the litter were extracted from scanned image data using the LeafArea R package [20]. Litter samples were put into a drying oven for 24 h to obtain the oven-dry weight. Similar to the characteristics of most *Q. acutissima*, individual leaves in the deciduous samples tended to be broad and round-tooth shaped, and they were, on average, 155.1 ± 22.0 mm long and 41.2 ± 7.5 mm wide at their widest point. The leaves in the *P. koraiensis* litter samples had a noticeably needle-shaped surface, with an average length of 84.9 ± 16.1 mm. The oven-dry weight of *Q. acutissima* litter was lower than that of *P. koreiensis* litter.

Prior to weighing the lysimeters, all lysimeter containers were filled with fallen leaves, which were obtained from random locations in the study area with as little disturbance to their accumulation as possible. Litter load in each lysimeter approximately corresponded to the litter accumulated on a 1 m × 1 m area adjacent to each lysimeter location. The undecomposed litters were carefully collected on the ground surface, avoiding leaves with obvious symptoms of pathogen or herbivore attack or with a decomposed entity. After litter collection, decomposed organic matter and twigs were manually removed prior to piling the litter in the lysimeter container.

The water retention capacity of leaf litter can be assessed per unit of dry weight. The oven-drying method is the simplest and most commonly adopted technique for determining the oven-dried weight of the litter; however, as this direct measurement requires destructive sampling, we decided not to utilize it in this study to maintain an intact litter structure as much as possible. Alternatively, the dry weight of litter samples was indirectly estimated by multiplying the air-dried litter weight by the ratio of the weights of air-dried and oven-dried litter, which were obtained from the 10 litter samples collected near each lysimeter. Table 2 shows the variation in litter weight within the forest, which may be attributed to the varying thickness and consolidation of litter accumulation.

Table 2. Characteristics of rainfall and leaf load at each lysimeter location.

Lysimeter	Litter Load (kg/m^2) [1]		Rainfall Data			
	Air-Dried	Oven-Dried	No. Event	Amount (mm)	Duration (h)	Intensity [2] (mm/h)
TBK1	1.42	1.26	99	9.19 ± 13.99	5.14 ± 5.46	3.54 ± 4.77
TBK2	0.62	0.55	70	9.68 ± 14.59	5.55 ± 5.64	3.55 ± 4.96
TBK3	1.52	1.36	85	11.76 ± 19.90	5.91 ± 6.07	4.17 ± 5.77
TCK1	0.94	0.58	91	13.06 ± 22.36	6.88 ± 7.72	3.70 ± 5.79
TCK2	0.75	0.46	80	11.57 ± 22.13	5.36 ± 5.58	4.22 ± 6.58
TCK3	1.40	0.86	87	12.55 ± 24.65	5.32 ± 6.02	4.49 ± 7.03

[1] Weight of litter piled in the lysimeter container; [2] 1-h maximum intensity for a rainfall event.

Gross rainfall was measured at KoFlux towers that were located in an open field adjacent to each site (approximately 100 m from each site). The characteristics of the rainfall events between November 2015 and November 2017 are presented in Table 2. Some of the observations were missing values owing to equipment failures, power outages, or severe weather conditions. Therefore, 99 rainfall events were selected for TBK1, 70 for TBK2, and 85 for TBK3. In the coniferous forest, 91, 80, and 87 rainfall events were used for the analyses of TCK1, TCK2, and TCK3, respectively. Evaporation was derived from the eddy covariance flux measurements of the KoFlux towers. A more detailed explanation of the eddy covariance method can be found in Kang et al. [21].

2.3. Estimation of Water Balance Parameters

To understand the hydrologic behavior of leaf litter, the water retention capacity and free drainage were estimated from lysimeter measurements. The term retention capacity refers to the ability of the leaf litter surface to store water and is expressed by the maximum and minimum retention [9,10]. The maximum retention is the maximum volume of water held on the litter during a rainfall event, and this value is obtained just before the rainfall stops. The minimum retention is quantified as the amount of water stored in the leaf litter after drainage (or gravitational flow) completely ceases. This value represents the inherent ability of litter to retain water owing to surface tension and adhesive force [10,13].

The water retained in the litter layer was later depleted via evaporation after rainfall, which was characterized by a litter moisture depletion curve [10]. The depletion curve is the lower part of the falling limb of a litter moisture curve and provides the length of time required to achieve a certain weight of litter moisture, assuming no further rainfall. Litter moisture depletion is a drying process that can be expressed as a simple exponential form [10,22],

$$S_t = S_{mx}e^{-at} = S_{mx}k^t \quad (1)$$

where S_t is the retention storage at time t (mm), S_{mx} is the maximum retention storage after the cessation of rainfall (mm), t represents the elapsed time (h), and k is the depletion constant (=e^{-a}).

The depletion constant k represents the time-dependent decline of litter moisture and determines the line of best fit through litter moisture measurements with time. The drying of litter did not occur in a uniform manner across the experiments. Thus, a representative k value for each litter type was obtained by averaging over all rainfall events.

As no drainage data were available in this study, drainage outflow from the litter layer was indirectly determined from the difference in litter weight over a duration of 30 min. During a rainfall event, evaporation typically occurs at a rate of 0.1 to 0.5 mm/h because the ambient air near the litter surface is closely saturated [23,24]. Therefore, evaporation loss can be disregarded in the calculation of free drainage during rainfall. During the rainfall period, the free drainage for a duration of 30 min can be calculated as follows:

$$D_t = R_t - \Delta S_t, \text{ where } \Delta S_t = S_t - S_{t-1} \quad (2)$$

where D_t and R_t are the amounts of water in the litter drainage and rainfall at time t (mm/30 min), respectively. ΔS_t implies the change in litter moisture for a duration of 30 min.

When the rain ceases, the evaporation process gradually recovers. Therefore, drainage can be estimated as follows:

$$D_t = \Delta S_t - E_t, \text{ for } \Delta S_t > E_t \tag{3}$$

$$D_t = 0, \qquad \text{for } \Delta S_t \leq E_t \tag{4}$$

where E_t is the amount of evaporative loss for a duration of 30 min (mm/30 min).

2.4. Statistical Analysis

Tests for normality must be checked before full statistical analysis can be conducted. Because the data didn't follow normal distribution due to high skewness and heterogeneous variance, differences among treatments were tested using the Kruskal–Wallis test. Bonferroni correction was also applied to minimize the alpha inflation when assessing the treatment effects. Two litter types were designed in this study to evaluate the effects of litter and rainfall on the water retention capacity with three lysimeters. Differences were considered statistically significant at $p < 0.05$. All statistical analyses were performed using the R statistical package, version 4.2.0 (22 April 2022) [25].

3. Results

3.1. Water Retention Capacity of Leaf Litter

The water retention capacity of the leaf litter was estimated from the litter lysimeter measurements and is presented in Table 3. For the entire period of measurement (Table 2), the maximum retention capacity (mean ± std. dev) of the *Q. acutissima* litter (1.69 ± 1.28 mm for TBK1, 1.46 ± 1.33 mm for TBK2, and 1.67 ± 1.57 mm for TBK3) was equivalent to 3.07 ± 2.23, 1.16 ± 1.06, and 1.23 ± 1.16 mm per unit litter mass (kg/m^2), respectively. The maximum retention capacity in the *P. koraiensis* litter (1.75 ± 1.79 mm for TCK1, 1.90 ± 1.79 for TCK2, and 2.01 ± 1.99 mm for TCK3) was equivalent to 2.04 ± 2.08 mm, 4.13 ± 3.89 mm, and 3.46 ± 3.42 mm per unit litter mass (kg/m^2), respectively, which was higher than that of the *Q. acutissima* leaf litter. This demonstrated that more rainfall was retained in needle litter than in broadleaf litter, regardless of rainfall amount, and this result was consistent with the finding of Li et al. [10]. Maximal water retention depends on the development of free drainage in the litter layer during rainfall events [9]. Needle litter forms an extremely dense accumulation that obstructs the dispersion and percolation of rainfall. However, in broadleaf litter such as that of *Q. acutissima*, biomat flow can be developed in the litter layer that allows water to move laterally and vertically [26].

Table 3. Water retention in leaf litter per rainfall event.

Lysimeter	Maximum Retention Storage (mm)	Minimum Retention Storage (mm)
TBK1	1.69 ± 1.28	1.46 ± 1.21
TBK2	1.46 ± 1.33	1.35 ± 1.32
TBK3	1.67 ± 1.57	1.56 ± 1.47
TCK1	1.75 ± 1.79	1.58 ± 1.75
TCK2	1.90 ± 1.79	1.89 ± 1.74
TCK3	2.01 ± 1.99	1.97 ± 1.90

In contrast to the previous laboratory experiments, where gravitational water was readily drained within a short period (approximately 30 min) after rainfall cessation [9,10], this in situ study demonstrated that water continued to drain out of litter samples beyond 30 min after rainfall ended. Therefore, the lowest, nearly asymptotic water retention was considered in this study as the minimum retention [10]. The minimum retention varied, ranging from 1.35 ± 1.32 mm (TBK2) to 1.56 ± 1.47 mm (TBK3) in broadleaf litter and from 1.58 ± 1.75 mm (TCK1) to 1.97 ± 1.90 mm (TCK3) in needle litter. The average values of minimum retention were 1.66 mm and 2.72 mm per unit mass for *Q. acutissima* and *P. koraiensis* litters, respectively. No significant differences were observed in retention among the groups with different litter weights ($p = 0.1912$); however, a statistically significant difference was observed between litter types ($p < 0.01$).

According to previous studies [9,10], the water retention capacity of litter depends on litter mass, regardless of its thickness. The relationship between retention and litter mass is shown in Figure 3. The coefficients of determination for linear relationship were also presented in Figure 3, implying a measure of how well linear regression represents the measurements. As expected, both the maximum and minimum retention values increased linearly as litter mass increased. The maximum retention value increased considerably in the *P. koraiensis* litter, whereas it tended to change only slightly in the *Q. acutissima* litter. As shown in Figure 3, all lysimeter measurements presented a linear relationship between minimum retention and litter mass, with slope coefficients of 0.187 for *Q. acutissima* and 0.399 for *P. koraiensis* litter. This implies that a stronger adhesive force, which can resist the vertical movement of pore water, exists among the elements of the needle litter compared to that of the broadleaf litter. The difference in maximum water retention between litter types was small because it is controlled by rainfall intensity. Unlike the maximum retention values, the minimum retention values exhibited high variability between litter types. There were significant differences in both maximum and minimum retention storages between broadleaf and needle litters ($p < 0.01$).

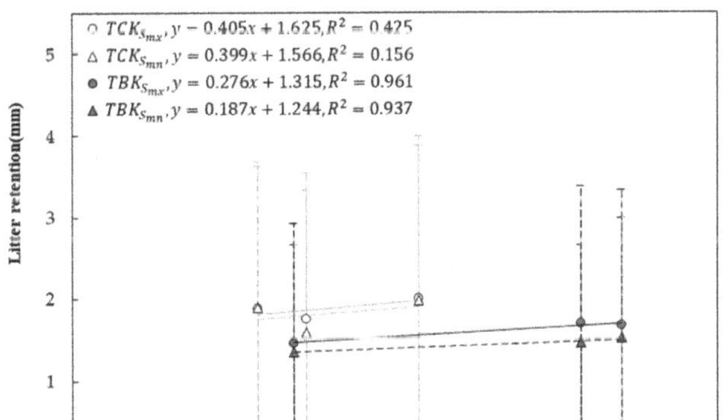

Figure 3. Variation in water retention with litter amount.

As shown in Figure 4, the high variability in the minimum retention capacity within litter type was likely due to natural variability in rainfall intensity and duration. The minimum retention capacity as a percentage of rainfall could be as high as 80% during light rainfalls (<10 mm), and it became no more than 20% in cases where rainfall amount was >30 mm. The minimum retention capacity of broadleaf litter was higher than that of needle litter for light rainfall events (<20 mm) but could be low for heavy rainfalls. Broadleaf litter can easily capture water with large, curve-shaped surfaces when rainfall does not exceed the storage capacity [9,10]. Compared to *Q. acutissima* litter, *P. koraiensis* litter formed a relatively dense barrier layer owing to its smaller physical dimensions and lower porosity. Thus, the strong adhesion and surface tension of water molecules in the *P. koraiensis* layer caused a reduction in water movement through the litter layer for heavy rainfalls [10].

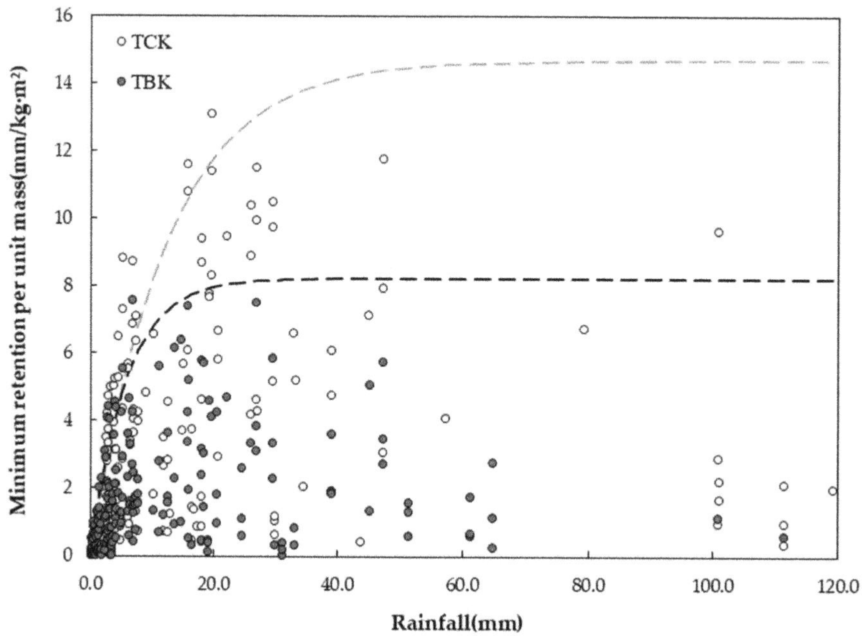

Figure 4. Variation in minimum water retention with rainfall and the corresponding envelope curve.

The percentage of rainfall retained by litter decreased as rainfall amount increased. The logarithmic relationship between retention capacity and rainfall for broadleaf (Equation (5)) and needle (Equation (6)) litter can be expressed as:

$$\mathrm{Log}(S_{mn}) = 3.461 - 0.043\ R_{tot}\ (\mathrm{mm})\ \left(R^2 = 0.600\right) \quad (5)$$

and

$$\mathrm{Log}(S_{mn}) = 3.173 - 0.027\ R_{tot}\ (\mathrm{mm})\ \left(R^2 = 0.407\right), \quad (6)$$

respectively, where $\log(S_{mn})$ is the natural log of the minimum water retention (%) and R_{tot} is the total amount of rainfall (mm). The derived relationships between water retention and rainfall amount were statistically significant ($p < 0.01$) and had residual standard errors of 1.159 and 0.604 for the broadleaf and needle litters, respectively.

For heavy rainfall with a long duration, the retention capacity reaches the potential value regardless of rainfall amount because the initial abstraction by litter is satisfied, and litter can be completely saturated. Figure 4 shows the upper boundary of the minimum retention value under natural conditions. The upper envelope curve (Figure 4) was fitted using the Aston curve [25]. The potential values of minimum retention per unit mass (kg/m^2) can be expressed as follows,

$$S_{mn,p} = 14.705\left(1 - e^{-0.081R_{tot}}\right) \text{ for needle litter} \quad (7)$$

$$S_{mn,p} = 8.226\left(1 - e^{-0.173R_{tot}}\right) \text{ for broadleaf litter} \quad (8)$$

respectively, where $S_{mn,p}$ is the potential (maximum) values of minimum retention per unit mass of litter (mm/kg/m^2).

The minimum retention value increased exponentially with light rainfall but approached asymptotic boundaries when rainfall exceeded 30 mm. This phenomenon was similar to the findings of Sato et al. [9], Li et al. [10], and others [27]. Light rainfall produced greater retention potential in broadleaf litter than in needle litter. The influence of rainfall on retention capacity was not significant in heavy rainfall events, although the capacity slightly increased as rainfall increased. Figure 4 indicates that the upper envelope for minimum retention was 8.226 mm per unit mass for *Q. acutissima* litter and 14.705 mm for *P. koraiensis* litter.

3.2. Litter Moisture Depletion Curve

In this study, a moisture depletion curve was plotted for each litter type. As shown in Figure 5, a simple exponential relationship was observed in the water depletion curve of the leaf litter, implying that average *k* values for broadleaf and needle litters were 0.975 and 0.991, respectively.

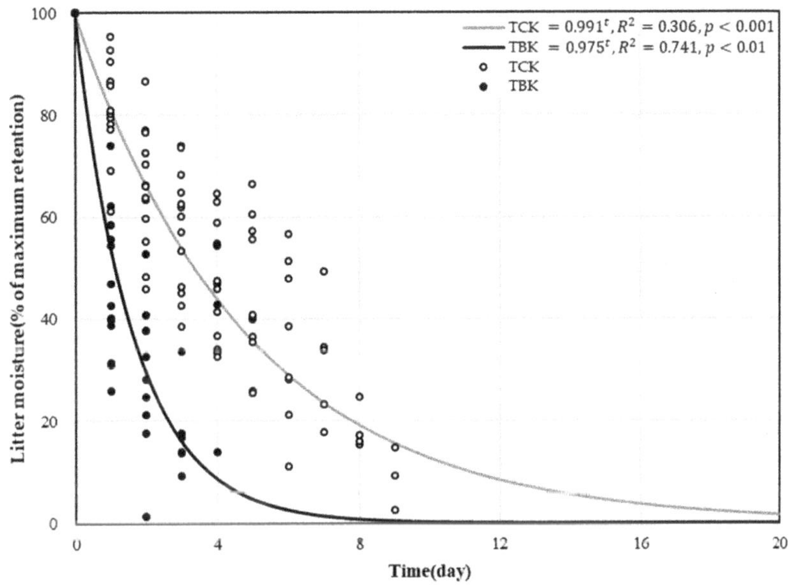

Figure 5. Depletion curve of litter moisture after the cessation of rainfall.

The litter drainage led to the decline in litter moisture content immediately after rain stopped, and as time elapsed, water absorbed by the litter asymptotically approached the minimum retention value of the litter. Figure 5 shows that half of the water retained in the *Q. acutissima* litter was depleted within approximately 1.1 days (27 h) after the cessation of rainfall. Over 3.8 days (90 h), the remaining moisture was slowly extracted from the *Q. acutissima* litter layer, which then reached approximately 10% of its maximum retention value. Water depletion in *P. koraiensis* litter followed a similar pattern to that in the broadleaf litter, depleting 50% and 90% of its maximum retention value within 3.4 days (80 h) and 11.2 days (268 h), respectively.

3.3. Water Balance Analysis for Rainfall Periods

This study revealed an increase in litter weight due to water retention and a decrease due to evaporation loss. The water balance for rainfall periods was analyzed, and the total rainfall was partitioned into litter retention, evaporation loss, and free drainage, as presented in Table 4. Because there were no significant differences among lysimeters within litter type, the average values across three lysimeters within each litter type were calculated. Regardless of rainfall amount, approximately 83.7% of incoming rainfall drained from the *Q. acutissima* litter into the uppermost soil layer, whereas 84.5% of rainfall percolated through the *P. koraiensis* litter.

Table 4. Water balance analysis for rainfall periods.

Lysimeter	Rainfall Amount (mm) [1]	Water Retention (mm)	Evaporation Loss (mm)	Litter Drainage (mm)
TBK1	836.0	103.3 (12.4%) [2]	51.9 (6.2%)	680.8 (81.4%)
TBK2	774.6	84.9 (11.0%)	33.5 (4.3%)	656.2 (84.7%)
TBK3	1023.3	103.2 (10.1%)	52.3 (5.1%)	867.8 (84.8%)
TCK1	1292.5	166.3 (12.9%)	25.5 (2.0%)	1100.7 (85.2%)
TCK2	809.8	120.8 (14.9%)	12.2 (1.5%)	676.8 (83.6%)
TCK2	1066.8	152.4 (14.3%)	14.4 (1.3%)	900.0 (84.4%)

[1] Rainfall events used for lysimeter measurements are included; [2] Parentheses indicate the portion of rainfall amount.

After litter drainage was complete, the remaining water retained in the litter layer was available for evaporation or retention. Evaporation losses during the depletion periods accounted for 5.2% of total rainfall in the broadleaf litter layer and 1.6% in the needle litter layer. Naturally, some retained water was not entirely lost via evaporation and contributed to the antecedent moisture of the litter for subsequent rainfall events. When lysimeter measurements were considered, the portion of total rainfall retained within the litter layer after the completion of litter drainage was 11.1% in the *Q. acutissima* litter and 14.0% in the *P. koraiensis* litter.

This was attributable to the physical properties of the litter, as reported in previous studies [10,28]. A horizontally oriented and densely compacted needle litter layer retains more water than curved and loosely compacted broadleaf litter layer. However, a relatively stronger adhesion and surface tension in needle litter can resist the evaporation of water from the litter surface and consequently enhance litter drainage, compared to broadleaf litter.

As shown in Figure 6, litter drainage varied with rainfall amount. For light rainfall of less than 10 mm, 25% and 32% of rainfall was intercepted or evaporated by the *Q. acutissima* and *P. koraiensis* litters, respectively, resulting in a decrease in the amount of rainfall reaching the mineral soil. The portion of rainfall that drained increased as the rainfall increased. These findings correspond with those of previous studies [10,23].

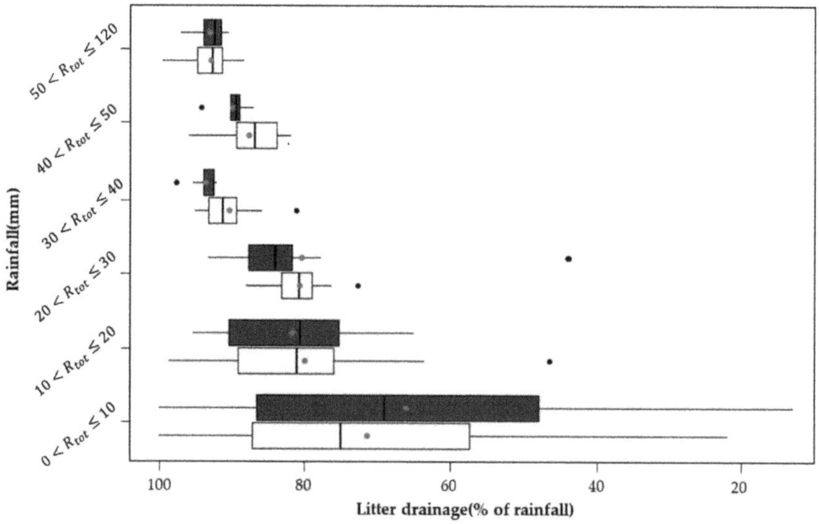

Figure 6. Variation in litter drainage with rainfall.

4. Discussion

The water balance of leaf litter was analyzed based on lysimeter measurements. In this study, total rainfall was partitioned into three components: retention, evaporation loss, and litter drainage. Water balance analysis revealed a proportionately greater amount of litter drainage due to lower water retention of leaf litter. The water retention values observed at the end of a rainfall period fell within the range of interception storage reported by Li et al. [10]. The amount of water retained is dependent on the initial moisture content of the litter prior to rainfall. Leaf litter that is very dry will retain a larger proportion of the rainfall arriving on the litter layer, and that which is close to the saturation point will retain very little additional moisture. The influence of antecedent moisture conditions on water dynamics in the litter layer was not quantitatively considered in this study.

Water retention per unit mass of leaf litter has been reported by previous studies based on laboratory experiments. Sato et al. [9] reported that the maximum water retention levels of *Lithocarpus edulis* (broadleaf litter) and *Cryptomeria japonica* (needle litter) were 1.56 mm and 1.59 mm, respectively, and the minimum retention of *L. edulis* and *C. japonica* were 1.53 mm and 0.81 mm, respectively. Li et al. [10] reported that the average maximum retention of broadleaf litter ranged from 3.96–6.56 mm for *Q. variabilis* and 5.26–6.25 mm for *Q. acutissima*; however, the values for needle litter ranged from 0.35–0.43 mm for *Abies holophylla* and 1.65–2.47 mm for *Pinus strobus*. Putuhena and Cordery [13] reported that the maximum and minimum water retention of pine litter was 1.25 mm and 0.97 mm, respectively. Compared to previous results measured in laboratories, the water retention values derived from the in situ lysimeter experiments in this study are within the same order of magnitude but exhibit wide variation due to the high variability of natural rainfall.

According to the lysimeter measurements in this study, water retention capacity was 10%–15% of the rainfall for storm periods. Other studies also published the water retention of gross rainfall or annual value for different litter types as follows: 8%–12% for mixed oak stands in India [29], 8%–16% for *Quercus petraea*, 12.1% for *Pinus patula*, 8.5% for *Eucalyptus grandis*, and 6.6% for *Acacia mearnsii* in South Africa [30]. Gerrits et al. [16] found litter interception to be as high as 22% in a beech forest and 18% in needle leaf litter in a cedar forest, while Helvey and Patric [11] found litter interception to be 15%–34% in a poplar stand in the USA. Comparison of the current study with these past studies is limited due to differences in climate conditions, tree species, litter mass, and methods of measurement.

Litter drainage was indirectly calculated as the difference between the amount of rainfall and the amount of water retained in the leaf litter. This concept is valid only if the evaporation loss is neglected for shorter durations. Even if this condition is not satisfied, it leads to a small discrepancy in almost all instances because the evaporation rate is too small to have a substantial impact on the drainage calculation for a very short period [24,31].

The gravimetric method is the most widely used technique for the moisture determination of litter [6,7], where litter samples are transported to the laboratory for experimental measurements. Although this is a very simple and accurate method, it disturbs the samples and requires 12–24 h to dry in an oven to weigh the litter sample under dry conditions. Therefore, the gravimetric method is not applicable to continuously measure the water partitioning in field conditions. A field lysimeter is the best solution to quantify litter interception and retention under natural conditions [12]. Because the experiment is executed in the original environment, litter samples are less disturbed compared to the gravimetric method.

Naturally, some biological and pathological changes could have occurred during the measurement period [32,33]. Decomposition can modify the physical characteristics of leaf litter, resulting in a change in its weight [34], which would influence the hydrologic processes in the litter by reducing its water retention capacity and resisting water penetration into the layer. In this study, litter decomposition was not considered because it was the intent of the experiment to obtain moisture measurements from relatively undisturbed litter samples. Litter-decomposing fungi are also known to have higher water resistance (hydrophobicity) due to fungal mycelia [33], although they have rarely been observed in litter samples.

5. Conclusions

The water retention capacity of leaf litter was estimated from lysimeter measurements taken over a duration of 30 min. Regardless of rainfall characteristics, needle litter had a greater capacity to retain rainfall than broadleaf litter did. During the experiment, approximately 1.66 ± 1.72 mm per rainfall event was stored per unit mass (kg/m^2) of broadleaf litter after rainfall, while 2.72 ± 2.82 mm of rainfall reaching needle litter was retained per unit mass (kg/m^2). Approximately 83.7% of incident rainfall drained into the uppermost soil layer below the *Q. acutissima* litter, whereas 84.5% of rainfall percolated through the *P. koraiensis* litter. The depletion curves indicated that the water retained in the *Q. acutissima* litter was more easily lost via evaporation than the *P. koraiensis* litter.

The duration and intensity of rainfall are known to affect litter water retention capacity. Most previous studies have been conducted under limited conditions, where the short duration and/or high intensity of artificial rainfall might lead to an underestimation of the retention capacity of litter, likely due to the litter accumulation not being thoroughly wetted prior to the simulated rainfall. On the contrary, as this study was conducted under natural rainfall conditions, it resulted in a higher variation in retention capacity, likely due to the variation in rainfall. Practically, the upper envelope curves derived from lysimetry experiments can provide the upper boundary of the retention capacity, reflecting the effect of rainfall characteristics under natural conditions.

This lysimetry experiment has some limitations. First, evaporation from the litter layer has not been measured but can be derived in this study. It would require intensive field experiments where hydrological components such as rainfall, evaporation, percolation, and drainage are measured simultaneously. Second, litter decomposition as a result of fungal and soil faunal activity also occurs over time. Changes in the physical traits of leaf litter can affect hydrological function; however, this phenomenon was not considered in the current study. Therefore, long-term measurements are required to accurately understand the hydrological role of leaf litter, considering spatial and temporal variations in production, accumulation, and decomposition.

This study highlighted the litter retention and evaporation processes. However, relatively little is known on water dynamics occurring on the forest floor. Biomat flow likely follows preferential flow paths through the litter layer. Further research can be directed to the clear understanding of how biomat flow influences the residence time of water within the layer, time to litter saturation, and, in consequence, the water retention capacity of litter.

Author Contributions: Conceptualization, T.K. and S.I.; methodology, T.K. and S.I.; software, J.K. and J.L.; validation, T.K., J.K. and J.L.; investigation, J.K., J.P. and J.L.; resources, H.S.K.; data curation, T.K. and J.P.; writing—original draft preparation, T.K.; writing—review and editing, S.I. and H.S.K.; visualization, J.K. and J.L.; supervision, S.I.; project administration, S.I.; funding acquisition, S.I. All authors have read and agreed to the published version of the manuscript.

Funding: This study was carried out with the support of 'R&D Program for Forest Science Technology (Project No. 2021343C10-2323-CD01)' provided by Korea Forest Service (Korea Forestry Promotion Institute).

Data Availability Statement: Not applicable.

Acknowledgments: We acknowledge the Forest Ecophysiology Laboratory of Seoul National University, Korea, for providing the experimental data, and also thankful for the support of the Mt. Taehwa Seoul National University Forest.

Conflicts of Interest: The authors declare no conflict of interest.

References

1. Korea Forest Service. *Statistical Yearbook of Forestry 2021*; Korea Forest Service: Daejeon, Republic of Korea, 2021; pp. 40–41.
2. Dunkerley, D. Percolation through leaf litter: What happens during rainfall events of varying intensity? *J. Hydrol.* **2015**, *525*, 737–746. [CrossRef]
3. Gomyo, M.; Kuraji, K. Effect of the litter layer on runoff and evapotranspiration using the paired watershed method. *J. For. Res.* **2016**, *21*, 306–313. [CrossRef]
4. Gavazzi, M.J.; Sun, G.; McNulty, S.G.; Treasure, E.A.; Wightman, M.G. Canopy rainfall interception measured over ten years in a coastal plain loblolly pine (*Pinus taeda* L.) plantation. *Trans. ASABE* **2016**, *59*, 601–610.
5. Guevara-Escobar, A.; González-Sosa, E.; Véliz-Chávez, C.; Ventura-Ramos, E.; Ramos-Salinas, M. Rainfall interception and distribution patterns of gross precipitation around an isolated Ficus benjamina tree in an urban area. *J. Hydrol.* **2007**, *333*, 532–541. [CrossRef]
6. Bernard, J.M. Forest floor moisture capacity of the New Jersey pine barrens. *J. Ecol.* **1963**, *44*, 574–576. [CrossRef]
7. Reynolds, J.F.; Knight, D.H. The magnitude of snowmelt and rainfall interception by litter in lodgepole pine and spruce-fir forests in Wyoming. *Northwest Sci.* **1973**, *47*, 50–60.
8. Rosalem, L.M.; Wendland, E.C.; Anache, J.A. Understanding the water dynamics on a tropical forest litter using a new device for interception measurement. *Ecohydrology* **2019**, *12*, e2058. [CrossRef]
9. Sato, Y.; Kumagai, T.O.; Kume, A.; Otsuki, K.; Ogawa, S. Experimental analysis of moisture dynamics of litter layers—The effects of rainfall conditions and leaf shapes. *Hydrol. Process.* **2004**, *18*, 3007–3018. [CrossRef]
10. Li, Q.; Lee, Y.E.; Im, S. Characterizing the interception capacity of floor litter with rainfall simulation experiments. *Water* **2020**, *12*, 3145. [CrossRef]
11. Helvey, J.; Patric, J. Canopy and litter interception of rainfall by hardwoods of eastern United States. *Water Resour. Res.* **1965**, *1*, 193–206. [CrossRef]
12. Gerrits, A.M.J.; Savenije, H.H.G. Forest floor interception. In *Forest Hydrology and Biogeochemistry*; Ecological Studies 216; Levia, D.F., Carlyle-Moses, D., Tanaka, T., Eds.; Springer: New York, NY, USA, 2011; p. 423.407.
13. Putuhena, W.M.; Cordery, I. Estimation of interception capacity of the forest floor. *J. Hydrol.* **1996**, *180*, 283–299. [CrossRef]
14. Li, X.; Niu, J.; Xie, B. Study on hydrological functions of litter layers in North China. *PLoS ONE* **2013**, *8*, e70328. [CrossRef]
15. Schaap, M.; Bouten, W. Forest floor evaporation in a dense Douglas fir stand. *J. Hydrol.* **1997**, *193*, 97–113. [CrossRef]
16. Gerrits, A.; Savenije, H.; Hoffmann, L.; Pfister, L. New technique to measure forest floor interception–an application in a beech forest in Luxembourg. *Hydrol. Earth Syst. Sci.* **2007**, *11*, 695–701. [CrossRef]
17. Beeson, R.C. Weighing lysimeter systems for quantifying water use and studies of controlled water stress for crops grown in low bulk density substrates. *Agric. Water Manag.* **2011**, *98*, 967–976. [CrossRef]
18. Korea Meteorological Administration. Available online: https://data.kma.go.kr (accessed on 13 December 2022).
19. Im, S.; Lee, W.S.; Kim, H.; Seoul National University Forests. *Developing a Network of Long-Term Research Field Stations to Monitor Environmental Changes and Ecosystem Responses in Asian Forests*, 1st ed.; Kamata, N., Kuraji, K., Owari, T., Guan, B.T., Eds.; The University of Tokyo Forests Press: Tokyo, Japan, 2019; Volume 1, pp. 39–50.
20. Katabuchi, M. LeafArea: An R package for rapid digital image analysis of leafarea. *Ecol. Res.* **2015**, *30*, 1073–1077. [CrossRef]
21. Kang, M.; Kim, J.; Lee, S.-H.; Kim, J.; Chun, J.-H.; Cho, S. Changes and improvements of the standardized eddy covariance data processing in KoFlux. *Korean J. Agric. For. Meteorol.* **2018**, *20*, 5–17.
22. Yang, H.; Choi, H.T.; Lim, H. Applicability assessment of estimation methods for baseflow recession constants in small forest catchments. *Water* **2018**, *10*, 1074. [CrossRef]
23. Semago, W.T.; Nash, A.J. Interception of precipitation by a hardwood forest floor in the Missouri Ozarks. *Univ. Mo. Agr. Exp. Stan. Res. Bull.* **1962**, *796*, 1–31.
24. Gash, J.H.; Lloyd, C.; Lachaud, G. Estimating sparse forest rainfall interception with an analytical model. *J. Hydrol.* **1995**, *170*, 79–86. [CrossRef]
25. R Core Team. *R: A Language and Environment for Statistical Computing*; R Foundation for Statistical Computing: Vienna, Austria, 2022; Available online: https://www.R-project.org/ (accessed on 11 November 2022).
26. Gerke, K.; Sidle, R.; Mallants, D.; Valisyev, R.; Karsanine, M.; Skvortsova, E.; Korost, D. Biomat Flow: From Field Experiments with Dyes to Pore-scale Modelling of Transport Properties and Flow Models. In Proceedings of the AGU Fall Meeting, San Francisco, CA, USA, 9–13 December 2013; p. 274313.
27. Aston, A.R. Rainfall interception by eight small trees. *J. Hydrol.* **1979**, *42*, 383–396. [CrossRef]
28. Sato, Y.; Otsuki, K.; Ogawa, S. Estimation of the Litter interception Loss in the Evergreen Forest. *J. Jpn. Soc. Hydrol. Water Resour.* **2003**, *16*, 640–651. [CrossRef]
29. Pathak, P.C.; Pandey, A.N.; Singh, J.S. Apportion of rainfall in central Himalayan forests (India). *J. Hydrol.* **1985**, *76*, 319–332. [CrossRef]
30. Bulcock, H.H.; Jewitt, G.P.W. Field data collection and analysis of canopy and litter interception in commercial forest plantation in the KwaZulu-Natal Midlands, South Africa. *Hydrol. Earth Syst. Sci.* **2012**, *16*, 3717–3728. [CrossRef]
31. Pereira, F.; Gash, J.; David, J.; Valente, F. Evaporation of intercepted rainfall from isolated evergreen oak trees: Do the crowns behave as wet bulbs? *Agric. For. Meteorol.* **2009**, *149*, 667–679. [CrossRef]

32. Gosz, J.R.; Likens, G.E.; Bormann, F.H. Nutrient release from decomposing leaf and branch litter in the Hubbard Brook Forest, New Hampshire. *Ecol. Monogr.* **1973**, *43*, 173–191. [CrossRef]
33. Cha, J.Y.; Im, S.; Lee, S.Y.; Ohga, S. Diversity of Fungal Species Isolated from Litter-mycelial Mats in the Litter Layer of a Korean Deciduous Forest. *J. Fac. Agric. Kyushu Univ.* **2011**, *56*, 237–241. [CrossRef]
34. Matthews, S. The water vapour conductance of Eucalyptus litter layers. *Agric. For. Meteorol.* **2005**, *135*, 73–81. [CrossRef]

forests

MDPI

Article

Variation in Hydraulic Properties of Forest Soils in Temperate Climate Zones

Victoria Virano-Riquelme [1,*], Karl-Heinz Feger [1] and Stefan Julich [1,2]

1 Institute of Soil Science and Site Ecology, Faculty of Environmental Sciences, Technische Universität Dresden, 01737 Tharandt, Germany
2 Faculty of Landscape Management and Nature Conservation, Eberswalde University for Sustainable Development, 16225 Eberswalde, Germany
* Correspondence: victoria.virano@tu-dresden.de

Abstract: The structure of forests in temperate climates has been changing to ensure the resilience of trees. This change affects the local water balance. Knowledge of soil hydraulic properties (SHP) is essential to assess the water cycle in ecosystems. There is little knowledge about the impact of tree species on SHP and the water balance. Based on a compilation of 539 related studies we aimed at identifying the effects of tree species and age on SHP in temperate climates. However, most studies concentrated on soil biogeochemical properties, whereas only 256 studies focused on SHP. The literature presents no standard methods for assessing SHP and there is no knowledge of their variations in forests. We present a systematic overview of the current state of knowledge on variations in SHP based on forest type in temperate climates. We identify the gaps and weaknesses in the literature and the difficulties of evaluating the reviewed studies. More studies following standardised methodologies are needed to create a robust database for each forest type and soil texture. It would improve the assessment of the forest water balance through calibrated plot/site-scale process models. Such a database does not yet exist, but it would greatly improve the management and development of future forest ecosystems.

Keywords: soil hydraulic properties; saturated hydraulic conductivity; bulk density; temperate forests

Citation: Virano-Riquelme, V.; Feger, K.-H.; Julich, S. Variation in Hydraulic Properties of Forest Soils in Temperate Climate Zones. *Forests* **2022**, *13*, 1850. https://doi.org/10.3390/f13111850

Received: 1 September 2022
Accepted: 29 October 2022
Published: 4 November 2022

1. Introduction

In recent decades, forest ecosystems have shown a decline in productivity, exhibiting tree species with negative physiological processes undergoing high and abrupt mortality [1]. According to Allen et al. [2], the increasing tree death rate is caused by climate change with concomitant drought and heat as the main factors compromising the prosperity of the forests. The severity of droughts and the high stress in temperate forest ecosystems create conditions favouring pests and pathogens [3], creating an environment of high risk, especially in areas with low water availability [4].

In addition, future scenarios foresee increasing drought frequency, intensity, and duration as well as additional impacts such as wildfires and severe storms, leading to elevated tree mortality rates [1,2,4]. Long-term droughts, coupled with rapid tree decline, represent considerable challenges to the management and policy-making communities. In response to changing conditions, forests of the temperate zone are transforming, being managed to withstand these adverse constraints.

In the past, monocultures of needle-leaved trees (i.e., of Norway spruce—*Picea abies*) were widely used in Central Europe as a promising economic resource. However, in the context of climate change, spruce monocultural stands have weakened and are rapidly dying out. Therefore, current forest management considers the use of native broad-leaved tree species as mitigation to endure drought stress and preserve the vitality of trees. As reported by Pretzsch et al. [5], the process of transformation and adaptation from non-natural and widespread coniferous monocultures towards mixed forests including deciduous tree species will be of great ecological interest considering different rates and species.

This strategy plays a key role in overcoming the current climatic conditions and in the development of future forest ecosystems. The introduction of broad-leaved and hardwood species in needle-leaved stands should be performed carefully as it changes several soil properties (i.e., humus, litter deposition), likely affecting the soil hydraulic properties [6–8]. Therefore, the assessment of possible consequences on the water balance is essential for efficient forest productivity [9].

The evaluation of the water balance of forest stands in terms of different species can be achieved by using calibrated plot/site-scale process models such as LWF-Brook90 [10,11], which depend on soil hydraulic properties (SHP). SHP are essential to assess the water cycle in ecosystems. The evaluation of these parameters allows the description of several hydrological processes within the soil profile, such as water infiltration, surface runoff and water storage, the availability of water for root uptake, percolation, and the generation of surface runoff as a function of the soil management practices [12,13]. Furthermore, the soil hydraulic conductivity is particularly relevant since it allows the estimation of water movement towards the root zone and the recharge of underground aquifers, as well as the transport of minerals and nutrients into the deeper layers of the soil profile [14,15].

Some studies compare SHP between forested and non-forested sites, but only a few focus on the different management practices within forest ecosystems. In contrast to other land uses, studies reveal that forest soils usually show higher hydraulic conductivity [14,16,17], improved water retention [17], decreased runoff and increased soil moisture [17,18], with often lower density [19]. Moreover, the infiltration capacity and soil water retention in forested soils control the formation of surface runoff acting as a natural flood regulator [16,20,21]. However, most studies addressing the variation in soil properties as a result of forest management focus on aboveground characteristics including biomass [22–24], vegetation traits [25–27], and soil biochemical properties [28–32].

Articles assessing SHP in temperate forests are extremely scarce and present a great variety in findings. Differing values for soil bulk density, hydraulic conductivity, and the accumulation of soil organic carbon in relation to forest type and tree age have been observed [8,14,21,33–35]. In addition, the relief of the sites has a noticeable implication for the assessment of the SHP. On steep slopes, needle-leaved forests have revealed enhanced hydraulic conductivity in contrast to smoother surfaces [36]. However, under similar conditions, broad-leaved forests create conditions less prone to surface runoff generation compared to needle-leaved forests [34]. SHP may vary according to a vast combination of soil texture, tree species, root distribution, and the inherent complexity of the interacting site conditions, likely affecting the water dynamics in a catchment [6,16,33,37,38].

The objective of this study is to review the available literature on information for the SHP of forested sites in temperate zones. Our aim is to evaluate and assess the impact of tree species and age on soil hydraulic properties based on the reported values. We expect to find differences in SHP in terms of broad-leaved and needle-leaved tree species as reported by Archer et al. [14], Julich et al. [16], and Wahren et al. [17]. We hypothesize that in needle-leaved stands (at least for comparable soil geological settings), the overall hydraulic conductivity is lower, together with higher bulk density, compared to broad-leaved stands.

2. Materials and Methods

2.1. Nature and Classification of the Studies

For the review, a systematic search was conducted based on the PRISMA Statement [39]. The academic databases Scopus and Web of Science were used to search for information, with the last search performed on 24 March 2022. This review targeted peer-reviewed publications in English until the year 2021. Publications in other languages were excluded as a thorough analysis requires an adequate understanding of the study site descriptions.

Figure 1 shows the search steps we followed to create our database. Firstly, we performed a broad search ("Search 1", Figure 1), considering the terms "soil hydraulic properties" and "forests", obtaining a total of 462 articles. The results were classified according to "temperate forests" and "tropical forests (+others)", with the latter group presenting the highest number of studies regarding soil properties. In this step the articles that were not related to the study of soil properties (i.e., soil fauna, soil fungi, root development), that were not specifically carried out in forested areas, that did not specify the type of climate, or articles where it was not clear, were excluded. Then, the focus of our study was specifically directed to temperate forests as our first research revealed that the studies conducted on these forests were deficient.

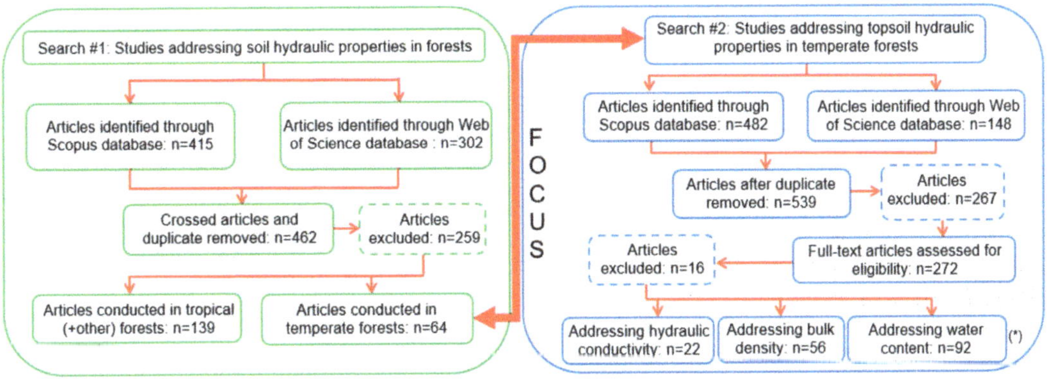

Figure 1. The flowchart on the left displays the steps performed in a first broad search, which shows the current state of the literature in relation to the variation in soil hydraulic properties in different climatic zones. The flow chart on the right illustrates the steps followed to filter the studies related to soil hydraulic properties in temperate forests (the focus) and how the final database based on current literature was compiled. The number of papers found (n) and the studies that were discarded at each step along the literature review are included. (*) Some articles are repeated between the subdivisions.

After the first literature review, we carried out a complementary search using the same academic databases to aim for studies addressing SHP in temperate forests, including the words "hydraulic conductivity", "water retention", and "water content" to make certain of covering relevant soil hydraulic properties ("Search 2", Figure 1). For each of the beforementioned words, different combinations of keywords such as "temperate forest", "temperate climate", "coniferous", "deciduous", "stand", "species", "mixed forest(s)", and "forest conversion" were used.

Although we collected a total of 539 articles up to 2021, several of these studies referred to the term "soil properties" in a general context. Therefore, 267 studies were discarded before the assessment as they focused specifically on soil biogeochemical properties, traits, and root water uptake, rather than on soil hydraulic properties. Moreover, some of those studies were discarded since they concerned other land uses. After neglecting the studies that were not within the scope of this review, the compilation consisted of 272 studies. Since our analysis aimed at soil hydraulic properties obtained between 0 and 30 cm depth, 16 additional studies not reporting depth of measurement or performed at greater depths were excluded from the database. In addition, studies that used units deviating from the standard were discarded to avoid confusion and misinterpretation over converted values. Once all these steps were completed, the final database was subdivided into those articles containing studies related to soil bulk density, hydraulic conductivity, and soil water content.

To assess the effect of the different forest ecosystems on the soil properties, we defined a classification for this study according to needle-leaved, broad-leaved, and mixed forests. Within each forest group, we attempted to classify the studies in terms of site parameters such as soil texture, organic matter concentration, tree age, and tree mixing rate (in the case of mixed forests). Still, due to the limited information, we were able to classify the articles only in terms of soil texture and tree age. The classification of soil texture was performed based on fine for clay-dominated soils, medium for mostly loamy soils, and coarse for mostly sandy soils. Given the range in ages of the trees in the reviewed studies, and for the purpose of our study, we decided to group the values collected according to the following criteria: young forest for trees < 50 years old, mature stand for those containing trees of 50–100 years old, and old forest for trees older than 100 years.

When evaluating the articles that assess soil water content, we observed that most of the studies were concerned with variations in water content over a given period of time. Such results are of interest for assessing temporal fluctuations in soil moisture content; however, they are relative to seasonal and climatic factors. Therefore, we focused on measurements of water holding capacity, as this property is relevant to assess the effect of forest type. From the 92 articles collected that addressed water content, only 15 could be used for our research.

In the case of saturated hydraulic conductivity, 3 out of 22 articles performed their measurements under laboratory conditions, which were excluded since such measurements likely neglect the site characteristics. Regarding the experiments performed in the field, one article did not specify the measurement device, whereas 10 studies used permeameters and nine used infiltrometers. Although both methods work under different techniques, we considered all the 20 articles in our analysis. From these studies, we compiled a total of 145 values for analysis. Regarding soil bulk density, from a total of 56 studies, we collected 145 values to be further assessed.

2.2. Statistical Analysis

As most of the results came from diverse studies and were performed under specific conditions, the collected data could not be assumed to be uniformly distributed. Even when some of the reviewed papers presented comparable site parameters, they frequently differed in other aspects, such as measurement approaches, devices, and sampling designs. Therefore, the Kruskal–Wallis test was used to assess whether the data presented significant differences between the forest types under the assumption that all data points were obtained independently under the same experimental conditions.

3. Results

The compilation of studies was screened for values of SHP in terms of water holding capacity, saturated hydraulic conductivity, and soil porosity. However, the available information on the aforementioned parameters was limited. Thus, it was not possible to add most of these hydraulic parameters to the analysis. Only for soil hydraulic conductivity could sufficient information be extracted for further assessment. Since soil bulk density was present in almost all the reviewed articles, it was included in the analysis, as it provides relevant information regarding soil structure.

In order to appropriately compare each soil hydraulic property in terms of comparable site parameters, we classified the data according to forest type (i.e., tree species and age of the trees) and the soil texture where the parameters were measured. We found that not all the collected articles reported these relevant parameters, hence several studies were unsuitable for further analysis. As a result, we created two datasets containing all values found in the pool of collected articles. Database A contains all results obtained for a soil depth of 0–30 cm (independent of the site description) and database B contains the filtered information of database A, comprising only those articles reporting forest type and soil texture. Furthermore, all values were converted to the same units for further comparison.

Table 1 presents the number of studies included in databases A and B, the latter containing the results that are suitable to assess the collected bulk density and hydraulic conductivity values. To visualize the information, the distribution was performed by forest type, and each field was additionally sorted by soil texture and tree age. It can be seen that most studies comprised information regarding bulk density and fewer studies measured hydraulic conductivity. It is also evident that the majority of the studies were conducted in needle-leaved and broad-leaved forests.

Table 1. Classification of values found in the reviewed literature filtered by forest type as main field and further divisions by soil texture and tree age, which were defined by this study. The left side of the table (a) corresponds to the hydraulic conductivity (K) and right side (b) refers to the soil bulk density (BD). All values are measurements between 0 and 30 cm soil depth.

					Needle-Leaved Forest				
(a) K	**Young**	**Mature**	**Old**	**Total**	**(b) BD**	**Young**	**Mature**	**Old**	**Total**
Coarse	10	1	0	11	Coarse	4	11	2	17
Medium	0	1	0	1	Medium	2	1	2	5
Fine	0	0	0	0	Fine	0	0	0	0
Total	10	2	0	12	Total	6	12	4	22
					Broad-Leaved Forest				
Coarse	0	6	9	15	Coarse	2	12	2	16
Medium	1	4	0	5	Medium	3	1	12	16
Fine	0	0	0	0	Fine	11	0	0	11
Total	1	10	9	20	Total	16	13	14	43
					Mixed Forest				
Coarse	0	12	2	14	Coarse	0	14	0	14
Medium	0	0	0	0	Medium	0	0	0	0
Fine	0	0	0	0	Fine	0	0	0	0
Total	0	12	2	14	Total	0	14	0	14

3.1. Soil Water Retention

Table 2 summarises the information regarding soil water retention characteristics found in the screened articles. From 15 studies a total of 30 values were found, yet not all studies referred to the same measurements of soil water holding capacity, resulting in limited information for the individual soil water types. Field capacity (FC) ranged from 0.12 m^3 m^{-3} for sandy soils up to 0.44 m^3 m^{-3} for silty clay soils. Regarding plant available water (PAW), only a little amount of information was available with values ranging from 0.02 to 0.33 m^3 m^{-3}. For water content at the permanent wilting point (PWP), values between 0.05 and 0.20 m^3 m^{-3} were reported.

3.2. Soil Bulk Density

Figure 2 compares the values for soil bulk density of the topsoil by using dataset A, which contains all values found in the literature screening, and dataset B containing those that were filtered regarding the available information on tree age and soil texture. Dataset A comprised a total of 163 bulk density values, whereas dataset B was reduced to 79, excluding about 52% of the information. For dataset A, the highest and lowest means were observed in the broad-leaved and needle-leaved forests, respectively, whereas dataset B showed comparable mean values between forest types (Figure 2).

Table 2. Detailed information on soil water holding capacity from the reviewed literature. In the division Forest Type, the abbreviations Nee, Broa, and Mix stand for needle-leaved, broad-leaved, and mixed forest type, respectively.

Author References	FC [1]	PAW [2]	PWP [3]	Forest Type	Soil Texture Class	Tree Age
		(m^3 m^{-3})				
Bittner et al. [40]	-	-	0.06	Broa	Silty clay	Old
	-	-	0.06	Broa	Silty clay	Old
	-	-	0.09	Broa	Silty clay	Old
Grant et al. [41]	0.28	0.09	-	Broa	Silty	Mature
Metzger et al. [42]	0.44	-	-	Broa	Fine textured	-
	0.42	-	-	Broa	Fine textured	-
Baldocchi et al. [43]	0.16	-	-	Nee	Sandy	Old
Archer et al. [14]	-	0.290	-	Nee	Sandy	Young
	-	0.240	-	Nee	Sandy	Mature
	-	0.220	-	Nee	Sandy	Old
	-	0.180	-	Nee	Sandy	Old
Dusek and Vogel [44]	-	-	0.20	Nee	Sandy	-
	-	-	0.20	Nee	Sandy	-
Tor-Ngern et al. [9]	0.15	-	-	Nee	Sandy	Young
	0.15	-	0.05	Nee	Sandy	Mature
Almahayni and Houska [45]	0.12	0.02	-	Nee	Sandy	-
Duarte et al. [46]	0.30	-	0.14	Nee	-	Old
Kirchen et al. [47]	0.40	-	0.18	Broa	Fine textured	Young
	0.40	-	0.28	Broa	Fine textured	Young
	0.31	-	0.15	Broa	Fine textured	Young
	0.34	-	0.17	Broa	Fine textured	Young
Leuschner [48]	-	0.32	0.13	Broa	Sandy	-
	-	0.14	0.13	Nee	-	Young
Curiel Yuste et al. [49]	0.12	-	-	Mix	Sandy	-
Julich et al. [16]	0.43	0.29	0.14	Broa	Silty	Young
	0.40	0.29	0.11	Broa	Silty	Old
	0.40	0.30	0.10	Nee	Silty	Mature
Wahren et al. [17]	-	0.22	-	Mix	Sandy loam	Mature
Schwärzel et al. [50]	-	0.27	-	Broa	Silty	Mature
	-	0.33	-	Nee	Silty	Old

[1] Soil water at field capacity. [2] Soil plant available water. [3] Soil water at permanent wilting point.

Table 3 shows the number of studies and descriptive statistics of bulk density for datasets A and B. The p-values suggest that in dataset A there is an effect of forest type on the soil bulk density (p-values < 0.05); however, the p-values in dataset B show no effect of forest type on this soil property. The comparison with mixed forest is not adequate in both cases since the number of studies on this forest type is not sufficient in contrast to the others (Table 3).

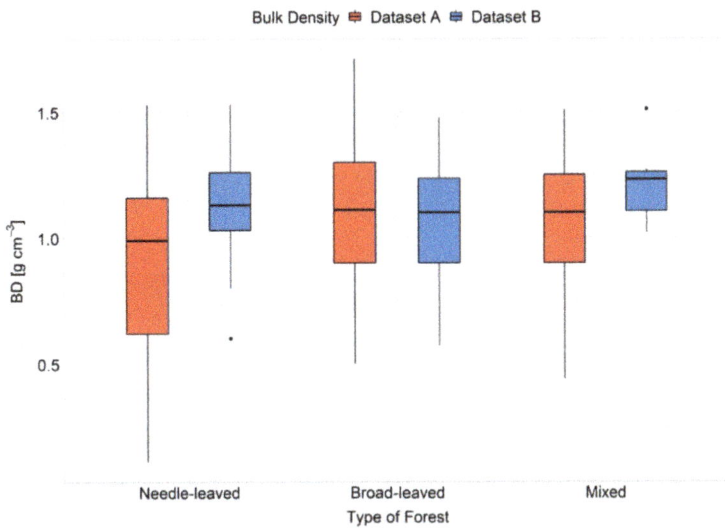

Figure 2. Descriptive statistics of top mineral soil bulk density by forest type considering the two data sets: dataset A containing all values found and dataset B filtered values based on available information on tree age and the soil particle size. All values refer to the topsoil (0–30 cm depth). Black dots represent the outliers.

Table 3. Summary of statistics for soil bulk density and saturated hydraulic conductivity sorted according to forest type. Dataset A contains all results found in the screened articles; dataset B considers only those values reported with additional information on soil particle size and tree age. All values refer to the topsoil (0–30 cm depth). G Mean refers to the geometric mean. (*) indicates the group of forest with significantly different BD values.

			Bulk Density (g cm^{-3})			
	Forest Type	*n*	**Mean**	**Min**	**Max**	***p*-Value**
A	Needle-leaved	45	0.89	0.11	1.53	0.0098 (*)
	Broad-leaved	83	1.09	0.50	1.71	0.2396
	Mixed	17	1.02	0.44	1.51	0.4682
B	Needle-leaved	19	1.12	0.60	1.53	0.5807
	Broad-leaved	40	1.10	0.57	1.48	0.3541
	Mixed	11	1.20	1.02	1.51	0.1241
			Hydraulic Conductivity (cm d^{-1})			
	Forest Type	*n*	**G. Mean**	**Min**	**Max**	***p*-Value**
A	Needle-leaved	20	132	6	1200	0.9806
	Broad-leaved	23	160	12	1400	0.4261
	Mixed	16	224	27	1201	0.3460
B	Needle-leaved	12	116	17	538	0.9379
	Broad-leaved	20	127	12	1096	0.1498
	Mixed	14	239	27	1201	0.1415

Figure 3 and Table 4 provide the descriptive statistics of dataset B regarding the bulk density of the topsoil. Most of the reviewed studies with information regarding tree age were conducted on broad-leaved forests, whereas for mixed forests only a few studies were carried out. Values for bulk density do not follow any trend (in terms of forest type or tree age) and reveal that mature and young stands are the dominant stages in the reviewed studies.

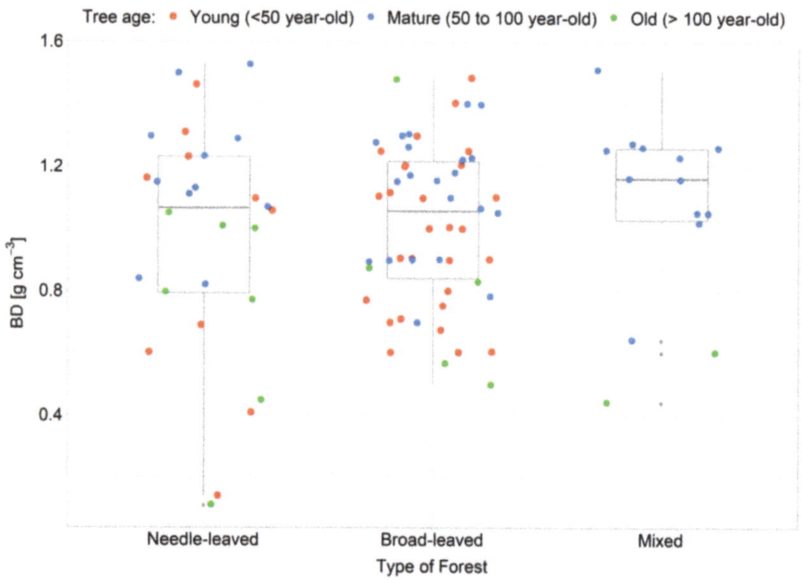

Figure 3. Descriptive statistics of the top mineral soil bulk density in relation to the tree age and forest type for dataset B. Scattered points represent the individual values classified by the different forest stages. Values correspond to the 0–30 cm soil depth. Grey points indicate the outliers.

Table 4. Summary of reported values on topsoil bulk density and hydraulic conductivity for the different forest types, data are subdivided based on the information on tree age: young (Y) less than 50 years, mature (M) 50 to 100 years, and old (O) more than 100 years.

Bulk Density (g cm^{-3})	**Needle-Leaved**			**Broad-Leaved**			**Mixed**		
Tree age	Y	M	O	Y	M	O	Y	M	O
mean values	0.92	1.18	0.74	0.98	1.11	0.85	-	1.16	0.52
n	10	11	7	28	21	5	-	12	2
max	1.46	1.53	1.05	1.48	1.40	1.48	-	1.51	0.60
min	0.14	0.82	0.11	0.60	0.70	0.50	-	0.64	0.44
Hydraulic Conductivity (cm d^{-1})	**Needle-Leaved**			**Broad-Leaved**			**Mixed**		
Tree age	Y	M	O	Y	M	O	Y	M	O
mean values	140	409	306	692	171	233	-	483	352
n	13	2	2	2	9	10	-	12	2
max	538	468	538	1096	312	751	-	1201	418
min	6	349	75	288	35	12	-	27	286

The topsoil bulk density for needle-leaved and broad-leaved forests presented the highest values in the mature stage; however, comparison with mixed forests was not possible since values for the young stage were not found. Most values for the old stages under needle-leaved and broad-leaved forests were lower in contrast to the other stages.

Figure 4 and Table 5 describe the reported values for bulk density in relation to soil texture and forest type. Most of the screened studies with information on soil texture were conducted in broad-leaved forests, whereas only a few were conducted in mixed stands. The distribution of values reveals no apparent relationship between bulk density and soil texture for the different forest types. Figure 4 also shows that very few reported values correspond to the Oh-horizon in contrast to those measured in the first 30 cm of soil.

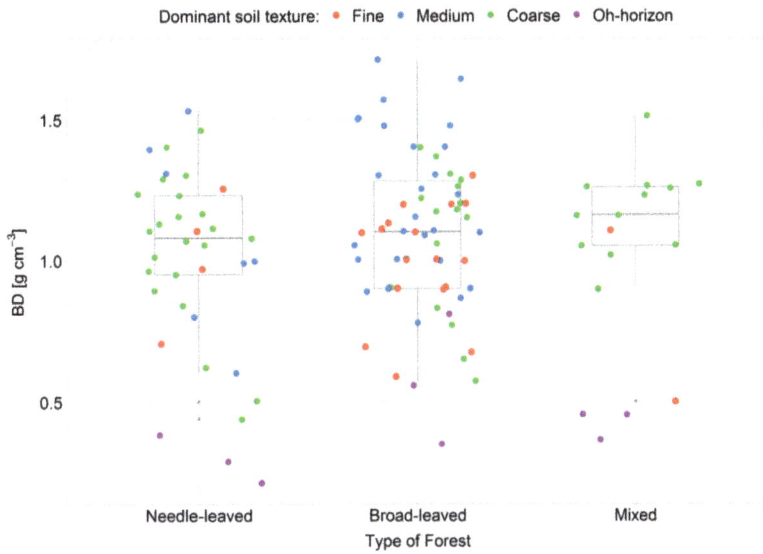

Figure 4. Descriptive statistics of the top mineral soil bulk density for dataset B in terms of forest type and soil texture: fine (clay-dominated), medium (mostly loamy), and coarse (mostly sandy). Scattered points represent the individual values between 0 and 30 cm soil depth, including the Oh-horizon. Grey points indicate the outliers.

Table 5. Topsoil bulk density and hydraulic conductivity values in terms of forest type, filtered by soil texture. Letters C, M, and F stand for soils with coarse (mostly sandy), medium (mostly loamy), and fine (clay-dominated) soil texture, respectively, between 0 and 30 cm soil depth. In the case of bulk density, the letter Oh represents the values for the Oh-horizon.

Bulk Density (g cm^{-3})	**Needle-Leaved**				**Broad-Leaved**				**Mixed**			
Soil texture	C	M	F	Oh	C	M	F	Oh	C	M	F	Oh
mean values	1.04	1.09	1.01	0.29	1.08	1.20	1.00	0.57	1.18	-	0.80	0.42
n	22	7	4	3	16	28	17	3	12	-	2	3
max	1.46	1.53	1.25	0.38	1.40	1.71	1.30	0.81	1.51	-	1.10	0.45
min	0.44	0.60	0.70	0.21	0.57	0.78	0.59	0.35	0.90	-	0.50	0.36

Hydraulic Conductivity (cm d^{-1})	**Needle-Leaved**				**Broad-Leaved**				**Mixed**			
Soil texture	C	M	F	Oh	C	M	F	Oh	C	M	F	Oh
mean values	204	261	875	-	242	286	1225	-	464	320	64	-
n	13	2	2	-	17	5	2	-	14	1	1	-
max	567	468	1200	-	751	1096	1400	-	1201	-	-	-
min	17	53	550	-	35	12	1050	-	27	-	-	-

For needle-leaved and broad-leaved forests, the highest mean values for bulk density were observed in forest stands with a loam-dominated soil texture, whereas the lowest were found in sites with a fine soil texture. Overall bulk density in the Oh-horizon were in the order: broad-leaved > mixed > needle-leaved, with a limited number of values reported in the screened articles (Table 5). Moreover, it was often difficult to identify whether the soil density values corresponded to the mineral layer or the organic layer. Therefore, it is possible that there is a bias in the values reported in our study due to a lack of detail in terms of the depth of measurements.

3.3. Soil Saturated Hydraulic Conductivity

Figure 5 and Tables 3–5 summarise all the information regarding the topsoil saturated hydraulic conductivity obtained from the reviewed articles. Analogous to the bulk density, dataset A comprises 59 values, while dataset B contains 46 values. Between the two datasets, no change in the mean values in terms of forest type was noticeable: mixed > broad-leaved > needle-leaved (Table 3, Figure 5). Furthermore, no statistical effect of forest type on the soil hydraulic conductivity of the topsoil was observed (p-values > 0.05, Table 3).

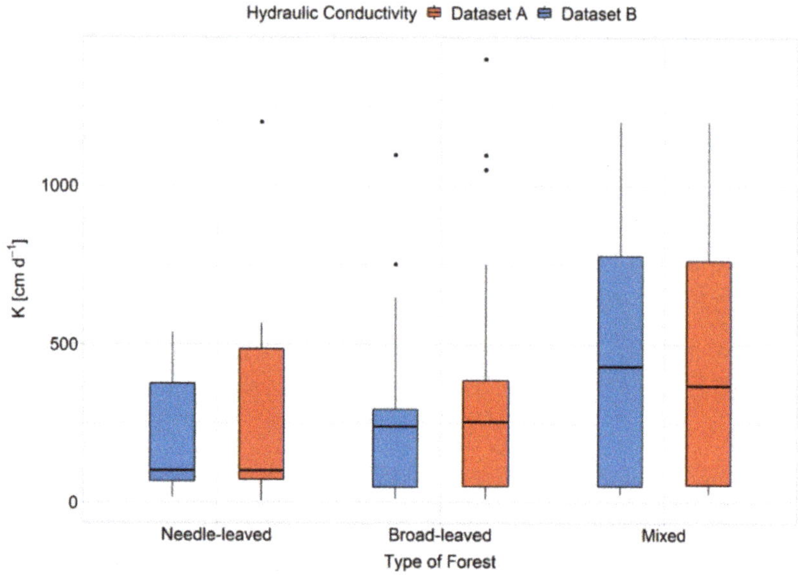

Figure 5. Descriptive statistics of topsoil hydraulic conductivity in terms of forest type considering two datasets: dataset A containing all values found and dataset B containing filtered values based on available information on tree age and soil particle size. All values refer to the topsoil (0–30 cm depth). Black dots represent the outliers.

For needle-leaved forests, most studies concentrated on the young stage, whereas the old stages of broad-leaved forests dominated the screened studies (Table 4, Figure 6). For mixed forests, there were no data on hydraulic conductivity at the young stage, while most studies occurred in the stands of the mature age (Figure 6). In the case of needle-leaved forests, the order of the mean values of hydraulic conductivity recorded was: mature > old > young, differing from that under broad-leaved forests: young > old > mature. However, the number of studies in the mature and old stages of needle-leaved forests was extremely low and therefore did not allow any statistical comparison.

Figure 7 and Table 5 describe the statistical distribution for dataset B with regard to the saturated hydraulic conductivity in relation to forest type and soil texture. Most of the reviewed studies with information on soil texture were conducted in broad-leaved forests and study sites were dominated by a coarse soil texture.

Under needle-leaved and broad-leaved forests, the order of magnitude of hydraulic conductivity was comparable between the coarse and medium soil textures, whereas soil dominated by fine texture presented higher values (Table 5). It has to be noted that the number of studies carried out in all forest types with regard to hydraulic conductivity in stands with fine-textured soils was very low.

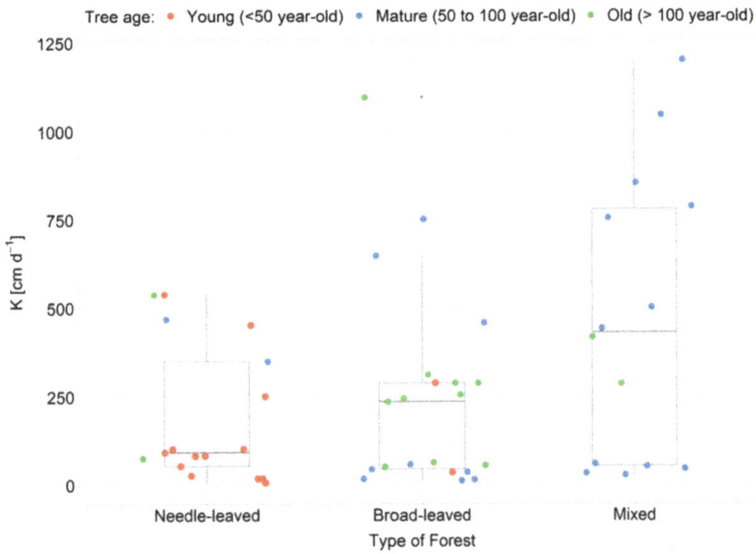

Figure 6. Descriptive statistics of top mineral soil hydraulic conductivity for dataset B in relation to tree age and forest type. Scattered points represent the individual values classified by the different forest stages considered in this study. Values correspond to the 0–30 cm soil depth. Grey points represent the outliers.

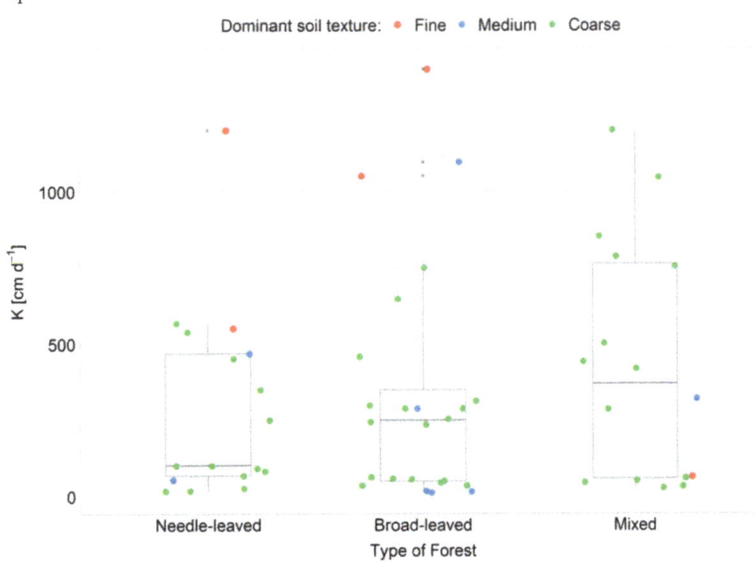

Figure 7. Descriptive statistics for dataset B top mineral soil hydraulic conductivity in terms of forest type and soil texture: fine (clay-dominated), medium (mostly loamy), and coarse (mostly sandy). Grey points represent the outliers.

4. Discussion

Our literature review revealed that with respect to water retention characteristics of soils under forests, only limited information is available, with fewer than 20 studies reporting values (Table 2). Thus, the reported information does not allow any conclusion on the influence of tree species and age on water retention characteristics (which are part of

SHP) for different site conditions (i.e., soil texture, tree species). Moreover, the reported studies differ in the applied methodology and experimental design, which allows only limited comparability among the studies. Dedicated studies such as Wahren et al. [17] and Julich et al. [16] indicate that for sites influenced by stagnant soil water conditions, differences in SHP for stands of different tree types and age can be expected. However, the transferability of those findings needs to be tested by more experiments on sites with different site conditions.

With respect to bulk density as an important indicator of soil structure, more information in the screened literature is available (Tables 1 and 3–5). Bulk density varies together with the changes in forest structure. Bakhshandeh-Navroud et al. [51] observed a soil bulk density of 1.33 g cm^{-3} in a broad-leaved monoculture, whereas a higher value was observed when eight different broad-leaved tree species were mixed (1.48 g cm^{-3}). Jost et al. [37] found a higher soil bulk density in a broad-leaved forest in contrast to a needle-leaved forest (1.13 vs. 0.97 g cm^{-3}). Błońska et al. [33] observed higher values in a broad-leaved forest (range: 0.35–0.81 g cm^{-3}) compared to needle-leaved stands (range: 0.21–0.28 g cm^{-3}), and in a mixed forest (0.36 g cm^{-3}); however, it has to be considered that in this study, the organic layer instead of the top mineral layer was sampled. This emphasizes the need for a clear description of the sampled depth and horizon. Moragues-Saitua et al. [52] reported a higher bulk density under needle-leaved forests (1.06 g cm^{-3}) and in broad-leaved forests (0.67 g cm^{-3}) in contrast to the beforementioned studies. The results of the presented literature review suggest an effect of forest type (tree species) on soil bulk density (Table 3a, p-value < 0.05). Moreover, it is not only soil texture that has been described in the literature as a factor that might explain variations in soil bulk density but tree age as well. Polláková et al. [53] observed that bulk density was higher (1.15 and 1.01 g cm^{-3}) in mature needle-leaved forests with comparable characteristics (110 years old, sandy soil), in contrast to old forests. Furthermore, they noticed that bulk density was higher in a silty soil young forest (50 years old, 1.31 g cm^{-3}). Rosenkranz et al. [54] reported that a 100-year-old mature needle-leaved forest presented a soil bulk density of 0.82 g cm^{-3}, but when mixed with a three-year-old broad-leaved tree species, it was reduced to 0.64 g cm^{-3}. Baldocchi et al. [43] found that soil bulk density was 1.05 g cm^{-3} in a 250-year-old needle-leaved forest with sandy soil, while Link et al. [55] reported lower values (0.8 g cm^{-3}) in an older forest (450 years old) containing needle-leaved tree species. Frêne et al. [29] observed a higher bulk density in a 110-year-old needle-leaved forest in contrast to an old (350/400 years old) mixed forest (0.77 vs. 0.44 g cm^{-3}, respectively); however, information on soil texture was not provided.

Regarding soil hydraulic conductivity, our literature review suggests that there is no apparent effect of forest type on soil hydraulic conductivity (Table 3). However, there are studies concluding that this parameter not only varies in terms of tree species, but also that site characteristics play a relevant role. Bauwe et al. [22] reported lower infiltration capacity under needle-leaved stands (349 cm d^{-1}) in contrast to broad-leaved forests (both with mature tree stands). Bens et al. [21] observed the highest hydraulic conductivity in mixed forests (76-year-old pine with 34-year-old beech) with 1045 cm d^{-1}, whereas needle-leaved and broad-leaved forests presented 57% and 38% lower infiltration rates, respectively. Similarly, Wahl et al. [7,8] reported the highest infiltration rates in mixed forests and the lowest in needle-leaved forests, both presenting comparable sandy soil. Julich et al. [16] reported a higher amount of well-connected macropores for a 170-year-old broad-leaved forest, and higher infiltration capacities, in contrast to a young 25-year-old needle-leaved forest, probably influenced by their deep-developed rooting system. Moreover, Frey et al. [56] found higher hydraulic conductivity in a mixed forest with loamy soil (320 cm d^{-1}) in contrast to a mixed forest with loamy soil but containing a much higher clay content (63 cm d^{-1}).

As reported by these authors, soil hydraulic conductivity is sensitive to other soil properties; still, many authors describe this parameter under varying conditions. For instance, Archer et al. [20] considered a wide range of parameters to assess the water infiltration rates such as the soil particle size, tree ages, and measurements at different depths. However, results were not able to be derived on the influence of such parameters since all forests differed in age and soil texture. They considered two mature stages (160/180 years old) and one old (500 years old) stage with broad-leaved stands and a young stage (~50 years old) with needle-leaved tree species. The sites differed in soil texture. In young needle-leaved forests the range was 82–101 cm d^{-1} (mean 90.9 cm d^{-1}), in mature broad-leaved sites the ranges were 12–17 cm d^{-1} (mean 14.3 cm d^{-1}) and 245–312 cm d^{-1} (mean 280.2 cm d^{-1}), and the range in the old broad-leaved site was 235–288 cm d^{-1} (mean: 258.3 cm d^{-1}). Similarly, Bens et al. [21] and Wahl et al. [7,8] reported higher values (648, 458, and 751 cm d^{-1}, respectively, by author) in broad-leaved forests in contrast to needle-leaved stands (451, 251, and 538 cm d^{-1}, respectively, by author). However, the study considered a young (<40 years old) needle-leaved and a mature (91 years old) broad-leaved forest, which also differed in terms of soil texture.

5. Conclusions

The various soils under forests present an extensive assortment of site conditions: i.e., soil texture, porosity, biotic activity (in particular that of the meso- and macrofauna), amount and vertical distribution of soil organic matter, and tree species and ages, among many others. Nevertheless, in our literature review we found considerable limitations in assessing the effect of forest type on soil hydraulic properties, which led us to the following conclusions:

- We found hardly any articles addressing SHP and describing an appropriate number of site parameters simultaneously in the reviewed studies, which made it difficult to determine the impact of forest type and age on soil hydraulic properties.
- The main purpose of many of the screened articles was not specifically to assess the SHP, which were measured as additional information; possibly one of the reasons why many parameters of interest for our analysis were poorly detailed.
- The great variability in results in the reviewed studies underlines the belief that SHP can be sensitive to a wide range of site parameters. Due to an incomplete description of the study site's conditions and differences in methodologies applied for each author (different measuring devices, diversity of sampling design, and inappropriate comparisons), there is not enough evidence to draw sophisticated conclusions in terms of the influence of specific forest characteristics (type and age) on SHP.
- Due to the large gap of knowledge in the literature and the abovementioned arguments, we were neither able to accept nor reject our hypothesis.

We suggest that the effect of forest type on soil water dynamics should be assessed based on studies with comparable site parameters to avoid overlooking the high sensitivity of SHP. For this reason, a much higher number of systematic studies following standardised methods is clearly needed. Such studies should include a minimum of information on tree age, mixture rate, depth of measurement, and soil texture in order to make future studies comparable and to create an understanding of the influence of forest type and forest structure on SHP under different site conditions.

Author Contributions: Conceptualisation S.J. and V.V.-R.; literature review, data analysis and writing-original draft preparation V.V.-R.; writing-review and editing S.J. and K.-H.F. All authors have read and agreed to the published version of the manuscript.

Funding: This research was supported by the scholarship programme of the Graduate Academy for the Promotion of Early-Career Female Scientists at the TU Dresden.

Data Availability Statement: The author, upon request, can provide information regarding the reviewed papers.

Conflicts of Interest: The authors declare no conflict of interest.

References

1. Allen, C.D.; Breshears, D.D.; McDowell, N.G. On underestimation of global vulnerability to tree mortality and forest die-off from hotter drought in the Anthropocene. *Ecosphere* **2015**, *6*, 129. [CrossRef]
2. Allen, C.D.; Macalady, A.K.; Chenchouni, H.; Bachelet, D.; McDowell, N.; Vennetier, M.; Kitzberger, T.; Rigling, A.; Breshears, D.D.; Hogg, E.H.; et al. A global overview of drought and heat-induced tree mortality reveals emerging climate change risks for forests. *For. Ecol. Manag.* **2010**, *259*, 660–684. [CrossRef]
3. European Commission, Eurostat. Forestry in the EU and the World: A Statistical Portrait 2011. Available online: https://data.europa.eu/doi/10.2785/13022 (accessed on 20 October 2021).
4. Lindner, M.; Maroschek, M.; Netherer, S.; Kremer, A.; Barbati, A.; Garcia-Gonzalo, J.; Seidl, R.; Delzon, S.; Corona, P.; Kolström, M.; et al. Climate change impacts, adaptive capacity, and vulnerability of European forest ecosystems. *For. Ecol. Manag.* **2010**, *259*, 698–709. [CrossRef]
5. Pretzsch, H.; Steckel, M.; Heym, M. Stand growth and structure of mixed-species and monospecific stands of Scot pine (*Pinus sylvestris* L.) and oak (*Q. robur* L., Quercus petraea (MATT.) LIEBL.) analysed along a productivity gradient through Europe. *Eur. J. For. Res.* **2020**, *139*, 349–367. [CrossRef]
6. Seiwa, K.; Kunii, D.; Masaka, K.; Hayashi, S.; Tada, C. Hardwood mixture enhances soil water infiltration in a conifer plantation. *For. Ecol. Manag.* **2021**, *498*, 119508. [CrossRef]
7. Wahl, N.A.; Bens, O.; Schäfer, B.; Hüttl, R.F. Impact of changes in land-use management on soil hydraulic properties: Hydraulic conductivity, water repellency and water retention. *Phys. Chem. Earth* **2003**, *28*, 1377–1387. [CrossRef]
8. Wahl, N.A.; Wöllecke, B.; Bens, O.; Hüttl, R.F. Can forest transformation help reducing floods in forested watersheds? Certain aspects on soil hydraulics and organic matter properties. *Phys. Chem. Earth* **2005**, *30*, 611–621. [CrossRef]
9. Tor-Ngern, P.; Oishi, A.C.; Uebelherr, J.M.; Palmroth, S.; Tarvainen, L.; Ottoson-Löfvenius, M.; Linder, S.; Domec, J.C.; Näsholm, T. Ecophysiological variation of transpiration of pine forests: Synthesis of new and published results. *Ecol. Appl.* **2017**, *27*, 118–133. Available online: http://www.jstor.org/stable/44132586 (accessed on 11 October 2021). [CrossRef]
10. Meusburger, K.; Trotsiuk, V.; Schmidt-Walter, P.; Baltensweiler, A.; Brun, P.; Bernhard, F.; Gharun, M.; Habel, R.; Hagedorn, F.; Köchli, R.; et al. Soil-plant interactions modulated water availability of Swiss forests during the 2015 and 2018 droughts. *Glob. Chang. Biol.* **2022**, *28*, 5928–5944. [CrossRef]
11. Schmidt-Walter, P.; Trotsiuk, V.; Meusburger, K.; Zacios, M.; Meesenburg, H. Advancing simulations of water fluxes, soil moisture and drought stress by using the LWF-Brook90 hydrological model in R. *Agric. For. Meteorol.* **2020**, *291*, 108023. [CrossRef]
12. Assouline, S.; Or, D. Conceptual and parametric representation of soil hydraulic properties: A review. *Vadose Zone J.* **2013**, *12*, 1–20. [CrossRef]
13. Weninger, T.; Bodner, G.; Kreiselmeier, J.; Chandrasekhar, P.; Julich, S.; Feger, K.-H.; Schwärzel, K.; Schwen, A. Combination of measurement methods for a wide range description of hydraulic soil properties. *Water* **2018**, *10*, 1021. [CrossRef]
14. Archer, N.A.L.; Otten, W.; Schmidt, S.; Bengough, A.G.; Shah, N.; Bonell, M. Rainfall infiltration and soil hydrological characteristics below ancient forest, planted Forest and Grassland in a Temperate Northern Climate. *Ecohydrology* **2016**, *9*, 585–600. [CrossRef]
15. Chandler, K.R.; Stevens, C.J.; Binley, A.; Keith, A.M. Influence of tree species and forest land use on soil hydraulic conductivity and implications for surface runoff generation. *Geoderma* **2018**, *310*, 120–127. [CrossRef]
16. Julich, S.; Kreiselmeier, J.; Scheibler, S.; Petzold, R.; Schwärzel, K.; Feger, K.H. Hydraulic properties of forest soils with stagnic conditions. *Forests* **2021**, *12*, 1113. [CrossRef]
17. Wahren, A.; Schwärzel, K.; Feger, K.H. Potentials and limitations of natural flood retention by forested land in headwater catchments: Evidence from experimental and model studies. *J. Flood Risk Manag.* **2012**, *5*, 321–335. [CrossRef]
18. Liu, Y. A numerical study on hydrological impacts of forest restoration in the southern United States. *Ecohydrology* **2011**, *4*, 299–314. [CrossRef]
19. Bogunovic, I.; Viduka, A.; Magdic, I.; Telak, L.J.; Francos, M.; Pereira, P. Agricultural and forest land-use impact on soil properties in Zagreb periurban area (Croatia). *Agronomy* **2020**, *10*, 1331. [CrossRef]
20. Archer, N.A.L.; Bonell, M.; MacDonald, A.M.; Coles, N. A Constant Head Well Permeameter Formula Comparison: Its Significance in the Estimation of Field-Saturated Hydraulic Conductivity in Heterogeneous Shallow Soils. *Hydrol. Res.* **2014**, *45*, 788–805. [CrossRef]
21. Bens, O.; Wahl, N.A.; Fischer, H.; Hüttl, R.F. Water infiltration and hydraulic conductivity in sandy cambisols: Impacts of forest transformation on soil hydrological properties. *Eur. J. For. Res.* **2007**, *126*, 101–109. [CrossRef]
22. Bauwe, A.; Koch, M.; Kallweit, R.; Konopatzky, A.; Strohbach, B.; Lennartz, B. Tree-ring growth response of Scots pine (*Pinus sylvestris* L.) to climate and soil water availability in the lowlands of north-eastern Germany. *Balt. For.* **2013**, *19*, 212–225. Available online: https://www.researchgate.net/publication/260435722 (accessed on 20 October 2021).
23. Gebauer, T.; Horna, V.; Leuschner, C. Canopy transpiration of pure and mixed forest stands with variable abundance of European beech. *J. Hydrol.* **2012**, *442–443*, 2–14. [CrossRef]
24. Meinzer, F.C.; Woodruff, D.R.; Eissenstat, D.M.; Lin, H.S.; Adams, T.S.; McCulloh, K.A. Above- and belowground controls on water use by trees of different wood types in an eastern US deciduous forest. *Tree Physiol.* **2013**, *33*, 345–356. [CrossRef]
25. Aranda, I.; Forner, A.; Cuesta, B.; Valladares, F. Species-Specific Water Use by Forest Tree Species: From the Tree to the Stand. *Agric. Water Manag.* **2012**, *114*, 67–77. [CrossRef]

26. Knutzen, F.; Meier, I.C.; Leuschner, C. Does reduced precipitation trigger physiological and morphological drought adaptations in European beech (*Fagus sylvatica* L.)? Comparing provenances across a precipitation gradient. *Tree Physiol.* **2015**, *35*, 949–963. [CrossRef]
27. Wang, Y.; Dong, X.; Wang, H.; Wang, Z.; Gu, J. Root trip morphology, anatomy, chemistry and potential hydraulic conductivity vary with soil depth in three temperate hardwood species. *Tree Physiol.* **2015**, *36*, 99–108. [CrossRef]
28. Christiansen, J.; Vesterdal, L.; Callesen, I.; Elberling, B.; Schmidt, I.K.; Gundersen, P. Role of six European tree species and land-use legacy for nitrogen and water budgets in forests. *Glob. Chang. Biol.* **2010**, *16*, 2224–2240. [CrossRef]
29. Frêne, C.; Dörner, J.; Zúñiga, F.; Cuevas, J.G.; Alfaro, F.D.; Armesto, J.J. Eco-hydrological functions in forested catchments of Southern Chile. *Ecosystems* **2020**, *23*, 307–323. [CrossRef]
30. Hansson, K.; Olsson, B.A.; Olsson, M.; Johansson, U.; Kleja, D.B. Differences in soil properties in adjacent stands of Scots pine, Norway spruce and Silver birch in SW Sweden. *For. Ecol. Manag.* **2011**, *262*, 522–530. [CrossRef]
31. Norman, J.S.; Barrett, J.E. Substrate and nutrient limitation of ammonia-oxidizing bacteria and archaea in temperate forest soil. *Soil Biol. Biochem.* **2014**, *69*, 141–146. [CrossRef]
32. Schiff, S.L.; Devito, K.J.; Elgood, R.J.; McCrindle, P.M.; Spoelstra, J.; Dillon, P. Two adjacent forested catchments: Dramatically different NO_{3-} export. *Water Resour. Res.* **2002**, *38*, 28-1–28-13. [CrossRef]
33. Błońska, E.; Klamerus-Iwan, A.; Lasota, J.; Gruba, P.; Pach, M.; Pretzsch, H. What characteristics of soil fertility can improve in mixed stands of Scots pine and European beech compared with monospecific stands? *Commun. Soil Sci. Plant Anal.* **2018**, *49*, 237–247. [CrossRef]
34. Kopler, I.; Burg, D.; Wittenberg, L.; Malkinson, D. Differences in soil moisture in two middle eastern oak forests: Comparing the effects of trees and soil composition. *Hydrol. Process.* **2019**, *33*, 86–100. [CrossRef]
35. Maes, S.L.; Blondeel, H.; Perring, M.P.; Depauw, L.; Brümelis, G.; Brunet, J.; Decocq, G.; den Ouden, J.; Härdtle, W.; Hédl, R.; et al. Litter quality, land-use history, and nitrogen deposition effects on topsoil conditions across European temperate deciduous forests. *For. Ecol. Manag.* **2019**, *433*, 405–418. [CrossRef]
36. Farahnak, M.; Mitsuyasu, K.; Jeong, S.; Otsuki, K.; Chiwa, M.; Sadeghi, S.M.M.; Kume, A. Soil hydraulic conductivity differences between upslope and downslope of two coniferous trees on a hillslope. *J. For. Res.* **2019**, *3*, 143–152. [CrossRef]
37. Jost, G.; Schume, H.; Hager, H.; Markart, G.; Kohl, B. A hillslope scale comparison of tree species influence on soil moisture dynamics and runoff processes during intense rainfall. *J. Hydrol.* **2012**, *420–421*, 112–124. [CrossRef]
38. Lange, B.; Germann, P.F.; Lüscher, P. Greater abundance of Fagus sylvatica in coniferous flood protection forests due to climate change: Impact of modified root densities on infiltration. *Eur. J. For. Res.* **2013**, *132*, 151–163. [CrossRef]
39. Moher, D.; Liberati, A.; Tetzlaff, J.; Altman, D.G.; The Prisma Group. Preferred reporting items for systematic reviews and meta-analyses: The PRISMA statement. *Int. J. Surg.* **2010**, *8*, 336–341. [CrossRef]
40. Bittner, S.; Talkner, U.; Krämer, I. Modeling stand water budgets of mixed temperate broad-leaved forest stands by considering variations in species specific drought response. *Agric. For. Meteorol.* **2010**, *150*, 1347–1357. [CrossRef]
41. Grant, R.F.; Zhang, Y.; Yuan, F.; Wang, S.; Hanson, P.J.; Gaumont-Guay, D.; Chen, J.; Black, T.A.; Barr, A.; Baldocchi, D.D.; et al. Intercomparison of techniques to model water stress effects on CO_2 and energy exchange in temperate and boreal deciduous forests. *Ecol. Model.* **2006**, *196*, 289–312. [CrossRef]
42. Metzger, J.C.; Wutzler, T.; Valle, N.D.; Filipzik, J.; Grauer, C.; Lehmann, R.; Roggenbuck, M.; Schelhorn, D.; Weckmüller, J.; Küsel, K.; et al. Vegetation impacts soil water content patterns by shaping canopy water fluxes and soil properties. *Hydrol. Process.* **2017**, *31*, 3783–3795. [CrossRef]
43. Baldocchi, D.D.; Law, B.E.; Anthoni, P.M. On measuring and modeling energy fluxes above the floor of a homogeneous and heterogeneous conifer forest. *Agric. For. Meteorol.* **2000**, *102*, 187–206. [CrossRef]
44. Dusek, J.; Vogel, T. Hillslope hydrograph separation: The effects of variable isotopic signatures and hydrodynamic mixing in macroporous soil. *J. Hydrol.* **2018**, *563*, 446–459. [CrossRef]
45. Almahayni, T.; Houska, T. Towards dynamic and process–based modelling of radionuclides cycling in terrestrial radioecology. *J. Environ. Radioact.* **2020**, *225*, 106380. [CrossRef] [PubMed]
46. Duarte, H.F.; Raczka, B.M.; Ricciuto, D.M.; Lin, J.C.; Koven, C.D.; Thornton, P.E.; Bowling, D.R.; Lai, C.; Bible, K.J.; Ehleringer, J.R. Evaluating the Community Land Model (CLM4.5) at a coniferous forest site in northwestern United States using flux and carbon-isotope measurements. *Biogeosciences* **2017**, *14*, 4315–4340. [CrossRef]
47. Kirchen, G.; Calvaruso, C.; Granier, A.; Redon, P.O.; Van der Heijden, G.; Bréda, N.; Turpault, M.P. Local soil type variability controls the water budget and stand productivity in a beech forest. *For. Ecol. Manag.* **2017**, *390*, 89–103. [CrossRef]
48. Leuschner, C. Water extraction by tree fine roots in the forest floor of a temperate Fagus-Quercus forest. *Ann. For. Sci.* **1998**, *55*, 141–157. [CrossRef]
49. Curiel, J.; Janssens, I.A.; Carrara, A.; Ceulemans, R. Annual Q_{10} of soil respiration reflects plant phenological patterns as well as temperature sensitivity. *Glob. Chang. Biol.* **2004**, *10*, 161–169. [CrossRef]
50. Schwärzel, K.; Menzer, A.; Clausnitzer, F. Soil water content measurements deliver reliable estimates of water fluxes: A comparative study in a beech and a spruce stand in the Tharandt forest (Saxony, Germany). *Agric. For. Meteorol.* **2009**, *149*, 1994–2006. [CrossRef]

51. Bakhshandeh-Navroud, B.; Abrari, V.K.; Pilehvar, B.; Kooch, Y. The interactions between tree-herb layer diversity and soil properties in the oriental beech (Fagus Orientalis Lipsky) stands in Hyrcanian forest. *Environ. Monit. Assess.* **2018**, *190*, 425. [CrossRef]
52. Moragues-Saitua, L.; Arias-González, A.; Gartzia-Bengoetxea, N. Effects of biochar and wood ash on soil hydraulic properties: A field experiment involving contrasting temperate soils. *Geoderma* **2017**, *305*, 144–152. [CrossRef]
53. Polláková, N.; Šimanský, V.; Jonczak, J. Characteristics of physical properties in soil profiles under selected introduced trees in the nature reserve Arboretum Mlyňany, Slovakia. *Folia Oecologica* **2017**, *44*, 78–86. [CrossRef]
54. Rosenkranz, P.; Dannenmann, M.; Brüggemann, N.; Papen, H.; Berger, U.; Zumbusch, E.; Butterbach-Bahl, K. Gross rates of ammonification and nitrification at a nitrogen-saturated Spruce (*Picea abies* (L.) Karst.) stand in southern Germany. *Eur. J. Soil Sci.* **2010**, *61*, 745–758. [CrossRef]
55. Link, T.E.; Flerchinger, G.N.; Unsworth, M.; Marks, D. Simulation of water and energy fluxes in an old-growth seasonal temperate rain forest using the simultaneous heat and water (SHAW) model. *J. Hydrometeorol.* **2004**, *5*, 443–457. [CrossRef]
56. Frey, B.; Niklaus, P.A.; Kremer, J.; Lüscher, P.; Zimmermann, S. Heavy-machinery traffic impacts methane emissions as well as methanogen abundance and community structure in oxic forest soils. *Appl. Environ. Microbiol.* **2011**, *77*, 6060–6068. [CrossRef]

forests

MDPI

Article

Hydrological Properties of Soil and Litter Layers of Four Forest Types Restored in the Gully Erosion Area of Latosol in South China

Zhihua Tu [1,2,*], Suyi Chen [1,†], Zexian Chen [1], Dongshuo Ruan [1], Wei Zhang [1], Yujie Han [1], Lin Han [1], Kang Wang [1], Yanping Huang [1] and Jinhui Chen [1,2,3,†]

1 Key Laboratory of Genetics and Germplasm Innovation of Tropical Special Forest Trees and Ornamental Plants, Ministry of Education, School of Forestry, Hainan University, Haikou 570228, China
2 Engineering Research Center of Rare and Precious Tree Species in Hainan Province, Haikou 570228, China
3 Sanya Nanfan Research Institute, Hainan University, Sanya 572025, China
* Correspondence: tuzhihua@hainanu.edu.cn
† These authors contributed equally to this work.

Abstract: Litter and soil play an important role in influencing hydrological processes and the global water cycle. Artificial afforestation, as a part of vegetation restoration, was constructed in the gully erosion areas of latosol with the objective to prevent erosion. Variations in the hydrological properties in soils that have undergone vegetation restoration from gully erosion are not well understood. In this study, we examine the variations in the litter thickness and mass, soil structure and porosity, and hydrological properties of four forest types (eucalyptus–grass forest, bamboo–grass forest, acacia–grass forest, and shrub–grass forest). The results show that the total litter thickness varied from 1.71 to 3.74 cm and was highest in the acacia–grass forest. The total litter mass for the acacia–grass forest, 3.49 ± 0.06 t·ha^{-1}, was significantly higher than that for the other forest types. The mass of the undecomposed litter (UL) layer was significantly lower than that of the semi-decomposed litter (SL). (2) The maximum water-retention capacity (W_{max}) and effective water-retention capacity (W_{eff}) of the SL layer were greater than those of the UL layer. The W_{max} and W_{eff} for the acacia–grass forest were markedly larger than those of the eucalyptus–grass, bamboo–grass, and shrub–grass forests. The water absorption rates of the SL and UL layers were highest during the onset of the immersion experiment, declined exponentially with time, and declined rapidly in the first 2 h. (4) The soil bulk density ranged from 1.46 g·cm^{-3} to 1.54 g·cm^{-3}, and the total porosity ranged from 32.06% to 37.13%. The soil bulk density increased with the increasing soil depth, while the total porosity decreased gradually. The soil water-holding capacity of the soil layer of 0–60 cm in the acacia–grass forest (301.76 t·ha^{-1}) was greater than that of the other forest types. A comprehensive evaluation of the water conservation capacity by the entropy weight method showed that the water conservation capacity was greatest in the acacia–grass forest. The higher water-holding capacity of the acacia–grass forest may be more effective in enhancing rainfall interception, minimizing splash erosion, and decreasing surface runoff. Here, the results indicate that acacia–grass forest restoration can mitigate soil erosion by favoring soil and water conservation, improving the environment in the gully erosion area of latosol.

Keywords: vegetation restoration; hydrological properties; litter layer; soil layer; latosol region

Citation: Tu, Z.; Chen, S.; Chen, Z.; Ruan, D.; Zhang, W.; Han, Y.; Han, L.; Wang, K.; Huang, Y.; Chen, J. Hydrological Properties of Soil and Litter Layers of Four Forest Types Restored in the Gully Erosion Area of Latosol in South China. *Forests* **2023**, *14*, 360. https://doi.org/10.3390/f14020360

Academic Editors: Yanhui Wang, Karl-Heinz Feger and Lulu Zhang

Received: 21 November 2022
Revised: 5 February 2023
Accepted: 9 February 2023
Published: 11 February 2023

1. Introduction

Soil erosion can have significant influences on ecosystem function and services, and it is thus a serious threat to sustainable global development [1,2]. Consequences from soil erosion can lead to soil fertility degradation, water eutrophication, riverbed aggradation, vegetation degradation, and the acceleration of ecosystem dysfunction [3–5]. The red soils in southern China are ranked second, after the Loess Plateau, in soil erosion severity.

These severely degraded soils [6,7] are mainly distributed in the Jiangxi, Fujian, Hunan, Guangdong, and Hainan provinces [2,3,8] and have been encountering soil and water loss since the 1950s [8]. The red soil region accounts for 22% of China's total land area but makes up more than 50% of the nation's total soil loss [9]. This area is regarded as a dominant factor of ecosystem degradation in southern China and has received increasing attention from the government [2]. Vegetation restoration plays a key role in water and soil conservation [10–14], and increases in vegetation cover have been shown as an effective measure to reduce soil erosion [15]. Over the past few decades, vegetation restoration through the State Key Forestry Ecological Projects has been carried out in the red soil region of southern China. These efforts are recognized as the main measure for controlling soil and water loss, greatly improving the regional environment [16–19].

Gully erosion in the red soil region is generally formed through a combination of water and gravity effects [2,20]. The rainfall in this region is particularly heavy and concentrated, generating high-intensity rainfall events [2,15]. More and more studies are showing that gully erosion has a strong influence on soil quality [21–23]. Reports have demonstrated that gully erosion affected 1220 km^2 in the red soil region from 1950 to 2005, leading to the loss of more than 60 Mt of soil [15], which is now effectively controlled by vegetation restoration [15,23]. The success of vegetation restoration to control gully erosion is reliant on the location and climate of the red soil region. This area is located in the subtropical and tropical monsoon region, which has abundant rainfall, high temperatures, and high productivity potential and could be more favorable to vegetation growth [15]. Vegetation restoration plays a key role in mitigating soil erosion through a supply of improved hydrological regulation services.

The vegetation in soil-eroded areas influences the hydrologic processes and is critical for the terrestrial ecosystem water cycle [24,25]. The forest canopy, litter layer, and mineral soil layer of the root zone are regarded as the first, second, and third functioning layers regulating the hydrological behaviors in forest ecosystems [26–30]. These layers are often used to scientifically assess the water conservation function of forest ecosystems [31,32]. Several studies have reported a significant benefit of water conservation from the litter and soil layers when a thicker litter layer and more porous soil exist in forest ecosystems [10,30,33–36]. However, the hydrological properties and water conservation function of the litter and soil layer of plantations in the gully erosion area of latosol in tropical zones have not been well studied. The service of the water conservation of forest ecosystems that have undergone optimal vegetation restoration for controlling gully erosion should be further investigated.

Forest litter and soil layers are important ecosystem components for regulating hydrological processes [30,33,37–40]. The litter layer acts as a sponge to intercept rainfall, relieving rainfall splash, and delaying or reducing surface runoff by infiltrating it into soils [35,41–43]. The forest ecosystem water budget is intensely affected by this process [4,34,35]. Consequently, the litter layer thickness and mass, soil structure and porosity, and water-holding capacity depend on forest vegetation [10,33,37,44,45]. Previous studies have investigated the effects of soil erosion on broad-leaved forests, which form a broad canopy and enhance the interception capacity of rainfall [28,37,46,47]. The broad-leaved forest ecosystem is highly conducive to water conservation [27,42,48]. An *Acacia mangium* plantation in the eroded area of latosol was found to generally favor the water-holding capacity in the litter layer [41]. However, the differences in the hydrological properties of broad-leaved forests in the gully erosion areas of latosol's tropical zones are not well understood.

Eroded latosol areas in Danzhou County, Hainan Province are a consequence of extensive deforestation in the 1960s to 1980s, which made vegetation scarce, thus loosening the soil and intensifying erosion during the rainy season [41,49]. Starting in 2000, vegetation restoration was conducted by planting soil- and water-conservation tree species, such as *Acacia mangium* and *Eucalyptus robusta*. Water and soil loss have been effectively controlled. However, barren soil in the gully erosion region that underwent restoration was more challenging than expected. In 2012, vegetation restoration was initiated in the gully erosion areas. Four vegetative restoration types were used: eucalyptus–grass forest, bamboo–grass forest, acacia–grass forest, and shrub–grass forest. As a result, gully erosion was effectively controlled and the regional environment was improved. However, the hydrological properties of the litter and soil layers in these forests are not well understood. In this study, the four forest types in the gully erosion area of latosol in the Mahuangling Watershed were investigated. We quantified the litter thickness and mass, the soil structure and porosity, and the hydrological properties. The aim of this study was to illuminate the effects of vegetation restoration on water conservation from the litter and soil layers to provide a theoretical and practical basis for vegetation restoration in the areas of latosol gully erosion.

2. Materials and Methods

2.1. Study Sites

This study was conducted in the Mahuangling Watershed at the Mahuangling Soil and Water Conservation Monitoring Station (19°41′~19°47′ N, 109°24′~109°30′ E) (Figure 1), located in Danzhou City, Hainan Province, China. The region has a tropical monsoon climate, with a mean annual temperature of 23.5 °C. The mean annual precipitation is approximately 1815 mm, with about 80%~85% occurring from May to October. The soils are typical latosol soils that originated from granitic parent materials. Due to vegetation over-harvesting and deforestation from the 1960s to 1980s, soil erosion intensified in the already eroded area of latosol. In the 2000s, as a part of the State Key Forestry Ecological Projects, vegetation restoration was conducted within the gully erosion area, specifically consisting of eucalyptus–grass forest (*E. robusta* and *P. distichum*), bamboo–grass forest (*Bambusa oldhamii* and *P. distichum*), acacia–grass forest (*A. mangium* and *P. distichum*), and shrub–grass forest (*Melastoma candidum*, *U. lobata*, and *P. distichum*). These restoration efforts have proven to be successful in controlling soil and water loss, greatly improving the local environment. Moreover, this study site is now primarily covered by plantation forests. The understory consists of species of *Urena lobata*, *Melastoma malabathricum*, *Chromolaena odorata*, *Lantana camara*, *Digitaria sanguinalis*, and *Paspalum distichum*, and the forest coverage is more than 75% [41].

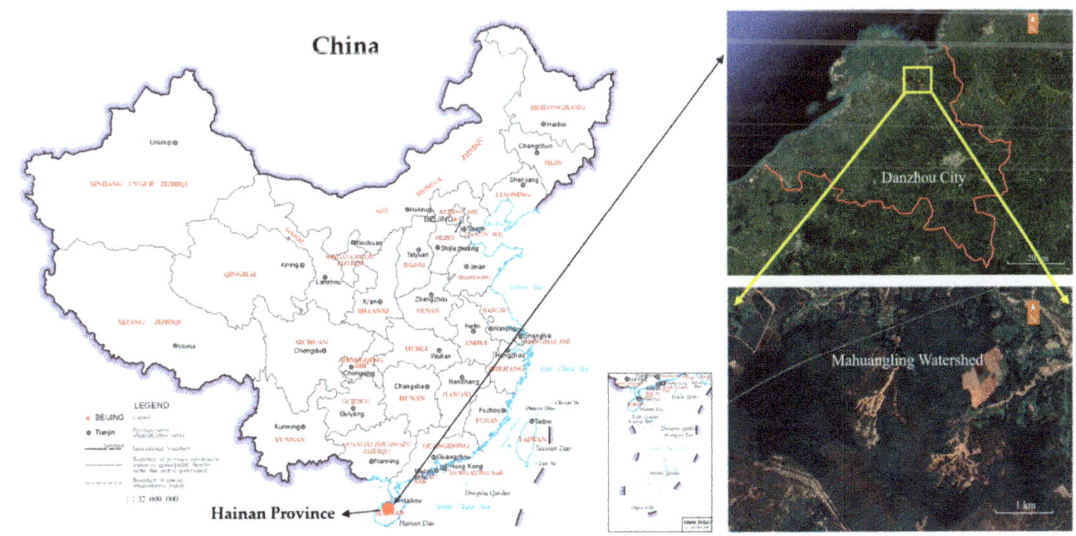

Figure 1. The locations of the sampling sites at the Mahuangling Soil and Water Conservation Monitoring Station.

2.2. Litter and Soil Samples Collection

In mid-January 2021, field sampling from the four forest types was conducted. Three 10 m × 40 m plots were established within each forest type, located in areas away from the road and relatively free from human disturbance. Each plot was selected following the methodology for field long-term observation of forest ecosystem, the National Standards of the People's Republic of China (GB/T 33027-2016). At each plot, we measured the tree height, the trunk diameter at breast height (1.3 m), and the canopy density. The basic information of the sampling plots is summarized in Table 1.

Table 1. The basic characteristics of the sampling plots.

Forest Type	Dominant Tree Species	Stand Age (Years)	Average Tree Height (m)	Average Diameter at Breast Height (cm)	Tree Density (Trees·ha^{-1})	Canopy Density	Slope Gradient (°)
Eucalyptus–grass	*E. robusta*	10	10.95 ± 2.15	10.10 ± 1.31	2500	0.53	2~5
Bamboo–grass	*Bambusa oldhamii*	10	3.52 ± 0.71	6.90 ± 0.32	2500	0.42	2~5
Acacia–grass	*A. mangium*	10	8.51 ± 1.13	8.70 ± 0.52	2500	0.75	2~5
Shrub–grass	*Melastoma candidum*	10	1.52 ± 0.25	1.70 ± 0.21 ★	2500	0.35	2~5

Note: Data are presented as mean ± S.D. ★: Ground diameter of shrub.

Within each forest type, we randomly selected five 0.5 m × 0.5 m quadrats for litter sampling. Intact litter layers were collected, and the semi-decomposed litter (SL layer) and undecomposed litter (UL layer) were graded and bagged separately following the method described in the National Standards. A total of 120 litter bags (2 litter layers × 5 random quadrats × 3 plots × 4 species) were collected. Around each quadrat, we randomly selected four points to measure the average thickness of the SL and UL layers representing that quadrat. The thickness of the total litter layer, as well as its SL and UL layer components, was recorded. The soil samples were collected at depths of 0 to 10 cm, 10 to 20 cm, 20 to 40 cm, and 40 to 60 cm soil layers from soil profiles using the cutting ring (100 cm^3) method. Three soil profiles were randomly selected for each standard sample plot and soil was extracted from each layer with two cutting rings. A total of 288 soil samples (4 forest types × 3 plots × 3 soil profiles × 4 soil layers × 2 cutting rings) were collected.

2.3. Laboratory Analyses

The mass of fresh litter was determined after the litter samples were brought back to the laboratory, and the litter was left to air-dry in the lab. The amount of litter mass (m_0) was determined by oven-drying samples at 75 °C [5]. The water-holding capacity of the litter was evaluated via the indoor water soaking method [28,37], where the litter was soaked in water for a predetermined time (0.25, 0.5, 1, 2, 4, 6, 8, 12, or 24 h). The percentage of water held in the litter samples after soaking was calculated as the water absorption rate. The amount and the percentage of water-holding during a 24 h soaking period were considered the maximum water-holding capacity of the litter (R_m) and calculated as follows [28,37]:

$$R_i = (m_i - m_0)/m_0 \times 100 \quad (1)$$

where R_i is the water-holding capacity of the litter at the immersion time i (i = 0.25, 0.5, 1, 2, 4, 6, 8, 12, or 24 h), m_0 is the dry litter mass, and m_i is the litter mass at the immersion time i after free drainage.

$$R_0 = (m_f - m_0)/m_0 \times 100 \quad (2)$$

Here, R_0 is the water-holding capacity of the litter under ambient conditions (%), m_0 is the dry litter mass, and m_f is the fresh litter mass.

$$R_m = (m_{24} - m_0)/m_0 \times 100 \quad (3)$$

Here, R_m is the maximum water-holding capacity of the litter (%), m_0 is the dry litter mass, and m_{24} is the litter mass soaked for 24 h after free drainage. The R_m was defined as the maximum levels of the respective parameters relative to the amount and the percentage of water-holding during a 24 h soaking period [28,37].

The litter effective water-retention capacity and maximum water-retention capacity were calculated as follows [28,37,47]:

$$W_{eff} = (0.85R_m - R_0)M \tag{4}$$

$$W_{max} = (R_m - R_0)M \tag{5}$$

where W_{eff} and W_{max} are the effective water-retention capacity (t·ha^{-1}) and maximum water-retention capacity (t·ha^{-1}), respectively; M is the unit litter mass (t·ha^{-1}).

Soil samples were brought back to the laboratory to determine physical properties of the soil. The oven-drying method was used to determine the soil water content. The cutting ring method was used to determine the soil bulk density, capillary porosity, and total porosity [30,50,51]. The oil water-holding capacity in different vegetation restoration types was calculated using the following equation:

$$S = 10{,}000 \times hp \tag{6}$$

where S is the water-holding capacity (t·ha^{-1}), h is the depth of soil layer (m), and p is the non-capillary porosity (%).

2.4. Comprehensive Evaluation of Water Conservation Capacity

To compare the hydrological properties of the different forest types more intuitively, we quantified the factors of the hydrological properties of the soil layer and litter layer using the entropy weight method (EWM). This was calculated under a standard system, and then we comprehensively evaluated the water conservation capacity. Based on the EWM, the calculation procedures were conducted via the following steps [30,33,52]:

Step 1. Construct decision matrix.

The set of evaluation indicators and objects are defined as $I = (I_1, I_2, I_3, \cdots, In)$, and $O = (O_1, O_2, O_3, \cdots, Om)$, respectively. The evaluation state value of each object set O_i against the indicator set I_i is expressed as x_{ij} ($i = 1,2,\cdots$, m; $j = 1,2,\cdots$, n). Thus, the decision matrix **X** of original data can be expressed as

$$X = \begin{pmatrix} x_{11} & \cdots & x_{1n} \\ \vdots & \ddots & \vdots \\ x_{m1} & \cdots & x_{mn} \end{pmatrix} \tag{7}$$

Step 2. Raw data standardization.

The original data matrix was normalized. For the indicator, the larger the value, the better:

$$r_{ij} = \frac{x_{ij} - \min(x_{ij})}{\max(x_{ij}) - \min(x_{ij})} \tag{8}$$

By contrast, the smaller the value, the better:

$$r_{ij} = \frac{\max(x_{ij}) - x_{ij}}{\max(x_{ij}) - \min(x_{ij})} \tag{9}$$

where r_{ij} is the normalized value of x_{ij}. x_{ij} is the value of the j-th indicator on the i-th object. Max(x_{ij}) and min(x_{ij}) are the maximum and minimum values of the j-th indicator, respectively.

Step 3. Calculate the feature weight.

The feature weight of the j-th indicator of the i-th object is p_{ij}, defined as

$$p_{ij} = \frac{r_{ij}}{\sum_{i=1}^{m} r_{ij}} \tag{10}$$

where p_{ij} is the feature weight of the j-th indicator of the i-th object.

Step 4. Calculate the entropy:

$$e_j = -\frac{1}{\ln m} \sum_{i=1}^{m} p_{ij} \ln p_{ij} \tag{11}$$

where $p_{ij} = 0$, $p_{ij} \ln p_{ij}$ is defined as 0. e_j is the entropy of the j-th indicator of the i-th object.

Step 5. Calculate the weight parameters.

As a result, the weight parameters are calculated as follows:

$$W_i = \frac{1 - e_j}{n - \sum_{j=1}^{n} e_j} \tag{12}$$

where W_i is the weight parameters of the j-th indicator of the i-th object.

Step 6. Calculate the water conservation capacity.

The water conservation capacity was calculated as follows [30,33]:

$$WCCI = \sum_{I=1}^{m} W_i \, r_{ij} \tag{13}$$

where $WCCI$ is the water conservation capacity index.

2.5. Statistical Analysis

The differences in the hydrological properties of the litter layer and soil layer were analyzed using a one-way analysis of variance (ANOVA). The least significant differences (LSD) were determined for multiple comparisons. The significance level was set at $p < 0.05$. The above analyses were completed with SPSS v. 18.0 (SPSS Inc., Chicago, IL, USA). Origin 2021 software (Origin, Origin Lab, Farmington, ME, USA) was used to plot the figures.

3. Results

3.1. Litter Thickness and Mass

Significant differences in the total litter thickness among the different forest types ($p < 0.001$) were found and are shown in Table 2. The litter thickness of the UL layer and SL layer differed significantly among the four forest types ($p < 0.05$), and these differences were dependent on the forest types ($p < 0.001$). The total litter thickness was the greatest in the acacia–grass forest (3.74 ± 0.38 cm), followed by the bamboo–grass forest (3.54 ± 0.31 cm), eucalyptus–grass forest (2.19 ± 0.09 cm), and shrub–grass forest (1.71 ± 0.10 cm). The same pattern was found for the SL layer thickness (Figure 2a). The bamboo–grass forest had the greatest SL thickness, followed by the eucalyptus–grass forest, acacia–grass forest, and shrub–grass forest. The thickness of the SL layers in the bamboo–grass forest and acacia–grass forest was larger than the thickness of their UL layers, but the SL layers of the eucalyptus–grass forest and shrub–grass forest were lower than the thickness of their UL layers ($p < 0.05$).

Table 2. The effects on the litter thickness, mass, total litter thickness, and total litter mass by tree species and litter layer (SL and UL) based on the analysis by F-statistics from factorial ANOVA. df: degrees of freedom; SS: sum of squares; MS: mean square; ***: $p < 0.001$.

Variable	Factor	df	SS	MS	F	*p*
Litter thickness	tree	3 112	17.48	5.83	68.36	<0.001 ***
	layer	1 112	1.80	1.80	21.15	<0.001 ***
	tree × layer	3 112	15.94	5.31	62.33	<0.001 ***
Litter mass	tree	3 112	9.20	3.07	21.35	<0.001 ***
	layer	1 112	29.88	29.88	208.11	<0.001 ***
	tree × layer	3 112	0.87	0.29	2.01	0.117
Total thickness	tree	3 56	44.88	14.96	233.34	<0.001 ***
Total mass	tree	3 56	18.42	6.14	26.27	<0.001 ***

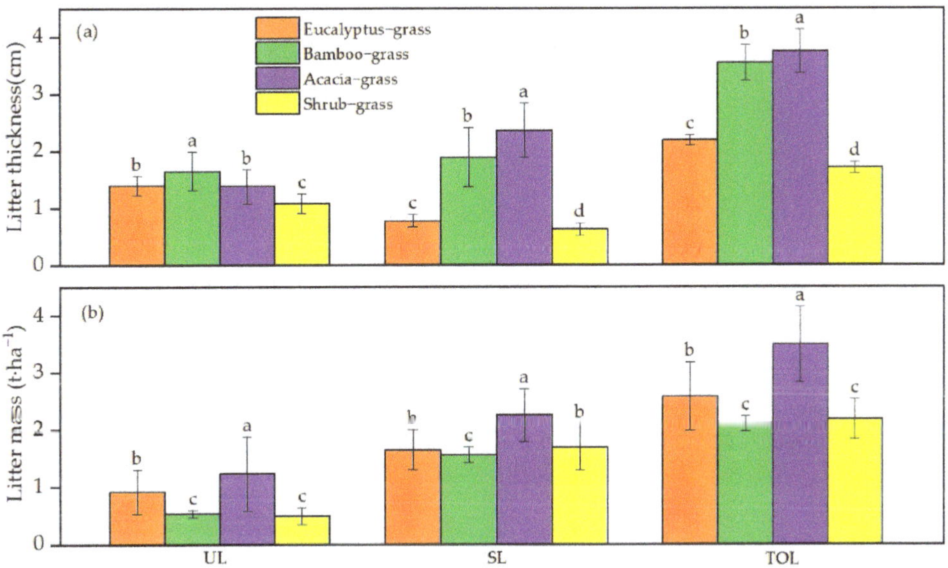

Figure 2. Variations in litter thickness (**a**) and litter mass (**b**) in the undecomposed litter (UL) layer and semi-decomposed litter (SL) layer of the forest type of eucalyptus–grass, bamboo–grass, acacia–grass, and shrub–grass. The different lowercase letters in each category indicate a significant difference among forest types.

Significant differences were observed in the litter mass among the different forest types and litter layers ($p < 0.001$) (Table 2). The acacia–grass forest had the highest total litter mass (3.49 ± 0.66 t·ha^{-1}), which was significantly higher than the eucalyptus–grass forest (2.58 ± 0.60 t·ha^{-1}), shrub–grass forest (2.18 ± 0.35 t·ha^{-1}), and bamboo–grass forest (2.10 ± 0.13 t·ha^{-1}) ($p < 0.001$) (Figure 2b). The litter mass of the SL layer was larger than the UL layer in all four forest types ($p < 0.05$) (Figure 2b). The litter mass of the SL layer was 2.26 ± 0.46 t·ha^{-1} for the acacia–grass forest, which was higher than the masses of the shrub–grass forest (1.69 ± 0.41 t·ha^{-1}), eucalyptus–grass forest (1.65 ± 0.36 t·ha^{-1}), and bamboo–grass (1.56 ± 0.14 t·ha^{-1}) ($p < 0.001$). There were significant differences in the mass of the UL layer, which decreased in the order of acacia–grass > eucalyptus–grass > bamboo–grass > shrub–grass ($p < 0.001$). The SL litter mass accounted for 64.97% ± 9.53% of the total litter mass in the eucalyptus–grass forest, 74.42% ± 3.61% in the bamboo–grass forest, 66.38% ± 14.18% in the acacia–grass forest, and 76.77% ± 8.07% in the shrub–grass forest.

3.2. R_m, W_{eff}, and W_{max}

The R_m in the SL layer was 292.88% ± 30.24% for the bamboo–grass forest and was not significantly different from the eucalyptus–grass forest (289.17% ± 33.21%). However, both were significantly greater than the acacia–grass forest (223.88% ± 37.78%) and shrub–grass forest (231.84% ± 38.20%) ($p < 0.05$) (Figure 3a). The R_m of the UL layer ranged from 232.94% ± 22.57% to 250.22% ± 20.05%, and no significant difference was found between the eucalyptus–grass forest and the bamboo–grass forest or between the acacia–grass forest and the shrub–grass forest ($p > 0.05$). The R_m of the bamboo–grass forest was 262.91% ± 21.46%, which is roughly equivalent to that of the eucalyptus–grass forest (261.51% ± 23.13%); both stands were significantly higher than those of the shrub-grass forest (241.03% ± 22.94%) and acacia-grass forest (232.95% ± 22.77%) ($p < 0.05$). The R_m of the SL layer was larger than that of the UL layer in the bamboo–grass forest and eucalyptus–grass forest, but the opposite trend was observed for the acacia–grass forest and shrub–grass forest.

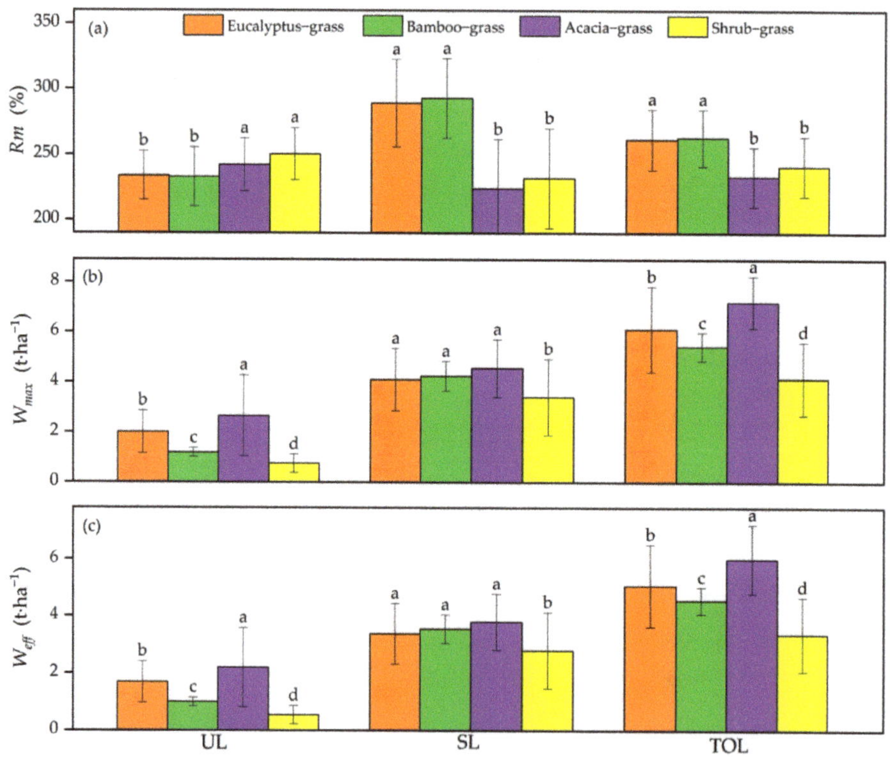

Figure 3. Variations in the R_m (**a**), W_{max} (**b**), and W_{eff} (**c**) in the undecomposed litter (UL) layer and semi-decomposed litter (SL) layer of the forest types of eucalyptus–grass, bamboo–grass, acacia–grass, and shrub–grass. The different lowercase letters in each category indicate a significant difference among the forest types.

The W_{max} differed significantly among the forest types ($p < 0.05$). The W_{max} of the acacia–grass forest was 7.20 ± 1.04 t·ha^{-1}, significantly greater than that of the eucalyptus–grass forest (6.11 ± 1.71 t·ha^{-1}), and higher than those of the bamboo–grass forest (5.42 ± 0.56 t·ha^{-1}) and shrub-grass forest (4.14 ± 1.46 t·ha^{-1}) ($p < 0.05$) (Figure 3b). The W_{max} of the SL layer was 4.65 ± 1.15 t·ha^{-1} for the acacia–grass forest, 4.24 ± 0.59 t·ha^{-1} for the bamboo–grass forest, and 4.10 ± 1.24 t·ha^{-1} for the eucalyptus–grass forest and was not observed to be significantly different. However, for all these vegetation restoration types, they were significantly larger than that for the shrub-grass forest (3.40 ± 1.52 t·ha^{-1}) ($p < 0.05$). Moreover, the W_{max} of the UL layer decreased in the following order: acacia–grass forest > eucalyptus–grass forest > bamboo–grass forest > shrub–grass forest ($p < 0.05$), ranging from 0.74 ± 0.37 t·ha^{-1} to 2.65 ± 1.62 t·ha^{-1}. The W_{max} of the SL layer was larger than the W_{max} of the UL layer in the four forest types ($p < 0.05$).

The W_{eff} was similar to the W_{max}. The W_{eff} significantly differed among the four forest types ($p < 0.05$). The W_{eff} of the acacia–grass forest was 6.00 ± 1.21 t·ha^{-1}, which was higher than the 5.07 ± 1.10 t·ha^{-1} of the eucalyptus–grass forest, the 4.55 ± 0.47 t·ha^{-1} of the bamboo–grass forest, and the 3.36 ± 1.30 t·ha^{-1} of the shrub–grass forest ($p < 0.05$) (Figure 3c). The acacia–grass forest had the greatest interception capacity (note that 1 mm of precipitation is equivalent to 1 t·ha^{-1}), which was equal to a 6.00 mm depth equivalent of rainfall, with 5.07 mm in the eucalyptus–grass forest, 4.55 mm in the bamboo–grass forest, and 3.36 mm in the shrub–grass forest. The SL layer W_{eff} did not differ significantly among the acacia–grass forest (3.80 ± 0.98 t·ha^{-1}), eucalyptus–grass forest (3.38 ± 1.06 t·ha^{-1}), and bamboo–grass forest (3.55 ± 0.50 t·ha^{-1}) ($p > 0.05$), but all were significantly higher than the 2.80 ± 1.33 t·ha^{-1} observed for the shrub–grass forest ($p < 0.05$). The UL layer W_{eff} significantly decreased in the following order: acacia–grass forest > eucalyptus–grass > bamboo–grass forest > shrub–grass forest ($p < 0.05$) and ranged from 0.55 ± 0.32 t·ha^{-1} to 2.20 ± 1.38 t·ha^{-1}. The SL layer W_{eff} was greater than the UL layer W_{eff} among the four forest types ($p < 0.05$).

3.3. Variations in Water-Holding Capacity of Litter

The water-holding capacity varied among the forest types. The water-holding ratio increased with increasing immersion time. The water-holding ratio of the SL layer litter was relatively higher than that of the UL layer at the same immersion time. After 0.25 h of water immersion, the water-holding ratio of the UL and SL layers reached 0.94 ± 0.40 and 2.62 ± 0.84 t·ha^{-1}, respectively, for the eucalyptus–grass forest, 0.80 ± 0.16 and 2.66 ± 0.51 t·ha^{-1} for the bamboo–grass forest, 1.26 ± 0.67 and 2.69 ± 0.77 t·ha^{-1} for the acacia–grass forest, and 0.54 ± 0.16 and 2.17 ± 0.77 t·ha^{-1} for the shrub–grass forest (Figure 4). The water-holding ratio slowly increased after two hours of water immersion and the UL and SL layers in the acacia–grass forest were higher than those in the other forest types (Figure 4). A logarithmic relationship was fitted between the water-holding ratio and the immersion times for both litter types of all four forest types.

3.4. Variations in Litter Water Absorption Rate

The water absorption rate was greatest at the beginning of the experiment and declined rapidly in the first 2 h. The water absorption rate of the SL layer was larger than the UL layer at the same immersion time, and the rate slowed until 12 h (Figure 5). With similar immersion times, a significant difference was observed in the water absorption rates among the forest types. The water absorption rate of the acacia–grass forest was larger than those of the eucalyptus–grass forest, bamboo–grass forest, and shrub–grass forest. After the first hour, the water absorption rates of the SL layer were 3.42 ± 0.55 t·ha^{-1}·h^{-1}, 3.55 ± 0.17 t·ha^{-1}·h^{-1}, 3.50 ± 0.98 t·ha^{-1}·h^{-1}, and 2.79 ± 0.25 t·ha^{-1}·h^{-1} for the eucalyptus–grass, bamboo–grass, acacia–grass, and shrub–grass forests, respectively (Figure 5b,d,f,h). Additionally, the water absorption rates of the UL layer were 1.34 ± 0.83 t·ha^{-1}·h^{-1}, 1.02 ± 0.51 t·ha^{-1}·h^{-1}, 1.83 ± 0.65 t·ha^{-1}·h^{-1}, and 0.78 ± 0.25 t·ha^{-1}·h^{-1} for the eucalyptus–grass, bamboo–grass, acacia–grass, and shrub–grass forests, respectively (Figure 5a,c,e,g). An exponential relationship was observed between the water absorption rate and the immersion time in both types of litter in the four forest types.

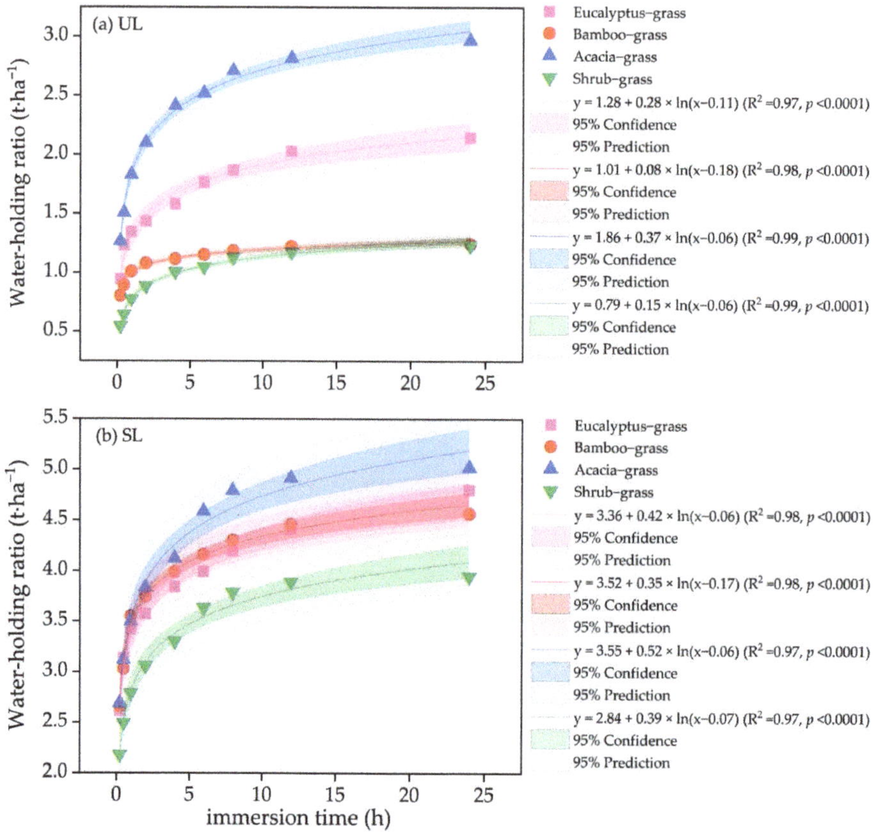

Figure 4. Logarithmic relationship between the water-holding ratio and immersion time of (**a**) undecomposed litter (UL) layer and (**b**) semi-decomposed litter (SL) layer of eucalyptus–grass, bamboo–grass, acacia–grass, and shrub–grass.

3.5. Variations in Soil Water-Holding Capacity

The soil bulk density increased with the soil depth in all forest types. The total porosity of the soil gradually decreased with increasing soil depth in all stands. No significant differences were observed for the soil bulk density or total porosity among the forest types ($p > 0.05$) (Table 3). At the 0 to 60 cm soil layer, the eucalyptus–grass forest had the highest soil bulk density (1.54 ± 0.08 g·cm^{-3}), followed by the shrub–grass forest, bamboo–grass forest, and acacia–grass forest (1.46 ± 0.05 g·cm^{-3}) (Table 3). The soil porosity did not significantly differ among the forest types. The non-capillary porosity was (5.13 ± 0.60)%, (4.36 ± 0.27)%, (3.99 ± 0.59)%, and (3.95 ± 0.99)% for the acacia–grass forest, eucalyptus–grass forest, shrub–grass forest, and bamboo–grass forest, respectively. The capillary porosity was (32.00 ± 1.15)%, (30.87 ± 5.21)%, (29.04 ± 4.66)%, and (28.06 ± 1.00)% for the acacia–grass forest, bamboo–grass forest, eucalyptus–grass forest, and shrub–grass forest, respectively. The total soil porosity was (37.13 ± 1.51)%, (34.82 ± 5.94)%, (33.41 ± 4.79)%, and (32.06 ± 1.35)% for the acacia–grass forest, bamboo–grass forest, eucalyptus–grass forest, and shrub–grass forest, respectively(Table 3). Additionally, at the 0 to 60 cm soil layer depth, the soil water-holding capacity was highest in the acacia–grass forest (301.76 t·ha^{-1}), followed by the eucalyptus–grass forest (263.53 t·ha^{-1}) and then the shrub–grass forest (233.58 t·ha^{-1}), and the bamboo–grass forest had the lowest value (220.78 t·ha^{-1}) (Table 3).

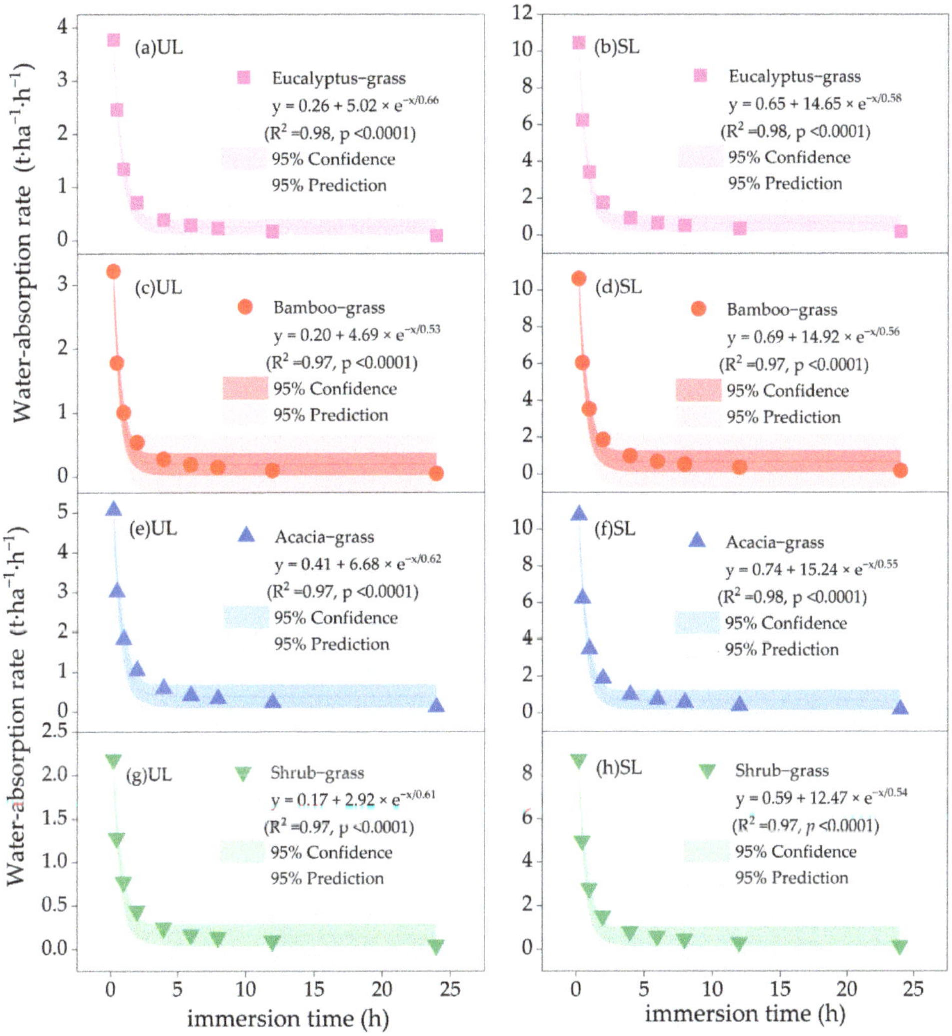

Figure 5. Exponential relationship between the water absorption rate and immersion time of the undecomposed litter (UL) layer for (**a**) eucalyptus–grass, (**c**) bamboo–grass, (**e**) acacia–grass, and (**g**) shrub–grass, and semi-decomposed litter (SL) layer for (**b**) eucalyptus–grass, (**d**) bamboo–grass, (**f**) acacia–grass, and (**h**) shrub–grass.

3.6. Comprehensive Evaluation of Water Conservation Capacity

The water conservation capacities of the four forest types were evaluated via EWM. The indexes of the litter layer (litter thickness, litter mass, R_m, W_{eff}, and W_{max}) and soil layer (bulk density, capillary porosity, non-capillary porosity, total porosity, soil water-holding capacity) were selected based on the principles of scientific hierarchy. The weighted value of each index was calculated, and they ranged from 0.0770 to 0.1579 (Table 4).

Table 3. Soil water-holding capacity in different forest types. The different lowercase letters in each category indicate a significant difference among the forest types.

Forest Type	Soil Depth (cm)	Bulk Density ($g \cdot cm^{-3}$)	Non-Capillary Porosity (%)	Capillary Porosity (%)	Total Porosity (%)	Soil Water-Holding Capacity ($t \cdot ha^{-1}$)
Eucalyptus–grass	0 to 10	1.43 ± 0.08 a	4.40 ± 2.10 a	32.26 ± 2.65 a	36.66 ± 1.91 a	44.00 ± 5.82 b
	10 to 20	1.53 ± 0.03 a	4.17 ± 2.36 a	31.32 ± 1.72 a	35.48 ± 0.98 a	41.70 ± 13.78 b
	20 to 40	1.59 ± 0.06 a	4.74 ± 1.38 a	30.46 ± 3.12 a	35.19 ± 2.83 a	94.80 ± 11.92 a
	40 to 60	1.60 ± 0.08 a	4.15 ± 0.58 a	22.14 ± 10.18 a	26.29 ± 9.74 b	83.03 ± 11.28 a
Bamboo–grass	0 to 10	1.45 ± 0.12 a	5.16 ± 0.52 a	34.33 ± 2.54 a	39.48 ± 2.74 a	51.60 ± 30.05 b
	10 to 20	1.50 ± 0.02 a	4.37 ± 2.52 a	34.15 ± 1.94 a	38.52 ± 2.00 a	43.77 ± 7.13 b
	20 to 40	1.51 ± 0.09 a	3.11 ± 0.71 a	31.75 ± 0.74 a	34.85 ± 0.29 a	62.21 ± 10.37 a
	40 to 60	1.45 ± 0.09 a	3.16 ± 3.01 a	23.26 ± 3.64 b	26.42 ± 4.30 b	63.20 ± 10.4 a
Acacia–grass	0 to 10	1.41 ± 0.10 a	5.96 ± 3.55 a	33.12 ± 0.53 a	39.07 ± 3.75 a	59.61 ± 10.64 c
	10 to 20	1.44 ± 0.06 a	4.95 ± 1.51 a	32.46 ± 4.24 a	37.40 ± 3.05 a	49.51 ± 9.17 c
	20 to 40	1.54 ± 0.04 a	4.54 ± 0.92 a	32.00 ± 1.84 a	36.54 ± 2.04 a	90.81 ± 10.16 b
	40 to 60	1.46 ± 0.13 a	5.09 ± 1.06 a	30.41 ± 3.12 a	35.50 ± 3.02 a	101.83 ± 11.07 a
Shrub–grass	0 to 10	1.44 ± 0.12 a	4.81 ± 3.55 a	28.74 ± 1.78 a	33.54 ± 5.11 a	48.11 ± 11.01 b
	10 to 20	1.47 ± 0.05 a	3.80 ± 1.51 a	29.06 ± 5.02 a	32.85 ± 4.01 a	38.03 ± 9.17 b
	20 to 40	1.56 ± 0.04 a	3.98 ± 1.10 a	26.93 ± 1.78 a	30.91 ± 1.56 a	79.63 ± 10.16 a
	40 to 60	1.50 ± 0.14 a	3.39 ± 0.92 a	27.53 ± 1.53 a	30.93 ± 0.67 a	67.81 ± 11.07 a

Table 4. Weighted values of water conservation capacity indexes in different forest types.

Grade I Index	Weight	Serial No.	Grade II Index	Weight
Litter layer	0.4826	X1	Litter thickness	0.0936
		X2	Litter mass	0.1467
		X3	R_m	0.0903
		X4	W_{eff}	0.0770
		X5	W_{max}	0.0751
Soil layer	0.5174	X6	Bulk density	0.0678
		X7	Non-capillary porosity	0.1579
		X8	Capillary porosity	0.0913
		X9	Total porosity	0.0918
		X10	Soil water-holding capacity	0.1085

The comprehensive evaluation value of the water conservation capacity of each forest type was calculated by combining the weight and the normalized index of each index. The WCCI was significantly the highest in the acacia–grass forest (0.5257), followed by the eucalyptus–grass forest (0.2310), bamboo–grass forest (0.1941), and shrub–grass forest, which had the lowest water conservation capacity (0.0492) (Table 5).

Table 5. Comprehensive evaluation of water conservation capacity in different forest types.

Forest Type	Water Conservation Capacity		Comprehensive Evaluation Value	Rank
	Litter Layer	Soil Layer		
Eucalyptus–grass	0.1218	0.1092	0.2310	2
Bamboo–grass	0.0938	0.1003	0.1941	3
Acacia–grass	0.2061	0.3197	0.5257	1
Shrub–grass	0.0352	0.0140	0.0492	4

4. Discussion

Broad-leaved trees as the dominant species in forest ecosystems significantly increase the interception capacity of rainfall [40]. A thick litter layer may be conducive to water retention in shallow soil [10,28,34], which is due to a reduction in evaporation from the surface soil [26,27]. The total litter thickness varied from 1.71 to 3.74 cm in this study and was the highest in the acacia–grass forest, suggesting that acacia–grass forest greatly reduces evaporation compared to other forests. This result may be attributed to the *A. mangium* tree having a larger leaf area. The lengths and widths of the leaves in the four forests are different; the leaves of *A. mangium* are 10.0–25.0 cm long and 5.0–10.0 mm wide, the leaves of *B. oldhamii* are 12.0–30.0 cm long and 2.5–5.0 mm wide, the leaves of *E. robusta* are 8.0–17.0 cm long and 2.5–6.0 mm wide, and the leaves of *B. oldhamii* are 4.0–10.0 cm long and 3.5–8.0 mm wide. The litter thickness of the SL layer of the acacia–grass and bamboo–grass forests was higher than the thickness of the UL, but the eucalyptus–grass and shrub–grass forests demonstrated the opposite. This suggests that the litter of *H. brasiliensis*, *M. candidum*, and *U. lobata* decomposes more easily than that of *A. mangium* and *B. oldhamii* under field conditions. Previous studies have reported a decreased thickness of the SL litter when compared to the UL litter in broad-leaved forests [37,41]. Additionally, significant differences in the litter mass among the forest types and litter layers were found, and it was exceptionally larger in the acacia–grass forest. This indicates that acacia–grass forest is conducive to rainfall interception. The litter mass in the SL layer accounted for 64.97%~76.77% of the total litter mass, which was greater than the UL layer, and aligns with the previous reports [28,37,47]. However, more rainfall interception by acacia–grass litter may evaporate back into the atmosphere [40,42,53]. Therefore, long-term monitoring is needed to determine the net effect of the litter on the forest's surface water flux.

The R_m defines the effectiveness per unit of mass litter in retaining water [28,32,47]. Forest ecosystems differ in terms of their water-holding capacity due to the variations in their litter coverage and litter decomposition rates [10,36,53,54]. In this study, the eucalyptus–grass and bamboo–grass stands demonstrated a higher R_m than the acacia–grass and shrub–grass stands. This difference in hydrological properties could be explained by the differences in the physical and chemical properties of the litter composition, such as the leaves, dead branches, and seeds. Moreover, this difference could be explained by the litter defoliation period since the litter components of acacia–grass and shrub–grass have thicker cuticles than eucalyptus–grass and bamboo–grass. In addition, we found that the R_m of the SL layer in the eucalyptus–grass and bamboo–grass forest was larger than that of the UL layer. The opposite was true in the acacia–grass and shrub–grass forests. Our results show that eucalyptus–grass and bamboo–grass had a higher degree of fragmentation and fewer dead branches in general [27,47]. These results align with Li et al. [37] in 2015 and Chen et al. [28] in 2018. Further investigations focusing on litter decomposition rates and processes are needed to better understand the differences among different vegetation types.

W_{max} is a measure of rainfall absorption and, thus, a decrease in runoff [37], which is dependent on the litter water content under ambient conditions, the litter mass, and the nature of rainfall [42,55]. Our study shows that the W_{max} of the acacia–grass forest was greater than those of the other forest types. These results could be explained by the larger litter mass of the acacia–grass forest being more effective than the other forest types. This is consistent with the previous studies that have shown that the W_{max} depends on the litter storage [28,34,36]. Dong et al. [56] reported that the W_{max} of *A. mangium* was larger than that of *E. robusta*. Additionally, we found that the W_{max} of the SL layer was greater than the UL due to a larger litter mass and a more effective SL layer. Zhou et al. [47] and Chen et al. [28] have reported a higher water-retention capacity of the SL litter layer compared to the UL litter layer. In contrast, Li et al. [37] reported that the maximum water-retention capacity of the SL litter layer was lower than that of the UL litter layer. A comprehensive investigation into the litter characteristics across different forest ecosystems in different climate regions is needed.

The W_{eff} defines the effective interception of precipitation by litter, which is an important hydrological property that can be used to consistently evaluate the potential to absorb rainfall and reduce surface runoff [31,37,53]. The W_{eff} is also affected by the water content of litter, litter storage, and the nature of rainfall [27,40,45]. In this study, we found that the litter layer can theoretically intercept up to 6.00 mm of rainfall in the acacia–grass forest, 5.07 mm in the eucalyptus–grass forest, 4.55 mm in the bamboo–grass forest, and 3.36 mm in the shrub–grass forest. The results suggest that the acacia–grass forest had a higher capacity to minimize splash erosion and decrease runoff and had the greatest water-retention capacity, making it more effective at rainfall interception [26,29,57]. This is likely overestimated due to the large intact leaves of the litter layer [37]. Nevertheless, whether the W_{eff} could be reached in situ during rainfall is unknown [28]. Therefore, the measured litter water content in situ should be further investigated to provide a more realistic evaluation of rainfall interception by litter.

The rapid water absorption rates of the SL land UL layers at the beginning of the experiment slowed down until 12 h and then were nearly unchanged. These results are likely explained by the drier litter having a lower matrix water potential in the first two hours after onset and becoming moist after 12 h. Previous studies have found the same trend in the litter water-holding ratio [28,30,33,56,57]. Additionally, the water-holding ratio of the SL layer was relatively larger than that of the UL layer. This may be attributed to the SL layer having a greater degree of fragmentation [10,37,53].

The soil layer of the forest ecosystem is a key component for water conservation. The water-holding capacity of the soil layer directly affects the surface runoff, soil subsurface flow, and groundwater recharge [30,37,44,56]. Variation in the hydro-physical properties of the soil layers indicates a likely dependence on the soil structure and pore space size [29]. In this study, our results show that the soil bulk density ranged from 1.46 $g{\cdot}cm^{-3}$ to 1.54 $g{\cdot}cm^{-3}$, and the total porosity ranged from 32.06% to 37.13%. The soil bulk density increased with increasing soil depth, while the total porosity gradually decreased. The results are consistent with the previous research [30,33,37]. Lower soil bulk density was associated with greater soil porosity and was responsible for increasing the water-holding capacity in the acacia–grass forest (301.76 $t{\cdot}ha^{-1}$). This was 1.37, 1.29, and 1.14-fold higher than that of the bamboo–grass, shrub–grass, and eucalyptus–grass forests, respectively. The higher soil total porosity in the acacia–grass forest may be attributed to the larger litter storage on the forest floor, which led to soil structural changes [17,56], improving the soil structure and porosity [30,33]. Further studies are needed to focus on the soil chemical properties for a better understanding of the hydrological properties of the soil layers, specifically, the soil organic matter, soil enzyme activities, and microbe communities among the different forest types.

We observed that the water storage capacity of the soil layer was larger than the litter layer, which is consistent with Bai et al. [30] in 2021 and Cheng et al. [33] in 2021. The *WCCI* was highest in the acacia–grass forest compared to the other forest types. From the perspective of the water conservation ecological service after vegetation restoration, our results suggest that the acacia–grass forest had a dense litter layer suitable for intercepting rainfall, reducing runoff, and buffering the impact of rainfall. Consequently, this improves the soil structure and size of the pore spaces in the soil layer, creating suitable conditions for infiltration and water holding in the soil, and thus reducing the loss of surface water [30,33,37,56]. Therefore, the vegetation restoration type of acacia–grass can improve the hydrological effects in the gully erosion area of latosol. In our study, we conducted experiments only under laboratory conditions. Further research should consider the climate and environmental factors affecting the hydrological properties of the litter and soil layer by way of in situ and long-term location investigation.

5. Conclusions

In the gully erosion area of latosol in the Mahuangling Watershed of Hainan province in China, we studied the hydrologic properties of the litter and soil layers in four restored plantation types. It was found that the litter thickness and litter mass in the acacia–grass forest were higher than those of the other forests. The effective water-retention capacity of the UL layer was relatively larger than that of the SL layer, especially in the acacia–grass forest, and is likely more effective for intercepting rainfall and reducing surface runoff. Lower soil bulk density was associated with greater porosity in the acacia–grass forest, which increases the soil water-holding capacity, thus reducing the loss of surface water. The comprehensive evaluation value of the water conservation capacity shows that the water storage capacity of the soil layer was larger than the litter layer. The *WCCI* in the acacia–grass forest was greater than in the other vegetation restoration types. This indicates that acacia–grass forest generally has favorable hydrological properties and plays an important role in water conservation during vegetation restoration. We suggest acacia–grass forest for restoration to mitigate soil erosion and improve hydrological effects in the gully erosion area of latosol in the future.

Author Contributions: Investigation, Z.T., S.C., D.R., Z.C., W.Z., Y.H. (Yujie Han), L.H., S.C., D.R., K.W. and Y.H. (Yanping Huang); methodology, Z.T.; software, Z.T.; validation, Z.T.; formal analysis, Z.T. and J.C.; data curation, Z.T. and S.C.; writing—original draft preparation, Z.T. and J.C.; writing—review and editing, Z.T. and J.C. All authors have read and agreed to the published version of the manuscript.

Funding: This research was funded by the National Natural Science Foundation of China, grant number 42267048; the Hainan Provincial Natural Science Foundation of China, grant numbers 420RC532 and 319QN159; the Scientific Research Fund of Hainan University, grant number KYQD(ZR)1950; and the Fund of Department of Water Resources of Hainan Province, grant numbers HD-KYH-2022350, HD-KYH-2022238, HD-KYH-2022017, and HD-KYH-2021062.

Institutional Review Board Statement: Not applicable.

Informed Consent Statement: Not applicable.

Data Availability Statement: The data that support the findings of this study are available from the corresponding author upon reasonable request.

Acknowledgments: We thank the reviewers and the editor for their valuable work and comments.

Conflicts of Interest: The authors declare no conflict of interest.

References

1. Borrelli, P.; Robinson, D.A.; Fleischer, L.R.; Lugato, E.; Ballabio, C.; Alewell, C.; Meusburger, K.; Modugno, S.; Schütt, B.; Ferro, V. An assessment of the global impact of 21st century land use change on soil erosion. *Nat. Commun.* **2017**, *8*, 2013. [CrossRef]
2. Xia, J.; Zhang, L.; Ge, P.; Lu, X.; Wei, Y.; Cai, C.; Wang, J. Structure degradation induced by wetting and drying cycles for the hilly granitic soils in collapsing gully erosion areas. *Forests* **2022**, *13*, 1426. [CrossRef]
3. Zhang, L.; Xiao, T.; Liu, H.; Ge, P.; Xia, J.; Dai, C.; Zhang, W.; Zhao, X. Effects of AM fungi and grass strips on soil erosion characteristics in red sandstone erosion areas in Southern China. *Forests* **2022**, *13*, 1351. [CrossRef]
4. Ahmad, N.N.S.B.; Mustafa, F.B.; Yusoff, S.Y.M.; Didams, G. A systematic review of soil erosion control practices on the agri-cultural land in Asia. *Int. Soil Water Conserv. Res.* **2020**, *8*, 103–115. [CrossRef]
5. Li, N.; Zhang, Y.; Wang, T.W.; Li, J.W.; Yang, J.W.; Luo, M.Y. Have anthropogenic factors miti-gated or intensified soil erosion over the past three decades in South China? *J. Environ. Manag.* **2022**, *302*, 114093. [CrossRef]
6. Liu, Z.; Tian, D.; Huang, Z.; Fu, Z.; Liu, J.; Hu, Y. Characteristics of soil and foliar N and P concentrations and stoichiometric ratio along restoration ages of *Pinus massoniana* plantations in red soils erosion regions of southern China. *Chin. J. Appl. Environ. Biol.* **2019**, *25*, 768–775. (In Chinese) [CrossRef]
7. Zhu, P.; Zhang, Q.; Yang, W.; Zhao, J. Characteristics of soil ecological stoichiometry of different vegetation types in ephemeral gully of forestland in Red Soil Region. *Res. Soil Water Conserv.* **2020**, *27*, 60–65. (In Chinese) [CrossRef]
8. Liang, Y.; Li, D.; Lu, X.; Xuan, Y.; Pan, X.; Mu, H.; Shi, D.; Zhang, B. Soil erosion changes over the past five decades in the red soil region of southern China. *J. Mt. Sci.-Engl.* **2010**, *7*, 92–99. [CrossRef]
9. Mao, Y.T.; Hu, W.; Chau, H.W.; Lei, B.K.; Di, H.J.; Chen, A.Q.; Hou, M.T.; Whitley, S. Combined cultivation pattern reduces soil erosion and nutrient loss from sloping farmland on red soil in Southwestern China. *Agronomy* **2020**, *10*, 1071. [CrossRef]
10. Pereira, L.C.; Balbinot, L.; Lima, M.T.; Bramorski, J.; Tonello, K.C. Aspects of forest restoration and hydrology: The hydrological function of litter. *J. For. Res.* **2022**, *33*, 543–552. [CrossRef]
11. Crouzeilles, R.; Curran, M.; Ferreira, M.S.; Lindenmayer, D.B.; Grelle, C.E.V.; Rey Benayas, J.M. A global meta-analysis on the ecological drivers of forest restoration success. *Nat. Commun.* **2016**, *7*, 11666. [CrossRef] [PubMed]
12. Bonner, M.T.L.; Herbohn, J.; Gregorio, N.; Pasa, A.; Avela, M.S.; Solano, C.; Moreno, M.O.M.; Almendras-Ferraren, A.; Wills, J.; Shoo, L.P.; et al. Soil organic carbon recovery in tropical tree plantations may depend on restoration of soil microbial composition and function. *Geoderma* **2019**, *353*, 70–80. [CrossRef]
13. Bastin, J.-F.; Finegold, Y.; Garcia, C.; Mollicone, D.; Rezende, M.; Routh, D.; Zohner, C.M.; Crowther, T.W. The global tree restoration potential. *Science* **2019**, *365*, 76–79. [CrossRef]

14. Yan, M.; Fan, L.; Wang, L. Restoration of soil carbon with different tree species in a post-mining land in eastern Loess Plateau, China. *Ecol. Eng.* **2020**, *158*, 106025. [CrossRef]
15. Huang, J.; Jiang, D.; Deng, Y.; Ding, S.; Cai, C.; Huang, Z. Soil physicochemical properties and fertility evolution of permanent gully during ecological restoration in granite hilly region of South China. *Forests* **2021**, *12*, 510. [CrossRef]
16. Ran, S.; Jin, J. Evolvement and control of vulnerable ecological region—A case study in ongniud banner and aohan banner, inner mongolia. *Chinese Geogr. Sci.* **2004**, *14*, 135–141. [CrossRef]
17. Liu, G.; Shangguan, Z.; Yao, W.; Yang, Q.; Zhao, M.; Dang, X.; Guo, M.; Wang, G.; Wang, B. Ecological Effects of Soil Conservation in Loess Plateau. *Bull. Chin. Acad. Sci.* **2017**, *32*, 11–19. (In Chinese) [CrossRef]
18. Ma, X.; Zhao, C.; Zhu, J. Aggravated risk of soil erosion with global warming—A global meta-analysis. *Catena* **2021**, *200*, 105129. [CrossRef]
19. Kumar, R.; Bhardwaj, A.K.; Rao, B.K.; Vishwakarma, A.K.; Kakade, V.; Dinesh, D.; Singh, G.; Kumar, G.; Pande, V.C.; Bhatnagar, P.R.; et al. Soil loss hinders the restoration potential of tree plantations on highly eroded ravine slopes. *J. Soil. Sediment.* **2021**, *21*, 1232–1242. [CrossRef]
20. Poesen, J.; Nachtergaele, J.; Verstraeten, G.; Valentin, C. Gully erosion and environmental change: Importance and research needs. *Catena* **2003**, *50*, 91–133. [CrossRef]
21. Liu, H.; Zhang, T.; Liu, B.; Liu, G.; Wilson, G.V. Effects of gully erosion and gully filling on soil depth and crop production in the black soil region, northeast China. *Environ. Earth Sci.* **2013**, *68*, 1723–1732. [CrossRef]
22. Mbaya, L.A.; Ayuba, H.K.; Abdullahi, J. An assessment of gully erosion in gombe town, gombe state, Nigeria. *J. Geogr. Geol.* **2012**, *4*, 110–121. [CrossRef]
23. Tang, F.; Yao, Y.; Song, J.; Wang, C.; Liu, Y. Interactive influence of soil erosion and cropland revegetation on soil enzyme activities and microbial nutrient limitations in the Loess Hilly-Gully Region of China. *Agronomy* **2022**, *12*, 2796. [CrossRef]
24. Austin, A.T.; Vivanco, L. Plant litter decomposition in a semi-arid ecosystem controlled by photodegradation. *Nature* **2006**, *442*, 555–558. [CrossRef] [PubMed]
25. Mackay, D.S.; Band, L.E. Forest ecosystem processes at the watershed scale: Dynamic coupling of distributed hydrology and canopy growth. *Hydrol. Process.* **1997**, *11*, 1197–1217. [CrossRef]
26. Ilek, A.; Kucza, J.; Szostek, M. The effect of stand species composition on water storage capacity of the organic layers of forest soils. *Eur. J. Forest Res.* **2015**, *134*, 187–197. [CrossRef]
27. Gomyo, M.; Kuraji, K. Effect of the litter layer on runoff and evapotranspiration using the paired watershed method. *J. Forest Res.* **2016**, *21*, 306–313. [CrossRef]
28. Chen, S.; Cao, T.; Tanaka, N.; Gao, T.; Zhu, L.; Zou, C. Hydrological properties of litter layers in mixed forests in *Mt. Qinling*, China. *iForest* **2018**, *11*, 243–250. [CrossRef]
29. Ilek, A.; Szostek, M.; Mikołajczyk, A.; Rajtar, M. Does mixing tree species affect water storage capacity of the forest floor? Laboratory test of pine-oak and fir-beech litter layers. *Forests* **2021**, *12*, 1674. [CrossRef]
30. Bai, Y.X.; Zhou, Y.C.; Zhang, X.Y.; Du, J.J. Water conservation capacity of litter and soil in mixed plantation of *Pinus massoniana* and Broadleaved Trees. *Sci. Silvae Sin.* **2021**, *57*, 24–36. (In Chinese) [CrossRef]
31. Guevara-Escobar, A.; Gonzalez-Sosa, E.; Ramos-Salinas, M.; Hernandez-Delgado, G.D. Experimental analysis of drainage and water storage of litter layers. *Hydrol. Earth Syst. Sci.* **2007**, *11*, 1703–1716. [CrossRef]
32. Zagyvai-Kiss, K.A.; Kalicz, P.; Szilágyi, J.; Gribovszki, Z. On the specific water holding capacity of litter for three forest ecosystems in the eastern foothills of the Alps. *Agr. Forest Meteorol.* **2019**, *278*, 107656. [CrossRef]
33. Cheng, C.; He, K.N.; Yu, G.F.; Chai, S.X. Comparative study on water conservation capacity of different forest types of artificial forest in arid and semi-arid area. *Acta Ecol. Sin.* **2021**, *41*, 1979–1990. (In Chinese) [CrossRef]
34. Wang, C.; Zhao, C.; Xu, Z.; Wang, Y.; Peng, H. Effect of vegetation on soil water retention and storage in a semi-arid alpine forest catchment. *J. Arid Land* **2013**, *5*, 207–219. [CrossRef]
35. Dunkerley, D. Percolation through leaf litter: What happens during rainfall events of varying intensity? *J. Hydrol.* **2015**, *525*, 737–746. [CrossRef]
36. Pang, X.; Bao, W. Effect of substituting plantation species for native shrubs on the water-holding characteristics of the forest floor on the eastern Tibetan Plateau. *J Resour. Ecol.* **2011**, *2*, 217–224.
37. Li, Y.; Li, B.; Zhang, X.; Chen, J.J.; Zhan, F.D.; Guo, X.H.; Zu, Y.Q. Differential water and soil conservation capacity and associated processes in four forest ecosystems in Dianchi Watershed, Yunnan Province, China. *J. Soil Water Conserv.* **2015**, *70*, 198–206. [CrossRef]
38. Wang, Y.; Wang, Y.; Zhang, H.; Xia, Y.; Xiao, J.; Wu, Y. Research on litter hydrology characteristic of typical vegetation in Jinyun mountain in Chongqing City. *J. Soil Water Conserv.* **2004**, *4*, 41–44. (In Chinese) [CrossRef]
39. Keith, D.M.; Johnson, E.A.; Valeo, C. A hillslope forest floor (duff) water budget and the transition to local control. *Hydrol. Process.* **2010**, *24*, 2738–2751. [CrossRef]
40. Neris, J.; Tejedor, M.; Rodríguez, M.; Fuentes, J.; Jiménez, C. Effect of forest floor characteristics on water repellency, infiltration, runoff and soil loss in Andisols of Tenerife (Canary Islands, Spain). *Catena* **2013**, *108*, 50–57. [CrossRef]
41. Tu, Z.; Chen, S.; Ruan, D.; Chen, Z.; Huang, Y.; Chen, J. Differential hydrological properties of forest litter layers in artificial afforestation of eroded areas of latosol in China. *Sustainability* **2022**, *14*, 14869. [CrossRef]

42. Sato, Y.; Kumagai, T.O.; Kume, A.; Otsuki, K.; Ogawa, S. Experimental analysis of moisture dynamics of litter layers—The effects of rainfall conditions and leaf shapes. *Hydrol. Process.* **2004**, *18*, 3007–3018. [CrossRef]
43. Sun, J.; Yu, X.; Wang, H.; Jia, G.; Zhao, Y.; Tu, Z.; Deng, W.; Jia, J.; Chen, J. Effects of forest structure on hydrological processes in China. *J. Hydrol.* **2018**, *561*, 187–199. [CrossRef]
44. Farahnak, M.; Mitsuyasu, K.; Otsuki, K.; Shimizu, K.; Kume, A. Factors determining soil water repellency in two coniferous plantations on a hillslope. *Forests* **2019**, *10*, 730. [CrossRef]
45. Liu, X.; Feng, Y.; Liu, P.; Zhang, Q.; Njoroge, B.; Zhou, Q.; Gan, X.; Zhang, W.; Li, Y. Soil moisture dominated the temporal dynamics of litter moisture content in subtropical forests: A 7-year observation in south China. *J. Hydrol.-Reg. Stud.* **2022**, *41*, 101102. [CrossRef]
46. Hua, W.; Jianli, Z.; Lifei, Y.; Lingbin, Y.; Congjun, Y.; Tengyong, L. Study on water conservation capacity of litter from different types of forest in Caohai Basin. *Meteorol. Environ. Res.* **2013**, *4*, 17–22, 26.
47. Zhou, Q.; Keith, D.M.; Zhou, X.; Cai, M.; Cui, X.; Wei, X.; Luo, Y. Comparing the water-holding characteristics of broadleaved, coniferous, and mixed forest litter layers in a Karst Region. *Mt. Res. Dev.* **2018**, *38*, 220–229. [CrossRef]
48. Zhang, Z.; Wang, D.; Lei, Y.; Su, K.; Wang, G.; Ma, H. Hydrological characteristics of litter in different forest succession stages at Liuxihe Watershed, southern China. *Front. For. China* **2009**, *4*, 317–322. [CrossRef]
49. Lin, Y. Present situation and countermeasures of soil and water conservation in Hainan Province. *Soil Water Conserv. China* **2015**, *3*, 7–9. (In Chinese) [CrossRef]
50. Carnol, M.; Bazgir, M. Nutrient return to the forest floor through litter and throughfall under 7 forest species after conversion from Norway spruce. *For. Ecol. Manag.* **2013**, *309*, 66–75. [CrossRef]
51. Lentz, R.D.; Ippolito, J.A. Biochar and manure affect calcareous soil and corn silage nutrient concentrations and uptake. *J. Environ. Qual.* **2014**, *43*, 775. [CrossRef]
52. Shi, Q.; Zhang, F.; Chen, Y.; Hu, Z. Optimal kinematics design of macpherson suspension: Integrated use of grey relational analysis and improved entropy weight method. *J. Harbin Inst. Technol. (New Ser.)* **2022**, *29*, 41–51.
53. Acharya, B.S.; Stebler, E.; Zou, C.B. Monitoring litter interception of rainfall using leaf wetness sensor under controlled and field conditions. *Hydrol. Process.* **2017**, *31*, 240–249. [CrossRef]
54. Martin, W.K.E.; Timmer, V.R. Capturing spatial variability of soil and litter properties in a forest stand by landform segmentation procedures. *Geoderma* **2006**, *132*, 169–181. [CrossRef]
55. Bai, Y.; Zhou, Y.; Du, J.; Zhang, X.; Di, N. Effects of a broadleaf-oriented transformation of coniferous plantations on the hydrological characteristics of litter layers in subtropical China. *Glob. Ecol. Conserv.* **2021**, *25*, e01400. [CrossRef]
56. Dong, H.; Yang, C.; Su, C.; Cao, H. Litter and Soil Hydrological Effects of Five No-commercial Forests in Dongguan. *J. Soil Water Conserv.* **2021**, *35*, 144–149, 160. (In Chinese) [CrossRef]
57. Levia, D.F.; Bollinger, W.C.; Hrabik, R.A. Evaporation of intercepted precipitation from fruit litter of *Liquidambar styraciflua* L. (sweetgum) in a clearing as a function of meteorological conditions. *Int. J. Biometeorol.* **2005**, *49*, 325–331. [CrossRef] [PubMed]

forests

MDPI

Article

Effects of Topography and Social Position on the Solar Radiation of Individual Trees on a Hillslope in Northwest China

Jiamei Li, Pengtao Yu *, Yanfang Wan, Yanhui Wang, Bingbing Liu and Yipeng Yu

Key Laboratory of Forest Ecology and Environment of National Forestry and Grassland Administration, Liupan Mountains Forest Ecosystems National Positioning Observation and Research Station, Ecology and Nature Conservation Institute, Chinese Academy of Forestry, Beijing 100091, China; lijiam@caf.ac.cn (J.L.); wanyf@caf.ac.cn (Y.W.); wangyh@caf.ac.cn (Y.W.); bingbing-l@caf.ac.cn (B.L.); yuyp@caf.ac.cn (Y.Y.)
* Correspondence: yupt@caf.ac.cn; Tel.: +86-10-62889562

Abstract: Solar radiation is a key factor influencing the photosynthesis and transpiration of trees. In mountainous regions, solar radiation income exhibits strong spatial heterogeneity due to topographical variations and the structural complexity of the forest. However, how the solar radiation income of individual trees in different social positions varies with slope position remains unclear. In this study, the daily solar radiation of the horizontal ground (R_h), different slope positions (i.e., at different locations on a hillslope, R_s) and individual trees with different social positions in the forest (R_i) were monitored from May to October in 2020 and 2021. The daily solar radiation income of a single hillslope (R_f) was applied to quantify the R_s response to the slope and aspect (i.e., slope effect) and the shade from the opposite mountain (i.e., shaded terrain effect). Our results showed that the R_f was 27.8% lower than R_h due to the slope effect of the sample slope. In the different slope positions, 2.7%–46.9% of solar radiation was lost due to the shaded terrain effect. A stronger limitation of R_s by the shaded terrain effect was detected on the bottom slope compared to that of the upper slope. The better the social position of an individual tree (i.e., tree dominance (*Dom*) and the distance between trees (*D*)), the more solar radiation it received, ranging from 22.4 to 95.3%. The dominant factor contributing to changes in R_i was slope position followed by *D* and *Dom* and, finally, R_h. These results provide an important basis for understanding the role of topography and tree social positions in solar radiation income in mountainous regions. Forest management measures should be varied with slope positions in mountainous regions, and forest density (i.e., distance between trees) should be considered as a key factor to optimize the forest functions.

Keywords: solar radiation; individual trees; topography; social position

Citation: Li, J.; Yu, P.; Wan, Y.; Wang, Y.; Liu, B.; Yu, Y. Effects of Topography and Social Position on the Solar Radiation of Individual Trees on a Hillslope in Northwest China. *Forests* **2023**, *14*, 561. https://doi.org/10.3390/f14030561

Academic Editor: Francois Girard

Received: 23 February 2023
Revised: 9 March 2023
Accepted: 9 March 2023
Published: 12 March 2023

1. Introduction

Solar radiation is one of the main environmental signals affecting plant biology, controlling multiple physiological responses [1,2], such as photosynthesis, transpiration and morphogenesis [3,4], with consequences for tree growth [5–8], which in turn can have an impact on forest productivity [9]. Solar radiation income often shows a high degree of spatial heterogeneity as a result of complex topographical variations and the spatial structure of trees [7,10–14], especially in mountain forests [15,16]. The accurate assessment of the effect of topography and the vertical and lateral position of trees in the forest (i.e., the niche of a tree in a forest is called social position in plant ecology) [17] on solar radiation income is essential for the measurement of solar radiation in mountain forests. Topography variations (such as slope gradient and aspect and the shadows cast by the surrounding topography) have a direct effect on solar radiation income [10,18–21]. At present, many studies of solar radiation in mountainous regions often focus on the difference in solar radiation income on various hillslopes [11,12,22]. For example, in the Northern Hemisphere, south-facing slopes generally receive a greater amount of solar radiation [23]. Other studies have focused on topographic shading effects [24,25]. The shaded terrain effect by topography leads to a significant reduction in solar radiation income on hillslopes [16]. However, few studies have considered the differences in solar radiation income at different slope positions on hillslopes and whether the shaded terrain effect is consistent at different slope positions.

The solar radiation income of individual trees within forests on hillslopes is also affected by the tree's spatial structure (i.e., the spatial structure of the tree in both the vertical and horizontal orientation). The essence of the social position of individual trees can describe the spatial information of trees well, which is influenced by different environmental conditions for each tree (i.e., the relative size and distance between the reference tree and its neighbors) and, hence, individuals generate specific social position features [26,27]. The social position of trees determines the living space and available water and radiation resources. Trees with a dominant social position have a vast living space and plenty of water and radiation resources, causing less competition [17]. The degree of dominance and forest density can be used as a proxy indicator [9,28]. Some studies have suggested that the degree of tree dominance affects its ability to receive solar radiation [29], and dominant trees in a stand receive a greater amount of solar radiation [6,30]. Furthermore, decreasing the forest density increases the fraction of the absorbed solar radiation of trees [28,31,32]. However, in many studies, trees were simply divided into different dominance groups based on their relative height or diameter at breast height *(DBH)* in the forest [7,33–35], without considering the shading effects of the surrounding adjacent trees. Assessments of forest density also did not take into account the effect of the social position of individual trees on solar radiation income.

Forests are mainly located in mountainous regions [36]. The rugged mountainous terrain creates a major obstacle to forest management because of the difficult physical approach by men and machines [12,37]. Therefore, it is difficult to accurately quantify the effects of solar radiation income on mountain forests. In addition, forest management is currently focused on the stand scale, i.e., only the stand density is considered without considering the differences in the social location of trees, resulting in the forest management being invalid or only slightly valid [7,38].

To gain a deeper understanding of the effect of topography and social position on the solar radiation income of individual trees, we hypothesize that: (1) solar radiation income is similar at different slope positions; (2) solar radiation income is greater under a more dominant social position. Hence, we aimed to assess the effects of topography and social position on the solar radiation income in a mountainous region, which can help us understand the character of the solar radiation income of trees, so as to make targeted plans for forest management in mountainous regions.

2. Materials and Methods

2.1. Study Site

The study site was located in the small watershed of Xiangshuihe (*XSH*), Southern Liupan Mountains (106°12′–106°16′ E, 35°27′–35°33′ N) in Ningxia, Northwest China (Figure 1a). *XSH* has an elevation range of 2010 to 2942 m and a semi-humid climate, with a mean annual air temperature of 5.8 °C and an annual precipitation of 550.8 mm (1981 to 2010); 70% of precipitation occurs in the growing season from May to October. The soil is dominated by gray cinnamon with a sandy loam texture. *XSH* is a forested watershed with a forest cover of 82.4%, of which plantations account for 24.4%.

The plantations of larch *(Larix principis-rupprechtti)* are the main components of the plantations in *XSH*, accounting for 23.6% of the *XSH* area. *Larix principis–rupprechtii* is the main tree species used in afforestation in North and Northwest China. Such larch plantations play a very important role in producing timber and ecological services (e.g., soil protection, hydrological regulation and water supply) [39,40]. In larch plantations, the understory shrubs were scattered, covering only approximately 5% of the area. Herbs cover approximately 40% of this area, with the dominant species being *Pteridium aquilinum* and *Carex hancokiana*.

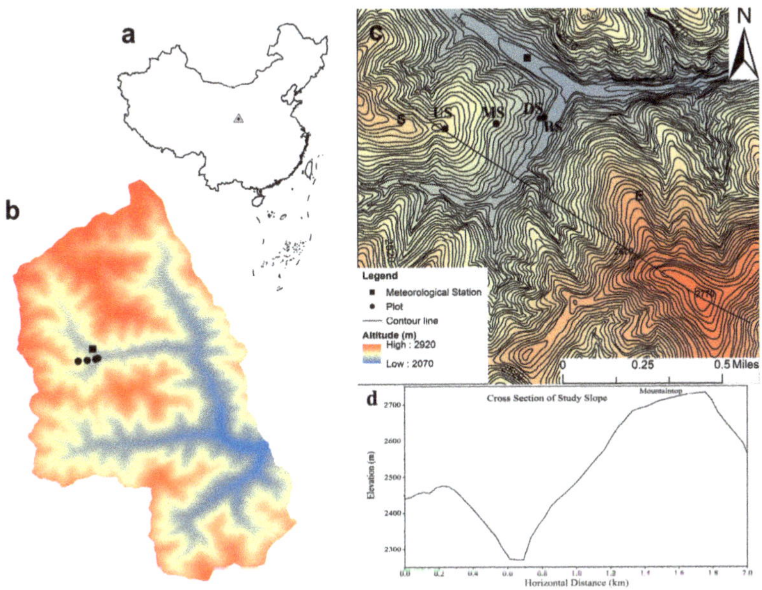

Figure 1. Information about the locations of the study sample plots and sample trees. (**a**) Location of the Xiangshuihe (*XSH*) study watershed in NW China. (**b**) The location of plots (*US*, *MS*, *DS* and *BS* represent the upper, middle, down and bottom slope) along slope positions on the sample hillslope. (**c**) The detailed terrain of the study slope and opposite slope. (**d**) The valley cross-section.

2.2. Study Slope and Plots

2.2.1. Study Slope

We selected a representative southeastern slope completely covered by an even-aged pure plantation of *Larix principis-rupprechtti*. The slope had a horizontal length of 425.1 m (corresponding to a slope length of 460.5 m), with a slope gradient of 22.6° and an elevation range of 2271 to 2480 m. The opposite mountain was the northwest slope with a horizontal length of 1052 m, a slope gradient of 23.5° and an elevation range of 2270 to 2731 m. The horizontal distance between the bottom slope of the opposite mountain and the bottom slope of the sample slope was 82 m (Figure 1b–d).

2.2.2. Study Plots

In 2020, four plots were selected at different slope positions on the sample slope, vertically ranging from 2273 to 2471 m a.s.l. The upper slope (*US*), middle slope (*MS*), down slope (*DS*) and bottom slope (*BS*) were selected with elevations of 2471 m, 2397 m, 2293 m and 2273 m, respectively (Figure 1). These elevation points contain the whole range of the hillslope and the solar radiation received through these elevations can represent the difference amounts of solar radiation received on the whole slope. Also, since the shading of the opposite mountain was more obvious nearer to the bottom of the slope, measurement points were set up on both the down slope and bottom slope. The location information of the *Larix principis-ruprechtii* plantation plots on the study slope were shown in Table 1. The stand characters of each plot were shown in Table 2.

Table 1. The location information of *Larix principis-ruprechtii* plantation plots on the study slope.

Plot Positions	Elevation (m)	Slope Gradient (°)	Elevation Difference (ΔH)	Horizontal Distance (Δd)
Upper slope (US)	2471	26.3	259	1501
Middle slope (MS)	2397	26.8	333	1339
Down slope (DS)	2293	29.0	437	1092
Bottom slope (BS)	2273	34.9	457	1050

The ΔH represents the elevation difference between the sample point and the opposite mountain top; the Δd represents the horizontal distance between the sample point and the opposite mountain top.

Table 2. The stand characteristics of *Larix principis-ruprechtii* plantation plots on the study slope.

Plot Positions	Stand Density (Trees ha^{-1})	Mean Diameter at Breast Height (cm)	Mean Tree Height (m)	Mean Canopy Diameter (m)
Upper slope (US)	856	21.83	19.4	4.5
Middle slope (MS)	844	22.29	20.0	3.7
Down slope (DS)	742	20.63	19.2	4.5
Bottom slope (BS)	711	21.40	19.5	4.4

2.3. Social Position of Trees in the Forest

The social position of trees can represent the trees' vertical and lateral position in the forest. The trees in the forest were divided into four social position levels according to tree dominance (*Dom*) and distance between trees (*D*), as shown by Equations (1) and (2).

$$Dom = \frac{H_i - \overline{H}}{\overline{H}} \tag{1}$$

where H_i (m) is the height of the sample tree *i* and $\overline{H}$ (m) is the average height of all the nearest neighbor trees around the sample tree [17]. The *Dom* represents the degree of dominance in the vertical direction of the sample tree from neighboring trees.

The distance among trees (*D*) is defined as the average horizontal distance between the sample tree and its neighboring trees, using Equation (2):

$$D_i = \frac{1}{n}\sum_{j=1}^{n} D_{ij} \tag{2}$$

where D_{ij} (m) represents the horizontal distance between the sample tree *i* and the neighbor tree *j*. The D_i represents the degree of superiority or inferiority in the horizontal direction of the sample tree from neighboring trees.

In order to better describe the differences in the social position of trees within the forest, the sample plots at the down slope (*DS*) were used as an example to illustrate the characteristics of the social position of trees. There were 91 trees in the down slope plot, with a mean tree height (*H*) of 19.2 m, a mean diameter at breast height (*DBH*) of 20.63 cm and a mean crown diameter of 4.5 m (Table 2). All trees in the plot were divided into four social position levels according to tree dominance (*Dom*) and the distance between trees (D_i). The vertical structure characteristics of the four social position levels are shown in Table 3.

Table 3. The vertical structure characteristics of *Larix principis-rupprechtii* plantation plot in the down slope (*DS*) (means ± standard deviations).

Tree Social Position Levels	Number of Trees	Mean Diameter at Breast Height (cm)	Mean Tree Height (m)	Mean Canopy Diameter (m)
Dominant	8	25.30 ± 1.7	21.8 ± 1.8	4.9 ± 0.5
Codominant	40	21.19 ± 2.2	19.8 ± 1.4	4.6 ± 0.4
Intermediate	31	19.16 ± 1.9	18.6 ± 1.7	4.3 ± 0.6
Suppressed	12	15.40 ± 2.5	17.1 ± 2.3	3.6 ± 0.5

2.4. Sample Tree Selection from the down Slope Plot (DS)

In the down slope plot (*DS*), one to two representative trees from each social position level were selected as sample trees using a stratified random sampling method (i.e., sample trees that can represent the average of this level). Therefore, we selected a total of seven sample trees, and the *DBH* varied in the range of 15.50 to 25.92 cm, the tree height ranged from 17.2 to 21.8 m and the canopy diameter ranged from 3.5 to 5.2 m (shown in Table 4).

Table 4. The characteristics of seven sample trees in the down slope plot (*DS*).

Tree Social Position Levels	Tree No.	Dom Value	D Value (m)	Diameter at Breast Height (cm)	Tree Height (m)	Canopy Diameter (m)
Dominant	29	0.12	4.8	25.92	21.8	5.2
	57	0.11	4.5	24.34	21.5	4.8
Codominant	61	0.01	3.9	21.50	19.8	4.4
	69	0.00	3.3	21.11	19.6	4.6
Intermediate	43	−0.04	3.0	19.53	18.5	4.2
	44	−0.04	2.9	18.82	18.4	4.0
Suppressed	81	−0.21	2.2	15.50	17.2	3.5

2.5. Solar Radiation Measurement

2.5.1. Solar Radiation Data Conversion

The standard for measuring solar radiation utilizes the units of watts per meter squared (W m^{-2}). However, pyranometers are both costly and limited in their ability to measure the solar radiation of individual trees in the slope field. With a lower cost, smaller size and more flexible installation, light meters measure luminous flux per unit area (illuminance), utilizing the units of lumens per meter squared or lux (lx). An effective conversion factor between watts per meter squared and lux would enable the use of light meters to evaluate the differences in the solar radiation in individual tree levels. Additionally, surveys of the literature have found no definitive and readily available "rule of thumb" conversion standard between solar radiation and illuminance data.

In this study, we converted the illuminance data (lx) into solar radiation data (W m^{-2}) using Equation (3) (Figure 2). This equation was established based on data recorded from a pyranometer in an automatic weather station (Weatherhawk 232; WeatherHawk Inc., Logan, UT, USA) and a light meter (HOBO MX2202; Onset Computer Corp., Bourne, MA, USA). An automatic weather station and a light meter were installed at the same position (i.e., 1.3 m above the ground and the distance between them was 1.0 m) on open and flat horizontal ground approximately 100 m from the sample slope. The data concerning total solar radiation arriving on horizontal ground were collected every 5 min by an automatic weather station (Weatherhawk 232; WeatherHawk Inc., Logan, UT, USA) from May to October in 2020. Radiation data were collected every 5 min by a light meter (HOBO MX2202; Onset Computer Corp., Bourne, MA, USA) from May to October in 2020.

$$y = 0.009x - 1.360 \quad (3)$$

where y (W m^{-2}) represents the solar radiation and x represents the illuminance (lx).

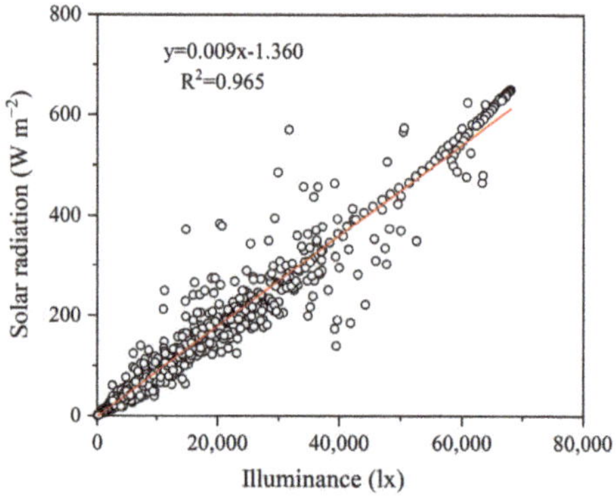

Figure 2. Relationship between solar radiation and illuminance data.

2.5.2. Solar Radiation of Individual Trees

The daily solar radiation income of individual trees (R_i, MJ m^{-2} day^{-1}) was measured with a light meter (HOBO MX2202; Onset Computer Corp., Bourne, MA, USA), which was installed at the top of the canopy of each of the seven sample trees. The sensor height of the sample trees is shown in the "Tree Height" column in Table 4. The data concerning the radiation arriving at the top of the canopy were collected every 5 min by the light meter (HOBO MX2202; Onset Computer Corp., Bourne, MA, USA) from July to October 2020 and May to October 2021.

2.5.3. Solar Radiation over the Forest Canopy

The daily solar radiation income over the forest canopy (i.e., outside the forest) on the sample hillslope (R_s, MJ m^{-2} day^{-1}) was measured in each of the four slope plots (i.e., *US*, *MS*, *DS* and *BS*) with four light meter loggers (HOBO MX2202; Onset Computer Corp., Bourne, MA, USA) at a height of 1.3 m above the ground. The data concerning the radiation arriving at different slope positions were collected every 5 min by the light meter (HOBO MX2202; Onset Computer Corp., Bourne, MA, USA) from July to October 2020.

2.5.4. Solar Radiation at the Horizontal Ground Measurement

The daily solar radiation income of the horizontal ground (R_h, MJ m^{-2} day^{-1}) was measured using the pyranometer of an automatic weather station (WeatherHawk 232, WeatherHawk Inc., USA), installed on open and flat horizontal ground approximately 100 m from the sample slope and at a height of 1.3 m above the ground [41]. The data concerning the total solar radiation arriving on horizontal ground were collected every 5 min by an automatic weather station (Weatherhawk 232; WeatherHawk Inc., Logan, UT, USA) from May to October in 2020 and 2021.

2.6. *The Calculation of Solar Radiation on a Single Hillslope*

The daily solar radiation income of the single hillslope (R_f) was introduced as a reference parameter (that is, the solar radiation of the slope is not affected by the surrounding mountains) and was calculated using the Fu (1958) model [10,42] as follows:

$$R_f = (\mu \sin\delta + \vartheta \cos\delta \cos\omega - \sin\beta \sin\alpha \cos\delta \sin\omega) R_h \tag{4}$$

$$\begin{aligned} \mu &= \sin\varphi \cos\alpha + \cos\varphi \sin\alpha \cos\beta \\ \vartheta &= \cos\alpha \cos\varphi - \sin\varphi \sin\alpha \cos\beta \end{aligned} \tag{5}$$

where R_f (MJ m^{-2} day^{-1}) represents the daily solar radiation income of the single hillslope, β (°) represents the slope direction (expressed as the azimuth clockwise from north to east) and the slope gradient α (°); δ (°) is the declination; ω (°) is the hour angle; and φ (°) is the latitude of the measurement point. The u and v are the composite parameters of the slope gradient α (°), the slope direction β (°) and the latitude of the measurement point φ (°) using Equation (5).

2.7. Relative Contribution of ΔH/Δd, R, Dom and D to Solar Radiation Received by an Individual Tree

The relative independent contributions of $\Delta H/\Delta d$ (the ratio of the height difference to the horizontal distance between the sample point and the top of the opposite mountain, i.e., different slope positions, Δ represents the difference value), R_h, *Dom* and *D* to R_i were quantified using the coupled R_i model, and this model was constructed in the Results Section 3.4. The method for determining the relative contribution of each factor to R_i involved a comparison of the difference in the response variable (R_i in this study) between the simulation with only one varying factor (e.g., $\Delta H/\Delta d$, R_h, *Dom* and *D*) and the reference R_i.

The relative independent contributions of $\Delta H/\Delta d$, R_h, *Dom* and *D* to R_i were calculated using Equations (6) and (7) [39]:

$$C_k = \frac{\Delta R_i(k)}{R_i(reference)} \times 100 \quad (6)$$

$$\Delta R_i(k) = R_i(k) - R_i(reference) \quad (7)$$

where C_k (%) is the independent contribution rate of each factor k ($\Delta H/\Delta d$, R_h, *Dom* and *D*) to the solar radiation of individual trees (R_i) compared to reference R_i. The reference R_i in this study was calculated using the developed R_i model and the value of $\Delta H/\Delta d$ (0.0, i.e., the sample point has no shaded terrain effect), the long-term means of the R_h (9.32 MJ m ^{2} day^{-1}, 2013–2021) and the means of *Dom* (0.001) and *D* (2.01 m) from all the sample trees (835 trees) on the hillslope. This reference R_i value could represent the long-term mean environmental conditions and social position in this study hillslope. R_i *(k)* was calculated by inputting the measured value of factor k and the reference values of other factors into the R_i model.

2.8. Statistical Analysis

To clarify how R_i responds to each single factor (*Dom* and *D*) and the appropriate function type between R_i and *Dom* and *D*, the upper boundary line method [43], which can present the real relation between R_i and a single factor by eliminating the interferences of other factors, was used to determine the R_i–*Dom* and R_i–*D* relation. The upper boundary line was widely used in the establishment of compound models with independent factors as indicators [39].

Statistics were analyzed by the Statistical Product and Service Solutions (SPSS), version 19.0 (IBM Inc., Chicago, IL, USA) and presented as the mean ± standard deviation (SD). The differences in the solar radiation income at the varying slope positions and social positions was analyzed by one-way analysis of variance (ANOVA) with a significance level of $p < 0.05$, and the letters a and b indicate significant differences [41]. The performance of the model was assessed using the coefficient of determination (R^2) [39].

3. Results

3.1. Variation of Solar Radiation at Different Slope Positions

The mean daily solar radiation income at different slope positions (R_s, slope gradient of 27.8°, slope aspect of 135°) was less than that of the horizontal ground (R_h) (Figure 3). The mean R_s at the upper slope (*US*) and the middle slope (*MS*) was slightly lower than the R_h (7.03 MJ m^{-2} day^{-1}) by 1.20 MJ m^{-2} day^{-1} and 1.60 MJ m^{-2} day^{-1}, respectively. However, R_s was even lower on the down slope (*DS*) and the bottom slope (*BS*), accounting for only 49.9% and 46.9% of R_h.

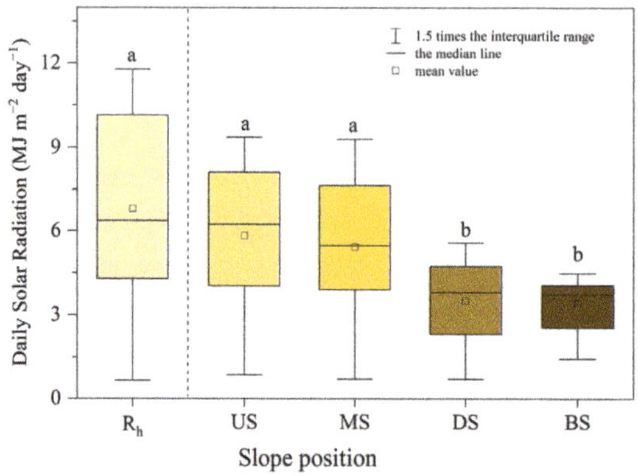

Figure 3. The differences in solar radiation income at various slope positions. (R_h represents the daily solar radiation income of horizontal ground; *US* represents the upper slope; *MS* represents the middle slope; *DS* represents the down slope; *BS* represents the bottom slope). 1.5 times IQR (interquartile range) was used to determine the anomalous values. Groups marked by different letters significantly differ from each other (i.e., a is significantly greater than b) ($p \leq 0.05$).

Daily solar radiation income of the single hillslope (R_f) was significantly lower than the R_h, accounting for 72.2% of the R_h (Figure 4a). In addition, the R_s was always less than the R_f at the four slope positions by 2.7%–46.9%. The response trend of the R_s to the R_f could be expressed by a saturated logarithmic function, and the threshold of R_s decreased gradually with decreasing slope position (Figure 4b). When R_f was 8.46 MJ m^{-2} day^{-1}, the difference in solar radiation income between *US* and *MS* was not significant, and the corresponding thresholds were 8.23 MJ m^{-2} day^{-1} and 8.10 MJ m^{-2} day^{-1}, respectively. However, the difference in R_s was more pronounced at the *DS* and *BS* than at the *US*, and the corresponding threshold was 5.58 MJ m^{-2} day^{-1} and 4.49 MJ m^{-2} day^{-1}, respectively.

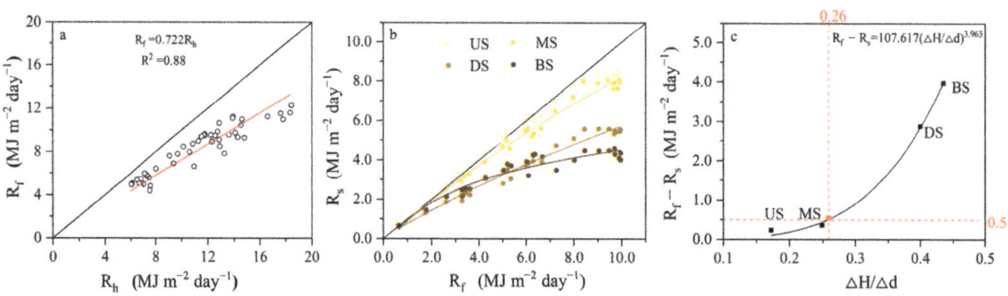

Figure 4. Effect of topographic change on solar radiation income on the hillslope. (**a**) The relationship between R_f and R_h. R_h represents the daily solar radiation income of the horizontal ground; R_f represents the daily solar radiation income of the single hillslope. (**b**) The relationship between R_s and R_f in different slope positions. R_s represents the daily solar radiation income outside the forest on the sample slope. The solid black line in (**b**) represents $R_f = R_i$. (**c**) The relationship between slope position and shade terrain effect. (R_f—R_s) represents the shaded terrain effect of the opposite mountain; $\Delta H/\Delta d$ represents the ratio of the height difference to the horizontal distance between the sample point and the top of the opposite mountain (i.e., the variation of the slope position), the ΔH value represents the elevation difference between the sample point and the opposite mountain top; the Δd value represents the horizontal distance between the sample point and the opposite mountain top; upper slope (*US*), middle slope (*MS*), down slope (*DS*) and bottom slope (*BS*). The ($\Delta H/\Delta d < 0.26$) was the threshold indicating that the sample point has no shaded terrain effect, that is, R_f—R_s < 0.5 MJ m^{-2} day^{-1}.

The response trend of R_f—R_s to $\Delta H/\Delta d$ (i.e., the ratio of the height difference to the horizontal distance between the sample point and the top of the opposite mountain) could be expressed by a power function, and R_f—R_s first increased slowly with increasing $\Delta H/\Delta d$ until 0.26 and then increased rapidly (Figure 4c). This indicated that the sample plot was not affected by the shaded terrain effect of the opposite mountain if the R_f—R_s was less than 0.5 MJ m^{-2} day^{-1}. In addition, the slope effect was always greater than the shaded terrain effect at *MS* and *US*, while the shaded terrain effect was more important than the slope effect at *DS* and *BS* (Figure 5).

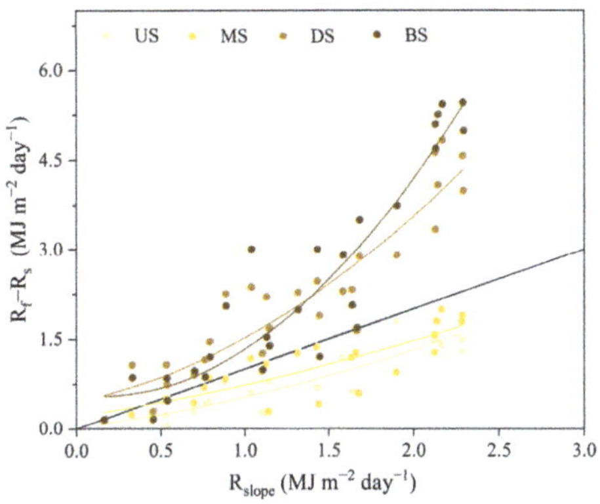

Figure 5. Comparison of the slope effect and the shaded terrain effect at different slope positions. R_{slope} represents the slope effect (i.e., R_h—R_f).

3.2. *Effect of Tree Social Position on Its Solar Radiation*

The better the social position of an individual tree, the more solar radiation (R_i) it received (Figure 6). R_i at the dominant level was the largest, followed by R_i at the codominant level and intermediate level and, finally, the suppressed level. The mean R_i of the dominant level (4.82 MJ m^{-2} day^{-1}) was 1.1 times, 2.0 times and 2.4 times more than the codominant level, the intermediate level and the suppressed level, respectively.

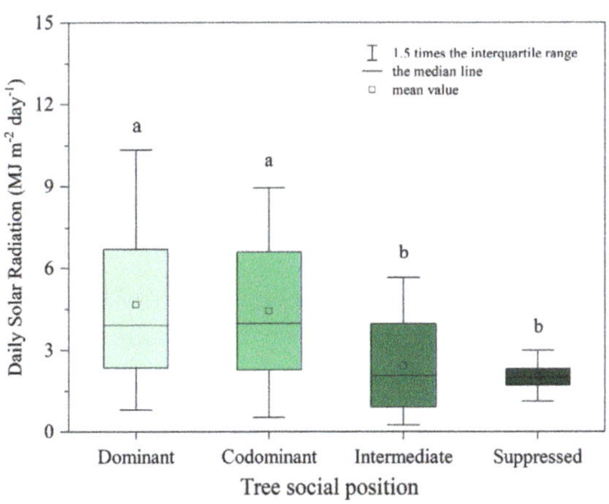

Figure 6. The differences in the solar radiation income of individual trees among social position levels in the down slope plot. Groups marked by different letters significantly differ from each other (i.e., *a* is significantly greater than *b*) ($p \leq 0.05$).

The R_i increased linearly with increasing daily solar radiation income from clearings outside of the forest on the sample slope (R_s) for the four social position levels, ranging from 22.4% to 95.3% (Figure 7a). The relationship between R_i at the dominant level and R_s was more sensitive than that of any other level. For example, the highest coefficient of determination (R^2) of the dominant levels was 0.978, which was 1.0, 1.1 and 1.5 times higher than the codominant, intermediate and suppressed levels, respectively. The *UBLs* in Figure 7b show that R_i gradually increased with the increasing dominance of trees (*Dom*) following a linear function, namely, $R_i = e \times Dom + f$. R_i first increased rapidly with increasing distance between the trees (*D*) until 4.5 m and then gradually stabilized around the maximum value as *D* continued to increase (Figure 7c). The relation between R_i and *D* could be expressed by a saturated logarithmic function, namely, $R_i = g \times \ln(D) + h$.

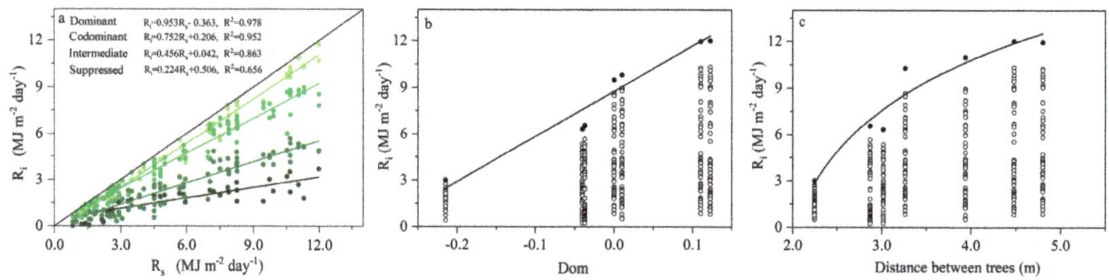

Figure 7. Effect of social positions on the solar radiation income of individual trees. (**a**) Response of R_i to R_s. (R_i represents the daily solar radiation income of individual trees; R_s represents the daily solar radiation income outside the forest of the sample slope). (**b**) Response of R_i to *Dom* (*Dom* represents the social position levels of the sample trees). (**c**) Response of R_i to *D*. (*D* represents the mean distance between the sample tree and neighboring trees). The upper boundary line (*UBL*) method [44] was used to determine the response function types of R_i to *Dom* and *D*.

3.3. The Construction of the Tree Solar Radiation Model and Its Validation

To understand the effect of topography and social position on the solar radiation income of individual trees (R_i), we built a solar radiation model of individual trees, which combined with the results of Sections 3.1 and 3.2. The R_i model was given as follows:

$$R_i = R_h \times f(topography) \times f(social\ position) \quad (8)$$

where R_i (MJ m^{-2} day^{-1}) represents the daily solar radiation income on the sample tree *i*, R_h (MJ m^{-2} day^{-1}) represents the daily solar radiation income of the horizontal ground, and *f* (*topography*) and *f* (*social position*) are the response functions of R_i to topography and social position factors, respectively. The response function of R_i to topography was obtained by the mathematical substitution method. The response functions of R_i to social position (i.e., dominance and distance) were obtained by concatenated multiplication since the solar radiation of individual trees showed a positive correlation with both dominance and the distance between trees (Figure 7b,c).

In addition, combining the methods in Section 2.6, the R_i model can be estimated by Equation (9):

$$R_i = (a \times \left((\mu \sin\delta + \vartheta \cos\delta\cos\omega - \sin\beta\sin\alpha\cos\delta\sin\omega)R_h - b \times \left(\frac{\Delta H}{\Delta d}\right)^c\right) + d) \times (e \times Dom + f) \times (g \times \ln(D) + h) \quad (9)$$

where R_i (MJ m^{-2} day^{-1}) represents the daily solar radiation income of individual trees, β (°) represents the slope direction (expressed as the azimuth clockwise from north to east) and the slope gradient α (°); δ (°) is the declination; w (°) is the hour angle; and ψ (°) is the latitude of the measurement point. u and v are the composite parameters of the slope gradient α (°), the slope direction β (°) and the latitude of the measurement point ψ (°) using Equation (5). R_h represents the daily solar radiation income of the horizontal ground, $\Delta H/\Delta d$ represents the ratio of the height difference to the horizontal distance between the sample point and the top of the opposite mountain,

Dom represents the social position levels of the sample trees, *D* represents the mean distance between the sample tree and neighboring trees and the *a*, *b*, *c*, *d*, *e*, *f*, *g* and *h* values represent the model parameters that need to be determined by the solar radiation data.

All parameters of the R_i model were newly fitted using observed data on the sample slope (slope 27.8°, aspect 135°) from July to October 2020, and the R_i model was further validated using the observed data from May to October 2021, Figure 8. The results indicated that the R_i model could still accurately estimate the varying R_i with a high R^2 value (0.83, n = 657). The simulated mean value (4.13 MJ m^{-2} day^{-1}) was 16.3% greater than the observed mean value (3.55 MJ m^{-2} day^{-1}). The R_i model of the sample slope was as follows:

$$R_i = (0.296 \times \left(0.722R_h - 107.617\left(\frac{\Delta H}{\Delta d}\right)^{3.963}\right) + 0.950) \times (2.132Dom + 2.032) \times (0.989\ln(D) - 0.426) \;\; R^2 = 0.83 \quad (10)$$

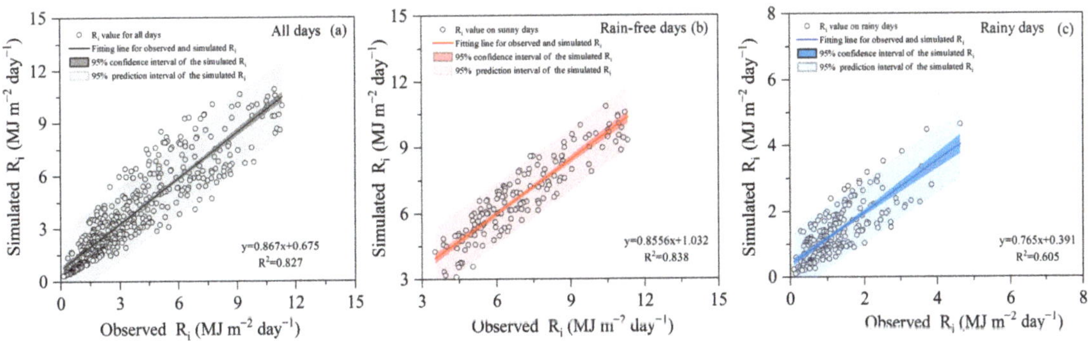

Figure 8. Comparison of the observed and simulated R_i values based on Equations (10)–(12) under the different weather conditions. (**a**) Comparison of the observed and simulated R_i values for all days. (**b**) Comparison of the observed and simulated R_i values for rain free days. (**b**) Comparison of the observed and simulated R_i values for all days. (**c**) Comparison of the observed and simulated R_i values for rainy days.

In addition, the R_i model accuracy of rain-free days was 0.01 greater than that of the weather-independent model (Figure 8a), with R^2 = 0.84, and the accuracy of the R_i model fit was low for rainy days with R^2 = 0.61.

$$R_i = (2.067 \times \left(0.722R_h - 107.617\left(\frac{\Delta H}{\Delta d}\right)^{3.963}\right) - 1.909) \times (3.786Dom + 2.667) \times (0.789\ln(D) - 0.593) \;\; R^2 = 0.84 \quad (11)$$

$$R_i = (0.760 \times \left(0.722R_h - 107.617\left(\frac{\Delta H}{\Delta d}\right)^{3.963}\right) + 0.314) \times (4.414Dom + 0.825) \times (1.078\ln(D) + 4.582) \;\; R^2 = 0.61 \quad (12)$$

3.4. Independent Contributions of $\Delta H/\Delta d$, Rh, Dom and D to R_i

The relatively independent contribution of $\Delta H/\Delta d$, R_h, *Dom* and *D* to R_i compared to the long-term mean of reference R_i is shown in Figure 9. The $\Delta H/\Delta d$, R_h and *Dom* exerted slight negative effects on R_i, with contribution rates of −18.4%, −4.5% and −6.9%, respectively, while *D* had an obviously positive effect on R_i, with a contribution rate of 16.0%. This indicated that variations in slope position had a greater effect on R_i than social position and R_h (i.e., weather variations), and the dominant social position factor affecting the change in R_i was *D*.

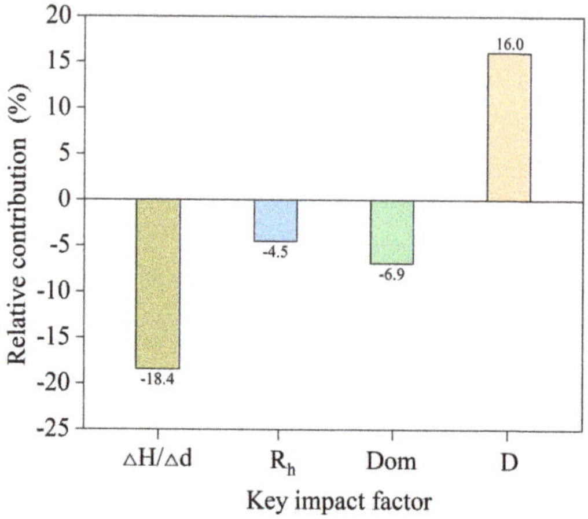

Figure 9. Relative contribution rate of R_h, *Dom* and D to the daily solar radiation income of individual trees (R_i) compared to a reference R_i. The reference R_i was modeled by Equation (10) with the long-term means of R_h, the $\Delta H/\Delta d$ value of 0 (Figure 5c) and the mean *Dom* and D of the 835 sample trees on the hillslope. ($\Delta H/\Delta d$ represents the ratio of the height difference to the horizontal distance between the sample point and the top of the opposite mountain (i.e., the variation of slope position); R_h represents the daily solar radiation income of the horizontal ground; *Dom* represents the social position levels of sample trees; D represents the mean distance between trees and neighboring trees).

The response of R_i to D exhibited the same trend at different slope positions and different *Dom* levels, and an obviously higher rate of increase occurred under higher *Dom* levels (Figure 10). For example, given the $\Delta H/\Delta d$ value of 0.1 and *Dom* value of 0.3, the effect of D on R_i only gradually decreased when D was greater than 6.0 m, while given the $\Delta H/\Delta d$ value of 0.1 and *Dom* value of −0.3, R_i stabilized when D was greater than 3.0 m.

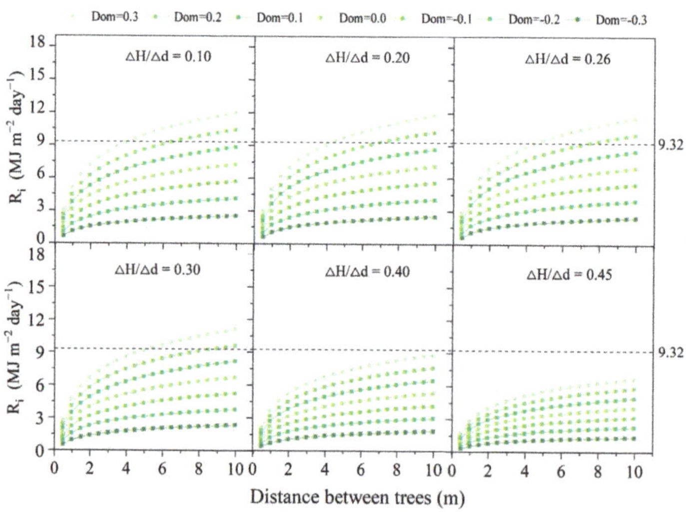

Figure 10. Variation in the simulated R_i with R_h at different *Dom* and slope position levels by the R_i model (Equation (10)). The value of R_h of 9.32 MJ m^{-2} day^{-1} used the long-term mean values from 2013 to 2021. $\Delta H/\Delta d$ represents the ratio of the height difference to the horizontal distance between the sample point and the top of the opposite mountain. $\Delta H/\Delta d$ = 0.26 was the threshold indicating that the sample point had no shaded terrain effect (Figure 4c).

There was a significant difference in the effect of D on R_i under different slope positions (i.e., different $\Delta H/\Delta d$ values) (Figure 10). The difference in R_i at different *Dom* levels gradually decreased as $\Delta H/\Delta d$ values increased, especially after $\Delta H/\Delta d > 0.26$. For example, given the $\Delta H/\Delta d$ value of 0.1 and *Dom* value of 0.3, the R_i was above the multi-year means (9.32 MJ m^{-2} day^{-1}) if the D values were higher than 4.0 m. However, given the $\Delta H/\Delta d$ value of 0.45, the R_i was always less than 9.32 MJ m^{-2} day^{-1} regardless of the value of *Dom*. This indicated that the higher the $\Delta H/\Delta d$ value (i.e., the lower the slope position), the weaker the effect of the social position of trees (both *Dom* and D).

4. Discussion

4.1. The Role of Topography on Solar Radiation: Slope Effect and Shaded Terrain Effect

In this study, the solar radiation income on the sample slope (slope 27.8°, aspect 135°) was lower than that of the horizontal ground (Figure 3). Similar results were obtained in earlier studies [45,46]. This was mainly because of the slope effect (i.e., slope and aspect) and the shaded terrain effect caused by the opposite mountain [25,47]. Since the slope effect of solar radiation income was the same on a hillslope [42,48], the difference in solar radiation income at different slope positions was mainly caused by different shaded terrain effects. Therefore, we rejected the first hypothesis that the solar radiation income was similar at different slope positions.

The difference in the shaded terrain effect was due to the different blocking effects of topographic features (that is, the distance and elevation difference from the sample slope) [21,49,50]. Our research showed that the shaded terrain effect did not limit solar radiation income on the slope when above the middle slope (i.e., the ratio of the elevation difference to the distance between the sample point and the top of the opposite mountain was satisfied ($\Delta H/\Delta d$) < 0.26) (Figure 4). This was mainly because the solar radiation income above the mid-slope was not affected by mountain shading due to the difference in height at *US* and *MS*, with the opposite mountain being reduced and the distance between the opposite mountain being increased. This was consistent with the result indicating that the forest above the middle slope was not affected by shaded terrain [37].

4.2. The Distance among Trees Was the Dominant Factor for Tree Solar Radiation

The social position of trees (including dominance of trees and distance between trees) was unique for each tree in the forest [14]. Our results suggested that the higher the social position of the individual tree, the more solar radiation it received (Figure 6). This result was completely consistent with our second hypothesis. This was primarily because the trees located in the dominant canopy, which were less shaded by neighboring trees because of a relatively higher tree height [51], could receive more solar radiation [6,7]. Solar radiation income increased as the distance between trees increased (Figure 7), which was consistent with the result indicating that an increase in the distance between trees resulted in greater intakes of solar radiation [52].

However, the contribution of the distance between trees to R_i was more important than that of dominance (Figure 9) because the vertical shading effect among trees weakened as the distance between trees increased [53]. Moreover, the solar radiation income of individual trees increased slowly when the distance between trees was above 4.5 m (Figure 7c), indicating that these trees obtained sufficient solar radiation and the distance between trees was no longer a key factor limiting solar radiation income. This was consistent with the finding that competition among neighbors within 5 m reduced the growth rate due to limited light resources [7,54].

4.3. Applications of the Newly Developed R_i Model

The R_i model developed in this study exhibited a satisfactory performance in estimating the daily solar radiation income of individual trees. The simulated value was 16.3% higher than the observed value since the impacts of leaf variation on solar radiation income were not considered [44]. When simulating only rainy days, the model accuracy was lower at $R^2 = 0.60$ (Figure 8c). This was mostly because on rainy days, cloud cover and ground humidity (changing albedo) in turn affected solar radiation income [55,56]. The developed R_i model couples the effects of topography (including latitude, longitude, elevation, slope and aspect) and the social position of trees (both horizontal distance and vertical dominance), which can easily be obtained in different mountain regions. Foresters could choose suitable measurements according to their actual conditions. For example, horizontal ground solar radiation data can be obtained from ground weather stations, topographic data can be obtained from remote sensing monitoring and the social position of trees can be obtained through field surveys [57].

The R_i model can be used to explain tree growth and thereby regulate forest productivity. For example, a study on this sample slope showed that the cumulative seasonal growth of the down slope plot (1.08 mm) was significantly less than that of the upper slope plot (1.54 mm) because high solar radiation facilitates photosynthesis of organic matter for tree growth and increases air temperature [58]. In addition, on this sample slope, previous transpiration studies focused on the stand scale because of the lack of environmental characteristics of individual trees [37,59,60]. The R_i model in this study accurately predicted the solar radiation received by individual trees, which can be beneficial in explaining the differences in the transpiration of trees. In summary, a series of mountain forest management plans (e.g., thinning) can be made based on the topographic conditions and social position of trees using the developed R_i model to improve the forest structure and alleviate the adverse effects of shade on the solar radiation income of trees.

4.4. Implications for Mountain Forest Management

The solar radiation income of individual trees in forests depended on topographic and social positions. Therefore, different management measures should be adopted for trees at different slope positions and different social positions to promote the receipt of more solar radiation. Our research suggests that the social position of trees (especially the distance between trees) has relevance for mountain forest management.

Different forest management measures should be implemented at different slope positions. When $\Delta H/\Delta d > 0.26$, the solar radiation income of these slope positions was significantly affected by the shaded terrain effect of the opposite mountain; therefore, the forest density should be reduced by increasing the distance between trees so that residual dominant trees can obtain more solar radiation during the hours of sunshine. For example, given a $\Delta H/\Delta d$ of 0.40 and a *Dom* of 0.3, the trees can receive a maximum R_i (8.79 MJ m^{-2} day^{-1}) at a distance between trees of 10.0 m (Figure 10).

However, when $\Delta H/\Delta d < 0.26$, the shaded terrain effect was almost negligible for the forest, and the shading of adjacent trees was the main factor affecting the solar radiation income of individual trees. The proportion of trees with different social positions can be adjusted by implementing thinning to reduce the light competition among trees. For example, given a $\Delta H/\Delta d$ of 0.10, the daily R_i will be above 9.32 MJ m^{-2} day^{-1} when the *Dom* values are 0.2 and 0.3 with D values of 6.5 m and 4.0 m, respectively (Figure 10).

4.5. Limitations of This Study

The R_i model established in this study can calculate the solar radiation income of trees at any location on the slope. However, due to the experimental conditions, this study only monitored the actual solar radiation income of trees on a sample down slope, so the differences in the solar radiation income of trees in different slope positions could not be verified. Therefore, the reliability of the model estimation will be improved when the model parameters are fitted using the data from more mountainous regions [61].

5. Conclusions

This study focused on the effects of topography and social position on the solar radiation income of individual trees. The slope effect resulted in lower solar radiation income on the hillslope, which was 27.8% lower than that of the horizontal ground. In the different slope positions, 2.7% to 46.9% of solar radiation was lost due to the shaded terrain effect. Solar radiation income above the middle slope was basically unaffected by the shading of the opposite mountain. The more dominant the tree's social position, the more solar radiation it received, ranging from 22.4% to 95.3%. The dominant factor contributing to the changes in R_i was the distance between trees, and R_i was basically stable when the distance between trees was greater than 4.5 m. The R_i model covered the effect of topography and the social position of individual trees, which explained 83.0% of the observed R_i variation.

These results advanced our understanding of the role of topography and tree social positions of solar radiation income in mountainous regions. Therefore, different forest management strategies should be developed at different slope positions, and the distance between trees should be selected as the main management measure to facilitate the precise adjustment of stand structure design, ensuring that residual trees receive more solar radiation.

Author Contributions: Formal analysis, J.L.; investigation, J.L. and B.L.; writing—original draft, J.L.; writing—review and editing, J.L., P.Y., Y.W. (Yanfang Wan), Y.W. (Yanhui Wang) and Y.Y. All authors have read and agreed to the published version of the manuscript.

Funding: This work was financially supported by the National Natural Science Foundation of China (U21A2005, 42161144008, U20A2085) and the National Key R & D Program of China (2022YFF0801803, 2022YFF1300404).

Data Availability Statement: Not applicable.

Acknowledgments: We thank Liu Hong, Songping Yu, Zhonghui Zhang and Xiaonan Huang for their assistance in the field work.

Conflicts of Interest: The authors declare no conflict of interest.

References

1. Smith, H. Light Quality, Photoperception, and Plant Strategy. *Annu. Rev. Plant Physiol.* **1982**, *33*, 481–518. [CrossRef]
2. Chiang, C.; Olsen, J.E.; Basler, D.; Bånkestad, D.; Hoch, G. Latitude and Weather Influences on Sun Light Quality and the Relationship to Tree Growth. *Forests* **2019**, *10*, 610. [CrossRef]
3. Sinoquet, H.; Rakocevic, M.; Varlet-Grancher, C. Comparison of Models for Daily Light Partitioning in Multispecies Canopies. *Agric. For. Meteorol.* **2000**, *101*, 251–263. [CrossRef]
4. Wang, H.; Tetzlaff, D.; Dick, J.J.; Soulsby, C. Assessing the Environmental Controls on Scots Pine Transpiration and the Implications for Water Partitioning in a Boreal Headwater Catchment. *Agric. For. Meteorol.* **2017**, *240–241*, 58–66. [CrossRef]
5. Lu, Y.; Lei, X.; Hong, L.; Ning, J.; Liu, X.; Meng, J. Demonstrative Application of the Close-to-Nature Forest Management Planning System to Forestry Practice. *J. Southwest For. Univ.* **2010**, *30*, 1–6.
6. Binkley, D.; Stape, J.L.; Bauerle, W.L.; Ryan, M.G. Explaining Growth of Individual Trees: Light Interception and Efficiency of Light Use by Eucalyptus at Four Sites in Brazil. *For. Ecol. Manag.* **2010**, *259*, 1704–1713. [CrossRef]
7. Binkley, D.; Campoe, O.C.; Gspaltl, M.; Forrester, D.I. Light Absorption and Use Efficiency in Forests: Why Patterns Differ for Trees and Stands. *For. Ecol. Manag.* **2013**, *288*, 5–13. [CrossRef]
8. Oogathoo, S.; Houle, D.; Duchesne, L.; Kneeshaw, D. Vapour Pressure Deficit and Solar Radiation Are the Major Drivers of Transpiration of Balsam Fir and Black Spruce Tree Species in Humid Boreal Regions, Even during a Short-Term Drought. *Agric. For. Meteorol.* **2020**, *291*, 108063. [CrossRef]
9. Rissanen, K.; Martin-Guay, M.-O.; Riopel-Bouvier, A.-S.; Paquette, A. Light Interception in Experimental Forests Affected by Tree Diversity and Structural Complexity of Dominant Canopy. *Agric. For. Meteorol.* **2019**, *278*, 107655. [CrossRef]
10. Fu, B. The total amount of solar radiation on the slope. *J. Nanjing Univ. (Nat. Sci.)* **1958**, *02*, 47–82.
11. Prévost, M.; Raymond, P. Effect of Gap Size, Aspect and Slope on Available Light and Soil Temperature after Patch-Selection Cutting in Yellow Birch–Conifer Stands, Quebec, Canada. *For. Ecol. Manag.* **2012**, *274*, 210–221. [CrossRef]
12. Baek, S.C.; Park, J.-H.; Na, S.I.; Park, J.-K. Distribution of Solar Radiation Including Slope Effect in South Korea. In Proceedings of the International Society for Optical Engineering; Entekhabi, D., Honda, Y., Sawada, H., Shi, J., Oki, T., Eds.; Kyoto, Japan, 21 November 2012; p. 85242K. Available online: https://www.spiedigitallibrary.org/conference-proceedings-of-spie/8524/1/Distribution-of-solar-radiation-including-slope-effect-in-South-Korea/10.1117/12.977420.short?SSO=1 (accessed on 22 February 2023).
13. Chen, J.M.; Chen, X.; Ju, W. Effects of Vegetation Heterogeneity and Surface Topography on Spatial Scaling of Net Primary Productivity. *Biogeosciences* **2013**, *10*, 4879–4896. [CrossRef]
14. Aakala, T.; Shimatani, K.; Abe, T.; Kubota, Y.; Kuuluvainen, T. Crown Asymmetry in High Latitude Forests: Disentangling the Directional Effects of Tree Competition and Solar Radiation. *Oikos* **2016**, *125*, 1035–1043. [CrossRef]
15. Fan, W.; Chen, J.M.; Weimin, J.; Nesbitt, N. Hybrid Geometric Optical–Radiative Transfer Model Suitable for Forests on Slopes. *IEEE Trans. Geosci. Remote Sens.* **2014**, *52*, 5579–5586. [CrossRef]
16. Nath, B.; Ni-Meister, W. The Interplay between Canopy Structure and Topography and Its Impacts on Seasonal Variations in Surface Reflectance Patterns in the Boreal Region of Alaska—Implications for Surface Radiation Budget. *Remote Sens.* **2021**, *13*, 3108. [CrossRef]
17. Zhang, X. Effects of Social Position and Competition on Tree Transpiration of a Natural Mixed Forest in Chongqing, China. *Trees* **2019**, *33*, 719–732. [CrossRef]
18. Garg, H.P.; Datta, G. Fundamentals and Characteristics of Solar Radiation. *Renew. Energy* **1993**, *3*, 305–319. [CrossRef]
19. Gracia, M.; Montané, F.; Piqué, J.; Retana, J. Overstory Structure and Topographic Gradients Determining Diversity and Abundance of Understory Shrub Species in Temperate Forests in Central Pyrenees (NE Spain). *For. Ecol. Manag.* **2007**, *242*, 391–397. [CrossRef]
20. Liu, B.; Yang, J.; Johnstone, J.F. Understory Vascular Plant Community Assembly in Relation to Time-since-Fire and Environmental Variables in a Chinese Boreal Forest. *J. Mt. Sci.* **2017**, *14*, 1317–1328. [CrossRef]
21. Díaz, F.; Montero, H.; Santana, D.; Montero, G.; Rodríguez, E.; Mazorra Aguiar, L.; Oliver, A. Improving Shadows Detection for Solar Radiation Numerical Models. *Appl. Math. Comput.* **2018**, *319*, 71–85. [CrossRef]
22. Courbaud, B.; de Coligny, F.; Cordonnier, T. Simulating Radiation Distribution in a Heterogeneous Norway Spruce Forest on a Slope. *Agric. For. Meteorol.* **2003**, *116*, 1–18. [CrossRef]

23. Warren, R.J. Mechanisms Driving Understory Evergreen Herb Distributions across Slope Aspects: As Derived from Landscape Position. *Plant Ecol.* **2008**, *198*, 297–308. [CrossRef]
24. Zhang, Y.; Chang, X.; Liang, J. Comparison of Different Algorithms for Calculating the Shading Effects of Topography on Solar Irradiance in a Mountainous Area. *Environ. Earth Sci.* **2017**, *76*, 295. [CrossRef]
25. Olson, M.; Rupper, S. Impacts of Topographic Shading on Direct Solar Radiation for Valley Glaciers in Complex Topography. *Cryosphere* **2019**, *13*, 29–40. [CrossRef]
26. Remeš, J.; Bílek, L.; Novák, J.; Vacek, Z.; Vacek, S.; Putalová, T.; Koubek, L. Diameter Increment of Beech in Relation to Social Position of Trees, Climate Characteristics and Thinning Intensity. *J. For. Sci.* **2015**, *61*, 456–464. [CrossRef]
27. Grote, R.; Gessler, A.; Hommel, R.; Poschenrieder, W.; Priesack, E. Importance of Tree Height and Social Position for Drought-Related Stress on Tree Growth and Mortality. *Trees* **2016**, *30*, 1467–1482. [CrossRef]
28. Zhang, S.Y.; Ren, H.; Jiang, Z. Wood Density and Wood Shrinkage in Relation to Initial Spacing and Tree Growth in Black Spruce (Picea Mariana). *J. Wood Sci.* **2021**, *67*, 30. [CrossRef]
29. Campoe, O.C.; Stape, J.L.; Albaugh, T.J.; Lee Allen, H.; Fox, T.R.; Rubilar, R.; Binkley, D. Fertilization and Irrigation Effects on Tree Level Aboveground Net Primary Production, Light Interception and Light Use Efficiency in a Loblolly Pine Plantation. *For. Ecol. Manag.* **2013**, *288*, 43–48. [CrossRef]
30. Gspaltl, M.; Bauerle, W.; Binkley, D.; Sterba, H. Leaf Area and Light Use Efficiency Patterns of Norway Spruce under Different Thinning Regimes and Age Classes. *For. Ecol. Manag.* **2013**, *288*, 49–59. [CrossRef]
31. Warren, C.R.; McGrath, J.F.; Adams, M.A. Water Availability and Carbon Isotope Discrimination in Conifers. *Oecologia* **2001**, *127*, 476–486. [CrossRef] [PubMed]
32. Hale, S.E. The Effect of Thinning Intensity on the Below-Canopy Light Environment in a Sitka Spruce Plantation. *For. Ecol. Manag.* **2003**, *179*, 341–349. [CrossRef]
33. Wan, Y.; Yu, P.; Wang, Y.; Wang, B.; Yu, Y.; Wang, X.; Liu, Z.; Liu, X.; Wang, S.; Xiong, W. The Variation in Water Consumption by Transpiration of Qinghai Spruce among Canopy Layers in the Qilian Mountains, Northwestern China. *Forests* **2020**, *11*, 845. [CrossRef]
34. Campoe, O.C.; Stape, J.L.; Nouvellon, Y.; Laclau, J.-P.; Bauerle, W.L.; Binkley, D.; Maire, G.L. Stem Production, Light Absorption and Light Use Efficiency between Dominant and Non-Dominant Trees of Eucalyptus Grandis across a Productivity Gradient in Brazil. *For. Ecol. Manag.* **2013**, *288*, 14–20. [CrossRef]
35. Fernández-Tschieder, E.; Binkley, D.; Bauerle, W. Production Ecology and Reverse Growth Dominance in an Old-Growth Ponderosa Pine Forest. *For. Ecol. Manag.* **2020**, *460*, 117891. [CrossRef]
36. Li, S. Forest in Mountainous Ecosystem and Its Relationship with Production. *Chin. J. Ecol.* **1985**, 35–38.
37. Tian, A.; Wang, Y.; Webb, A.A.; Liu, Z.; Yu, P.; Xiong, W.; Wang, X. Partitioning the Causes of Spatiotemporal Variation in the Sunny Day Sap Flux Density of a Larch Plantation on a Hillslope in Northwest China. *J. Hydrol.* **2019**, *571*, 503–515. [CrossRef]
38. Kübler, D.; Hildebrandt, P.; Günter, S.; Stimm, B.; Weber, M.; Muñoz, J.; Cabrera, O.; Zeilinger, J.; Silva, B.; Mosandl, R. Effects of Silvicultural Treatments and Topography on Individual Tree Growth in a Tropical Mountain Forest in Ecuador. *For. Ecol. Manag.* **2020**, *457*, 117726. [CrossRef]
39. Wang, L.; Liu, Z.; Guo, J.; Wang, Y.; Ma, J.; Yu, S.; Yu, P.; Xu, L. Estimate Canopy Transpiration in Larch Plantations via the Interactions among Reference Evapotranspiration, Leaf Area Index, and Soil Moisture. *For. Ecol. Manag.* **2021**, *481*, 118749. [CrossRef]
40. Liu, Z.; Wang, Y.; Yu, P.; Xu, L.; Yu, S. Environmental and Canopy Conditions Regulate the Forest Floor Evapotranspiration of Larch Plantations. *For. Ecosyst.* **2022**, *9*, 100058. [CrossRef]
41. Liu, Z.; Wang, Y.; Tian, A.; Yu, P.; Xiong, W.; Xu, L.; Wang, Y. Intra-Annual Variation of Stem Radius of Larix Principis-Rupprechtii and Its Response to Environmental Factors in Liupan Mountains of Northwest China. *Forests* **2017**, *8*, 382. [CrossRef]
42. Fu, B. The Influence of Slope on the Sunniness. *J. Nanjing Univ. (Nat. Sci.)* **1958**, *02*, 23–46.
43. Schmidt, U.; Thöni, H.; Kaupenjohann, M. Using a Boundary Line Approach to Analyze N_2O Flux Data from Agricultural Soils. *Nutr. Cycl. Agroecosystems* **2000**, *57*, 119–129. [CrossRef]
44. Liu, Z.; Wang, Y.; Tian, A.; Webb, A.A.; Yu, P.; Xiong, W.; Xu, L.; Wang, Y. Modeling the Response of Daily Evapotranspiration and Its Components of a Larch Plantation to the Variation of Weather, Soil Moisture, and Canopy Leaf Area Index. *J. Geophys. Res. Atmos.* **2018**, *123*, 7354–7374. [CrossRef]
45. Sun, Z.; Shi, B.; Weng, D. Distribution characteristics of direct solar radiation on mountain slopes in China. *Plateau Meteorol.* **1990**, *9*, 371–381.
46. Zeng, Y.; Qiu, X.; Pan, A.; Liu, C. Distributed Modeling of Global Solar Radiation over Rugged Terrain of the Yellow River Basin. *Adv. Earth Sci.* **2008**, *23*, 1185–1193.
47. Zhang, S.; Li, X.; She, J. Error Assessment of Grid-based Terrain Shading Algorithms for Solar Radiation Modeling over Complex Terrain. *Trans. GIS* **2020**, *24*, 230–252. [CrossRef]
48. Shi, G.; Qiu, X.; Zeng, Y. New Method for Estimating Daily Global Solar Radiation over Sloped Topography in China. *Adv. Atmos. Sci.* **2018**, *35*, 285–295. [CrossRef]
49. Aguilar, C.; Herrero, J.; Polo, M.J. Topographic Effects on Solar Radiation Distribution in Mountainous Watersheds and Their Influence on Reference Evapotranspiration Estimates at Watershed Scale. *Hydrol. Earth Syst. Sci.* **2010**, *14*, 2479–2494. [CrossRef]

50. Marsh, C.B.; Pomeroy, J.W.; Spiteri, R.J. Implications of Mountain Shading on Calculating Energy for Snowmelt Using Unstructured Triangular Meshes: Implications of Mountain Shading for Snowmelt. *Hydrol. Process.* **2012**, *26*, 1767–1778. [CrossRef]
51. Laurans, M.; Hérault, B.; Vieilledent, G.; Vincent, G. Vertical Stratification Reduces Competition for Light in Dense Tropical Forests. *For. Ecol. Manag.* **2014**, *329*, 79–88. [CrossRef]
52. Chase, C.W.; Kimsey, M.J.; Shaw, T.M.; Coleman, M.D. The Response of Light, Water, and Nutrient Availability to Pre-Commercial Thinning in Dry Inland Douglas-Fir Forests. *For. Ecol. Manag.* **2016**, *363*, 98–109. [CrossRef]
53. Strand, M.; Löfvenius, M.O.; Bergsten, U.; Lundmark, T.; Rosvall, O. Height Growth of Planted Conifer Seedlings in Relation to Solar Radiation and Position in Scots Pine Shelterwood. *For. Ecol. Manag.* **2006**, *224*, 258–265. [CrossRef]
54. Chi, X.; Tang, Z.; Xie, Z.; Guo, Q.; Zhang, M.; Ge, J.; Xiong, G.; Fang, J. Effects of Size, Neighbors, and Site Condition on Tree Growth in a Subtropical Evergreen and Deciduous Broad-leaved Mixed Forest, China. *Ecol. Evol.* **2015**, *5*, 5149–5161. [CrossRef]
55. Huang, D.; Qian, Y. The effects of the slope irradiance on different weather processes under different model resolutions. *Acta Meteorol. Sin.* **2008**, *01*, 90–100.
56. Choosakul, N.; Buddhakala, M.; Barnthip, N.; Muakngam, A.; Banglieng, C. Application of Solar Cells for Daytime Weather Study. *Energy Procedia* **2011**, *9*, 171–177. [CrossRef]
57. Hassan-Esfahani, L.; Torres-Rua, A.; Jensen, A.; McKee, M. Assessment of Surface Soil Moisture Using High-Resolution Multi-Spectral Imagery and Artificial Neural Networks. *Remote Sens.* **2015**, *7*, 2627–2646. [CrossRef]
58. Ma, J.; Guo, J.; Wang, Y.; Liu, Z.; Gao, D.; Hong, L.; Zhang, Z. Variations in Stem Radii of Larix Principis-Rupprechtii to Environmental Factors at Two Slope Locations in the Liupan Mountains, Northwest China. *J. For. Res.* **2021**, *32*, 513–527. [CrossRef]
59. Cao, G. Hydrological Impact and the Slpoe Scale Effect of the Vegetation Structure in the Xiangshuihe Watershed in Liupan Mountains. Ph.D. Thesis, Chinese Academy of Forestry, Beijing, China, 2014.
60. Liu, Z. Spatio-Temporal Variations and Scale Transition of Hydrological Impact of Larix Principis-Ruprechtii Plantation on a Slope of Liupan Mountains, China. Ph.D. Thesis, Chinese Academy of Forestry, Beijing, China, 2018.
61. Sawano, S.; Hotta, N.; Tanaka, N.; Tsuboyama, Y.; Suzuki, M. Development of a Simple Forest Evapotranspiration Model Using a Process-Oriented Model as a Reference to Parameterize Data from a Wide Range of Environmental Conditions. *Ecol. Model.* **2015**, *309–310*, 93–109. [CrossRef]

forests

MDPI

Article

Variability in Snowpack Isotopic Composition between Open and Forested Areas in the West Siberian Forest Steppe

Dmitry Pershin [1,2,*], Natalia Malygina [2], Dmitry Chernykh [2], Roman Biryukov [2], Dmitry Zolotov [2] and Lilia Lubenets [2]

1 Faculty of Geography, Lomonosov Moscow State University, 1 Leninskie Gory, Moscow 119991, Russia
2 Institute for Water and Environmental Problems SB RAS, 1, Molodezhnaya Street, Barnaul 656038, Russia
* Correspondence: pershindk@my.msu.ru

Abstract: The stable water isotopes in snow (primarily ^{18}O and ^{2}H) are widely used for tracing hydrological and ecological processes. However, isotopic signatures of snow can be significantly modified by topography and land cover. This study assesses spatial and temporal variability of the bulk snowpack isotopic composition ($\delta^{18}O$, $\delta^{2}H$, d-excess) between forested (pine and birch) and open areas in the West Siberian forest steppes. Isotopic samples were collected over the peak snow accumulation in 2017–2019. The snow isotopic composition within forested areas differed from open steppes, mainly in reducing d-excess (1.6‰ on average). We did not find a significant effect of canopy interception on snow enrichment in heavier isotopes. Snowpack in the pine forests was even lighter by 3.6‰ for $\delta^{2}H$ compared to open areas, probably, due to low energy inputs and interception capacity. Additionally, snow depth significantly influenced the isotopic composition spatial variability. As snow depth increased, $\delta^{18}O$ and $\delta^{2}H$ values decreased due to conservation within the snowpack and less influence of sublimation and moisture exchange with the soil. However, this pattern was only evident in winters with below-average snow depth. Therefore, taking into account snow depth spatial and seasonal variability is advisable when applying the isotopic methods.

Keywords: river basin; forest; grassland; interception; wind redistribution; stable water isotopes

Citation: Pershin, D.; Malygina, N.; Chernykh, D.; Biryukov, R.; Zolotov, D.; Lubenets, L. Variability in Snowpack Isotopic Composition between Open and Forested Areas in the West Siberian Forest Steppe. *Forests* **2023**, *14*, 160. https://doi.org/10.3390/f14010160

Academic Editors: Yanhui Wang, Karl-Heinz Feger and Lulu Zhang

Received: 23 November 2022
Revised: 12 January 2023
Accepted: 13 January 2023
Published: 16 January 2023

1. Introduction

Snow accumulation, storage, and melting dynamics affect multiple hydrological, ecological, and social processes in mountainous and high-latitude environments [1,2]. Tracking changes in snowpack accumulation and melt rates is challenging because the driving factors operate at multiple spatial and temporal scales [3–5]. Stable water isotopic composition of snow (primarily $\delta^{2}H$ and $\delta^{18}O$) has become a valuable tool for investigating various snow hydrological and ecohydrological processes [6]. Implementations include snow contribution to groundwater recharge [7,8], streamflow generation during rain-on-snow events [9,10], exploring vegetation water sources [11–14], etc. However, the application of isotopic methods is complicated by snow evolution processes that alter the isotopic signal [15–17].

Snow mass and energy balance is altered by complex processes such as sublimation, wind redistribution, forest canopy interception, melting, and metamorphism [18]. Since snow contains liquid, solid, and vapor phases, most of these processes are accompanied by phase transitions, changing the stable water isotopic composition [6,19,20]. The intensity of snow hydrological processes varies in space. Spatial factors affecting snow isotopic composition have included altitude [21–23], aspect [15,24], snow depth [21], and canopy interception [25–27]. Most of these factors affect snow sublimation fluxes, which leads to enrichment in heavier isotopes of the remaining snow cover [16,22,28,29].

Forest canopy interception affects the snow isotopic composition by increasing the sublimation of intercepted snow and making throughfall isotopically heavier [6]. The under-canopy snowpack in the north-western US was up to one-fourth smaller and isotopically heavier by roughly 2‰ in $\delta^{18}O$ compared with the snowpack in the clear-cut area [26]. Studies in Switzerland showed that the isotope ratios were higher in the snowpack under forest canopy than in open grasslands (by 13.4 ‰ in δ^2H and 2.3 ‰ in $\delta^{18}O$) [27]. However, a five-year study in the southwestern US has shown a much more significant influence of snowfall isotopic input and aspect on $\delta^{18}O$ than canopy density [28]. Additionally, several mechanisms of changes in the snow isotopic composition in complex landscapes remain poorly understood, such as the effect of wind redistribution [6].

Most of the works cited above were performed in mountain forests in relatively humid regions of Europe and North America. Over Siberia, studies have shown that the variability of the contribution of many precipitation sources to the snow results in large isotopic variability [30,31]. At the same time, more detailed catchment-scale studies investigating changes in snow isotopic composition have not been conducted either in the boreal forest area or in the forest steppe.

Snow in continental semi-arid regions is the major water source for ground and soil water recharge and streamflow generation [32]. Considering the significant differences in the snow isotopic signal compared with rainfall, implementation of the isotopic methods for tracing streamflow formation, groundwater recharge, and plant water use seems promising in these regions. However, the mechanisms of the snowpack isotopic composition spatial variability and post-depositional fractionation remain poorly understood. In this work, we focused on changes in the stable water isotopic composition of snow ($\delta^{18}O$, δ^2H, d-excess) between open and forested areas during peak snow accumulation over three years (March 2017–2019). The studies were conducted in the Kasmala River basin located in the forest steppe ecoregion in the south of Western Siberia. The basin structure consists of extensive arable lands, Scots pine forests, and small patches of deciduous forests. Basin landscape composition allowed us to study changes in the snow isotopic composition in open and forested areas and to evaluate factors influencing isotopic ratios considering the differences in topography and seasonal climate.

2. Materials and Methods

2.1. Study Area

We carried out the study in the 1768.7 km^2 Kasmala river basin (53°4′ N, 82°20′ E) in the south of the West Siberian Plain (Figure 1). The Kasmala basin is a snow-dominated watershed in the headwaters of the Ob River.

The study area belongs to the West Siberian forest steppe ecoregion [33]. This region is also part of the West-Siberian grain belt, an important agricultural region in southern Siberia. The dominant land cover type is arable land (59.7%). A unique characteristic of this part of the West Siberian Plain is the long strips of Scotch pine (*Pinus sylvestris* L.) forests [34–36]. This forest type (covers about 12%) is characterized by dense pine stands (canopy density 60–80 %) with a small proportion of mixed deciduous vegetation. The average diameter at breast height is about 25–30 cm, and tree height is between 20–25 m. The forests occupy the sandy massifs within the extended ancient flow depressions. Upland slopes oriented towards the depressions have small slope angles (1–3°) covered by arable land and steppe patches. Another type of forest presented here is the small patches of birch (*Betula pendula* Roth) and aspen (*Populus tremula* L.) stands (5.9%). The basin is naturally divided into three main parts: the northern (NP) and southern (SP) open steppe areas with deciduous forest patches and the central part, occupied mainly by pine forest (CP). Additionally, in the study, we considered arable land/open steppes, deciduous forests, and pine forests as three major land cover types.

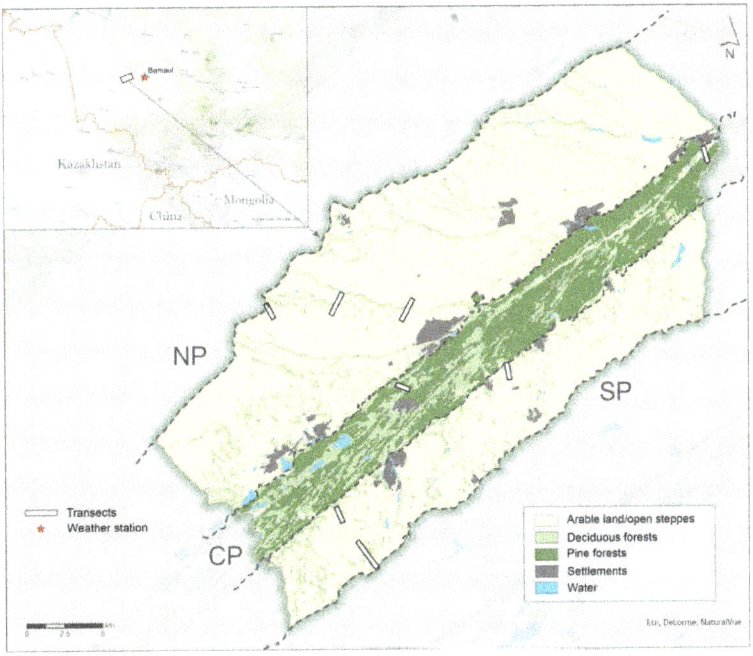

Figure 1. The Kasmala basin in the south of Western Siberia with sampling locations and land cover composition. NP—northern part, CP—central part, SP—southern part.

Continental climate with strong seasonality in the study region varies from sub-humid in the northeast to semi-arid (steppe) in the southwest. The area has cold winters and hot summers with a mean annual temperature (1966–2020) of 2.4 °C at the Barnaul weather station (53 km to the nearest transect) [37]. The average annual precipitation is 427 mm. The mean winter (November-March) precipitation is about 125 mm. Wind drift is the most crucial factor in snow redistribution. The average daily wind speed in November-March is 2.7 m/s [37].

2.2. Sampling and Laboratory Analysis

We conducted three field campaigns around 10–15 March, 2017–2019 during the peak snow accumulation. Sampling was performed along eight transects that varied in length from 500 m to 2 km. The distance between measurements was 200 m in open areas and 100 m in forested areas. NP and SP included three transects each, and CP included two transects (Figure 1). Each transect differed in terrain characteristics (ruggedness, slopes, and aspects) and land cover composition (open areas, deciduous, and coniferous forests). GPS location of each sampling point was recorded. During the field campaign, the bulk snowpack samples were collected at each location using a 60 cm snow coring sampler VS-43. The samples were weighed to obtain bulk density and calculate snow water equivalent (SWE). Then, the entire samples were immediately placed into sealable high-density polyethylene bags to prevent further evaporation and fractionation effects. The snow depth was measured at the same points using a special snow ruler. If the snow depth was higher than 60 cm, the snow was measured successively from deeper layers. A total of 192 samples were taken over three sampling years.

Snow samples were brought to the laboratory in sealed bags on the day of sampling and melted at room temperature. Once the samples completely melted, they were filtered through 0.45-μm filters (Minisart NML Plus) into 2-mL glass vials. The isotopic composition ($\delta^{18}O$, $\delta^{2}H$) of all samples was analyzed through laser spectroscopy (PICARRO L2130-i

(WS-CRDS). The measurement uncertainties were ±0.4‰ for $\delta^{18}O$ and ±0.1‰ for δ^2H. Water samples were calibrated against the international standards (V-SMOW, GRESP).

2.3. Data Processing and Analysis

We analyzed spatial and temporal variations in the bulk snowpack isotopic composition. In exploring interannual differences, we considered variations in temperature, precipitation, and snow accumulation relative to the interannual means from the Barnaul weather station [37]. The station is the closest to the study area (53 km to the nearest transect) and has the most consistent series of observations.

The $\delta^{18}O$ and δ^2H mean values among basin parts, land cover types, and sampling years were tested using non-parametric Kruskal–Wallis H test. Additionally, Wilcoxon test with the Bonferroni correction was applied to evaluate the differences between the group levels.

For assessing the site-specific covariation of hydrogen and oxygen stable isotope ratios, we created δ^2H vs. $\delta^{18}O$ plots (similar to local meteoric water lines) for the basin parts (NP, CP, and SP). Additionally, the second-order isotopic parameter deuterium excess was computed (d-excess = $\delta^2H - 8 \times \delta^{18}O$) [38]. D-excess helps distinguish equilibrium and nonequilibrium processes through the differences from the global meteoric water line (GMWL; $\delta^2H = \delta^{18}O + 10$) [39]. Values less than 10‰ often indicate kinetic fractionation due to evaporation or sublimation [6].

In order to assess the interactions between isotopic composition, topography, and land cover, we created a set of multiple regression (MLR) models. Stepwise regression analysis was performed separately for $\delta^{18}O$ and δ^2H as dependent variables. However, the relationships behaved similarly, and we used δ^2H for further analysis due to a smaller measurement error. All variables were first tested for normality of distribution, multicollinearity, and the presence of outliers. We also calculated Z-scores for the predictors and response variables of the final models. The selection of variables for each model was performed automatically by stepwise backward elimination according to the Akaike information criterion (AIC). Finally, each final model was checked for residual distribution and homoscedasticity.

Input variables were chosen based on their potential effect on the isotopic composition of a bulk snowpack. Snow depth and SWE were used as indicators of wind redistribution since wind is the main factor of depth and SWE spatial heterogeneity on the plains (no elevation gradient). Given that depth and SWE are correlated, they were included in the regression separately. The snow energy balance and, therefore, the processes of sublimation and melt depend on the local topography. The slope, aspect, general curvature, and Topographic Ruggedness Index (TRI) were calculated using the SRTM digital elevation model (DEM) [40]. The aspect was transformed with the cosine function, resulting in a value of −1 for north- and 1 for south-facing points. We used the land cover map based on the Landsat satellite images to calculate the forest ratio in the 200 m surrounding the sample points. This metric was chosen because forests in the study area affect snowpack accumulation not only via canopy interception but also through the obstruction of wind redistribution. The obstruction factor appears mainly in the deciduous forest patches within the steppes.

Statistical analysis was performed in R (http://www.r-project.org, accessed on 27 March 2021). Topographic variables were calculated using QGIS (www.qgis.org, accessed on 2 June 2021).

3. Results

3.1. Interannual Differences

Most seasonal climate parameters were close to average during winter seasons 2016/2017–2018/2019, excluding snow conditions. Mean winter temperatures did not differ much from the long-term mean (Table 1). Daily temperatures above 0 °C) were uncommon (for example, in December 2018). However, very low temperatures occurred more frequently, such as in November 2016, January 2018, and February 2019 (Figure 2). Wind speeds in 2016/2017 were close to the mean (2.3 m/s), 2017/2018 and 2018/2019 had lower average wind speeds (1.7–1.8 m/s).

Table 1. Winter summary statistics with standard deviations at Barnaul weather station calculated from daily data [37].

Winter Season	Mean Temperature (°C)	Mean Wind Speed (m/s)	Peak Snow Depth (cm)
2016/2017	−10.5 ± 7.8	2.3 ± 1.5	73 ± 22
2017/2018	−11.9 ± 8.8	1.7 ± 1.6	25 ± 6
2018/2019	−11.8 ± 9.4	1.9 ± 1.4	41 ± 10
Period of record (1966–2020) mean	−11.5 ± 4.7	2.7 ± 0.3	30 ± 13

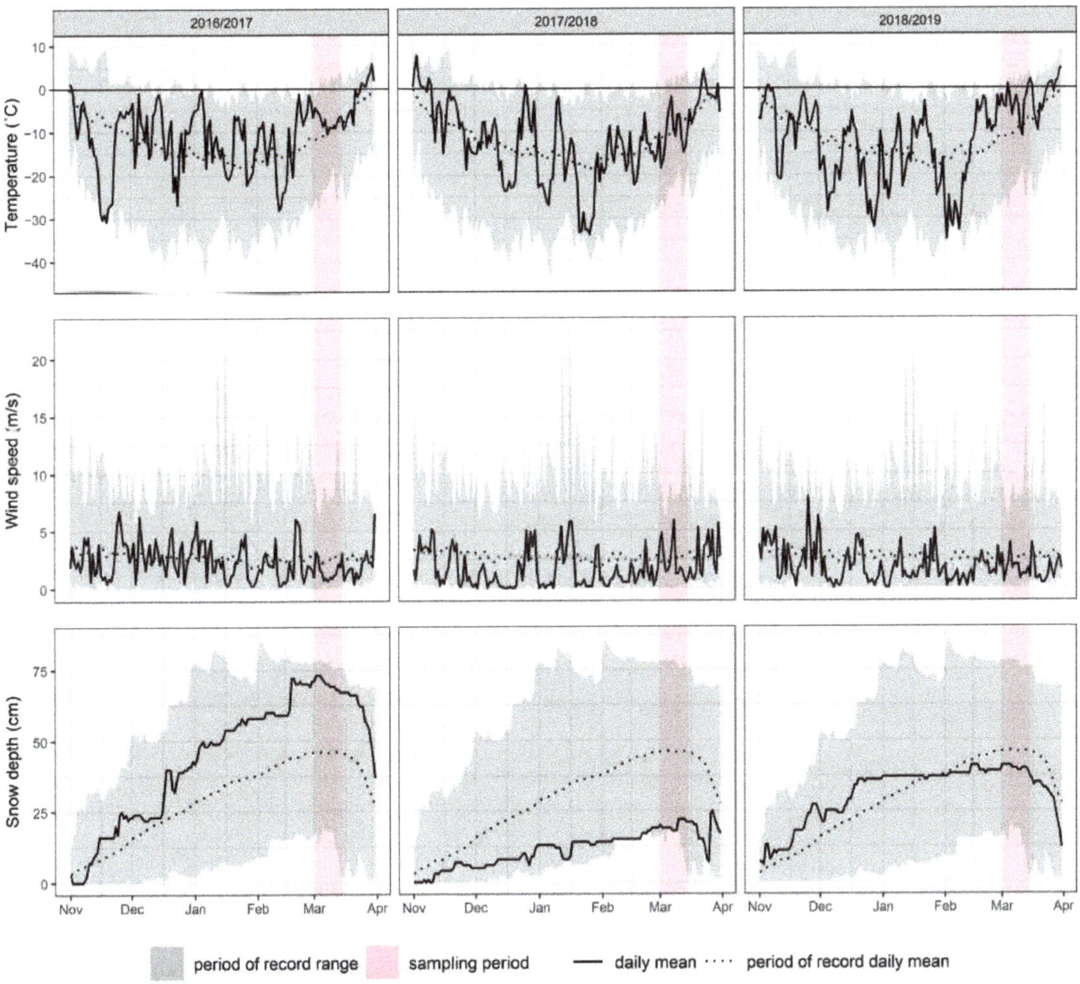

Figure 2. Barnaul weather station air temperature, wind speed, snow depth for the studied winter seasons as well as the station period of record (1966–2020).

The strongest differences were observed in snow cover dynamics. The maximum snow depth values were above average in 2016/2017 (73 cm) and extremely below average in 2017/2018 (25 cm). March 2018 had the lowest snow depth during the station period of record. Depth and SWE values also varied on our transects during the study period. The most considerable difference occurred between 2016/2017 and 2017/2018, when depth and SWE were 57 % and 70 % lower, respectively (Table 2). Therefore, the studied winter seasons had significantly different snow mass conditions but were relatively similar in other seasonal climate parameters.

Table 2. Means and standard deviations of the bulk snowpacks SWE, depth, and isotopic composition, averaged over 2017–2019 and northern (NP), central (CP), and southern (SP) parts.

Sampling Year	Basin Part	Depth (cm)	SWE (mm)	$\delta^{18}O$ (‰)	$\delta^{2}H$ (‰)	d-Excess (‰)
2017	NP	71.4 ± 22.1	196.6 ± 80.6	−19.8 ± 0.7	−151.2 ± 5.5	7.3 ± 1.7
	CP	69.5 ± 13.6	169.7 ± 28.6	−20.3 ± 0.5	−156.7 ± 3.1	5.8 ± 1.9
	SP	80.4 ± 17.3	212.2 ± 49.4	−20.2 ± 0.6	−154.6 ± 4.0	7.2 ± 1.9
2018	NP	27.8 ± 11.5	59.8 ± 40.9	−19.5 ± 1.3	−148.3 ± 9.3	7.4 ± 2.3
	CP	27.8 ± 8.9	41.6 ± 12.3	−20.1 ± 1	−154.9 ± 6.3	5.6 ± 2.4
	SP	36.8 ± 10	67.1 ± 19.9	−20 ± 1.1	−152.6 ± 7.3	7.6 ± 2.2
2019	NP	44.1 ± 8.7	103.6 ± 21.4	−19.4 ± 0.8	−148.1 ± 5.6	7.3 ± 1.0
	CP	40.8 ± 10	91.7 ± 46.7	−19.7 ± 0.7	−151.9 ± 4.7	5.6 ± 1.6
	SP	56.6 ± 11.1	108.7 ± 29	−19.7 ± 0.8	−150.6 ± 5.7	7 ± 1.8

According to Kruskal–Wallis test results (Table 3), differences in mean isotopic composition between sampling years were significant. The median levels varied only around 1‰ in $\delta^{18}O$ and 4‰ in $\delta^{2}H$ during sampling years.

Table 3. Kruskal–Wallis test results for snow isotopic composition ($\delta^{18}O$, $\delta^{2}H$, d-excess) among land cover types, basin parts, and sampling years.

	Factors	Df	H-Value	*p*-Value
$\delta^{18}O$	Sampling year	2	14.00	<0.001
	Basin part	2	9.74	0.008
	Land cover	2	7.13	0.029
$\delta^{2}H$	Sampling year	2	15.54	<0.001
	Basin part	2	18.2	<0.001
	Land cover	2	15.79	<0.001
d-excess	Sampling year	2	3.08	0.21
	Basin part	2	23.03	<0.001
	Land cover	2	19.38	<0.001

Variability within each sampling year was substantially higher (Figure 3). The isotopic variation had the highest rates in 2018 when the snowpack was extremely shallow. The range reached 6.3‰ in $\delta^{18}O$ and 42.6‰ in $\delta^{2}H$. In 2017, the range was approximately half and amounted to 3.2‰ and 26.0‰ in $\delta^{18}O$ and $\delta^{2}H$, respectively. The standard deviations (Table 2) of each sampling year also demonstrate differences in variability.

Differences in mean isotopic composition among the northern (NP), central (CP), and southern (SP) parts of the basin were also significant considering the whole study period (Table 3). Firstly, significant differences may indicate expected differences in the drivers of snow isotopic composition: primarily wind redistribution in open areas and canopy interception in forests. Looking at each year separately, the forested and open parts of the basin differed significantly only in $\delta^{2}H$ (Figure 3). Additionally, Wilcoxon test showed (p-value < 0.05) that $\delta^{18}O$ and $\delta^{2}H$ varied significantly between NP and CP-SP.

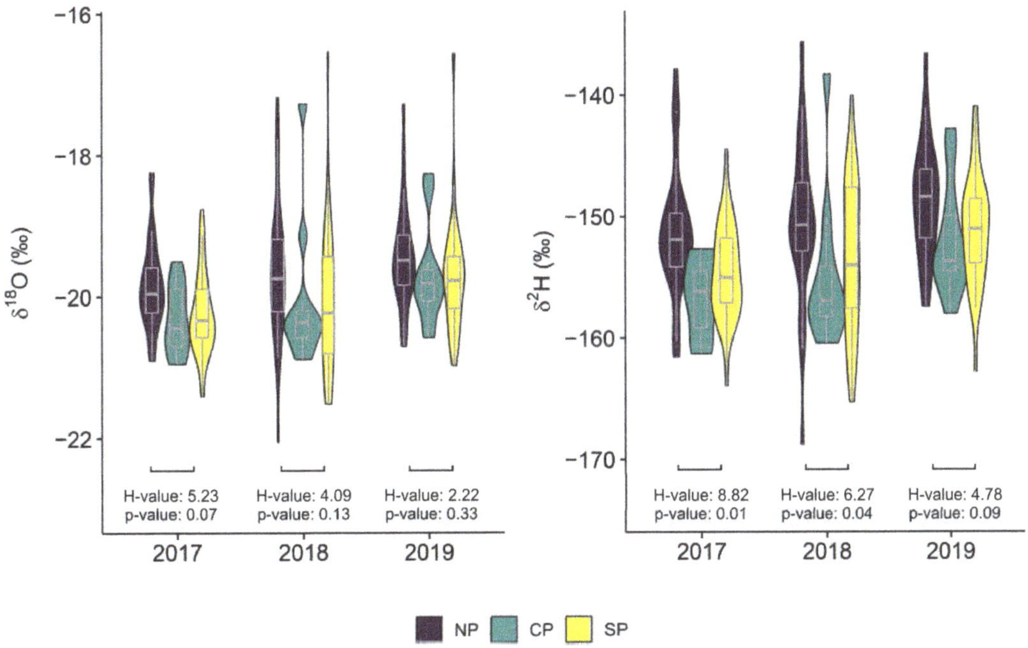

Figure 3. Temporal variability of the isotopic composition (δ^2H and δ^{18}O) of bulk snowpack over the basin parts (NP, CP, SP) with results of Kruskal–Wallis tests within sampling years. Violin and Tukey outlier box plots show the median value (horizontal line within the box), the 1st and 3rd quartile (ends of the box), minimum/maximum values (whiskers) and distribution.

Differences among land cover types were significant for δ^2H and d-excess mean values (Table 3). As in the case of the basin parts, δ^2H and d-excess varied significantly between open steppes, deciduous forests, and pine forests (Wilcoxon test *p*-value < 0.05). Pine forests and deciduous forests were not statistically different.

During the entire study period, the snowpack isotopic composition was slightly heavier within the open parts compared with the forested. The δ^{18}O and δ^2H values in CP were on average lower than in open NP and SP, but only by 0.2 and 3.6‰, respectively. We expected higher δ^{18}O and δ^2H values in the forested part due to canopy interception. However, no direct effect was observed.

3.2. Oxygen and Hydrogen Ratios

Figure 4 gives an overview on δ^2H vs. δ^{18}O relationships among the sampling years and basin parts. All samples lay lower than the GMWL, indicating the influence of nonequilibrium processes (Figure 4). In all equations, slopes were far below 8, and intercept values did not exceed 10, corresponding to the similar parameters of GMWL.

Slope and intercept values of δ^2H vs. δ^{18}O regressions showed clear spatial patterns. In the open SP and NP slopes ranged from 7.90 to 6.12. The slope values were significantly lower in the forested part (5.58 to 6.38). Furthermore, the CP intercepts also reached very low values (up to −43.3). This aspect indicated significantly higher sublimation rates and other nonequilibrium processes in the forested part.

Slopes changed little over time, especially depending on the amount of snow each winter season. Only the intercept values changed significantly (by more than 10–20) within the open and forest parts. On the one hand, this may indicate the constancy of fractionation processes on-ground (stable slope), and on the other hand, the influence of snow formation conditions during each winter season (unstable intercept).

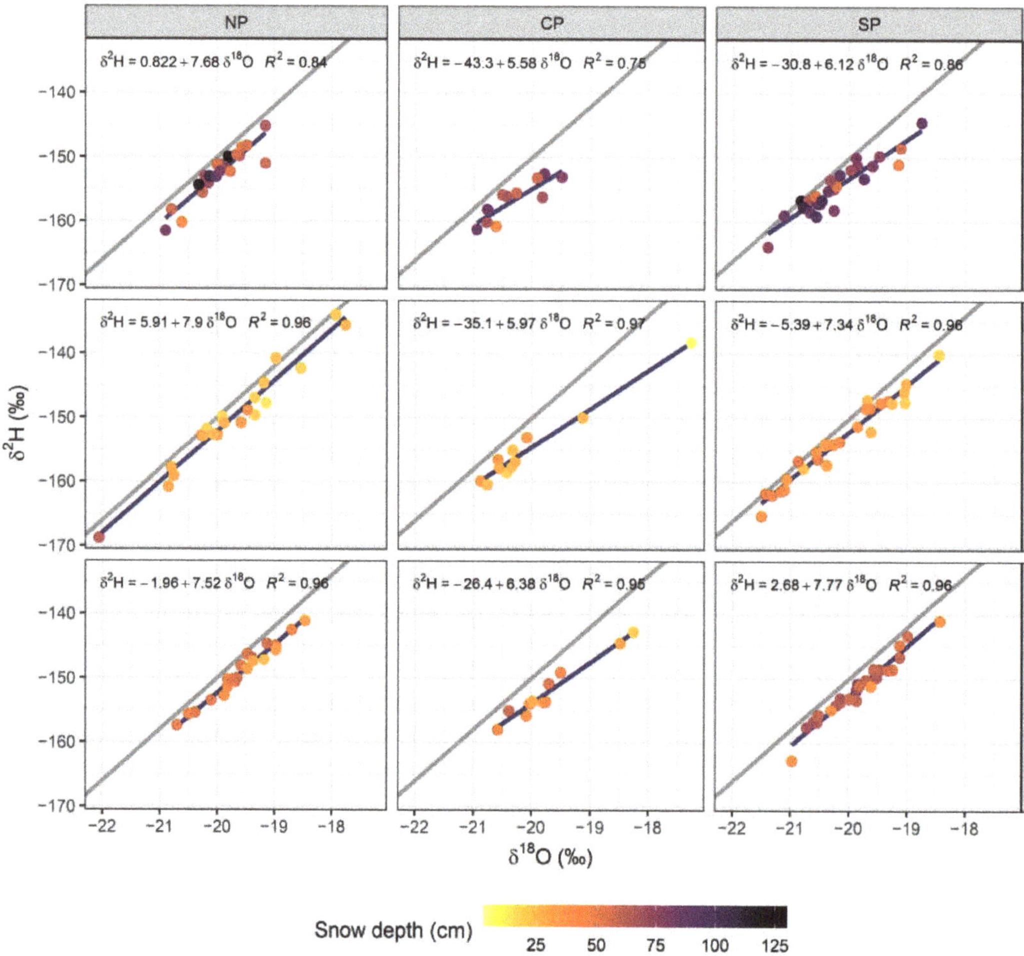

Figure 4. δ^2H vs. $\delta^{18}O$ plot over the basin parts (NP, CP, SP) and sampling years, including their regression lines (blue), global meteoric water line (gray), and snow depth gradient.

D-excess values also showed spatial patterns, remaining relatively stable interannually (Figure 5). CP's mean d-excess values (Table 2) were 1.6 ‰ lower than open SP and NP. Moreover, the d-excess fell below 10 (d-excess of GMWL) in all parts of the Kasmala basin. According to the Kruskal–Wallis test (Table 3), mean differences were significant across the basin parts during the study period. The major differences occurred between CP and NP/SP (Wilcoxon test p-value < 0.001). The contrast again referred to the high intensity of nonequilibrium processes in the forested part of the basin, which, however, did not lead to direct enrichment in heavier isotopes of the forest snowpack.

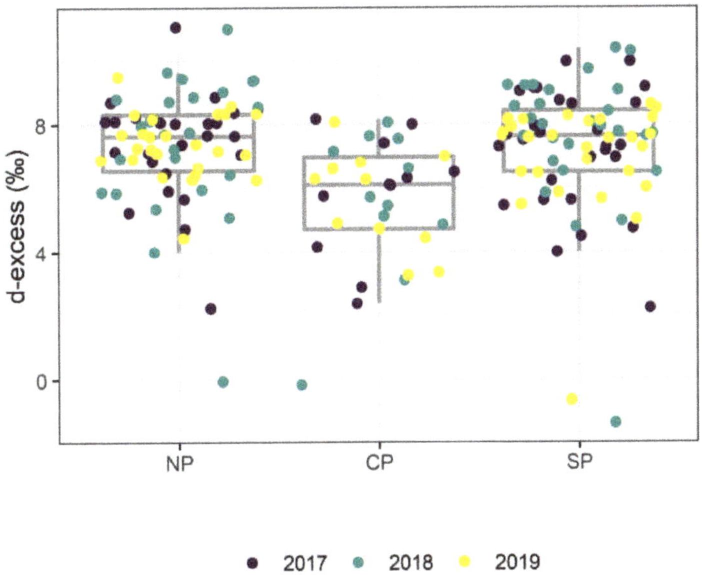

Figure 5. D-excess values over the basin parts (NP, CP, SP) and sampling years. Individual samples are shown using dots. Tukey outlier box plots show the median value (horizontal line within the box), the 1st and 3rd quartile (ends of the box), and the minimum/maximum values (whiskers).

3.3. *Influence of Topography and Land Cover Factors*

Only snow depth and TRI had a relatively stable effect on δ^2H. The influence was evident in two of the three years. The other predictors had no significant influence, except for a feeble impact of Aspect in 2017 (Table 4).

Table 4. Adjusted R^2, slopes, and intercepts for computed regression equations describing H^2 vs. land cover and topography predictors relationships (ns = predictor variable not selected).

Variables	2017	2018	2019
Snow Depth	ns	−0.608 ***	−0.224 *
SWE	ns	ns	ns
TRI	−0.445 ***	ns	−0.358 ***
General Curvature	ns	ns	ns
Forest ratio	ns	ns	ns
Slope	ns	ns	ns
Aspect (Cos)	−0.196 *	ns	ns
Intercept	0.019	0.025	0.013
R^2a	0.239 ***	0.325 ***	0.164 **

* $p < 0.1$; ** $p < 0.05$; *** $p < 0.01$.

The effect of snow depth was closely related to the snow amount over the sampling years. The influence of depth was not significant in 2017 (high snow). However, in 2018 (shallow snow), the influence of depth became considerable and described almost 32% of the variability. The slope was negative, which means that δ^2H values decrease with increasing snow depth. In moderate snow 2019, the influence of depth became weaker. The linear influence of snow depth is evident in the plot of partial residuals (Figure 6).

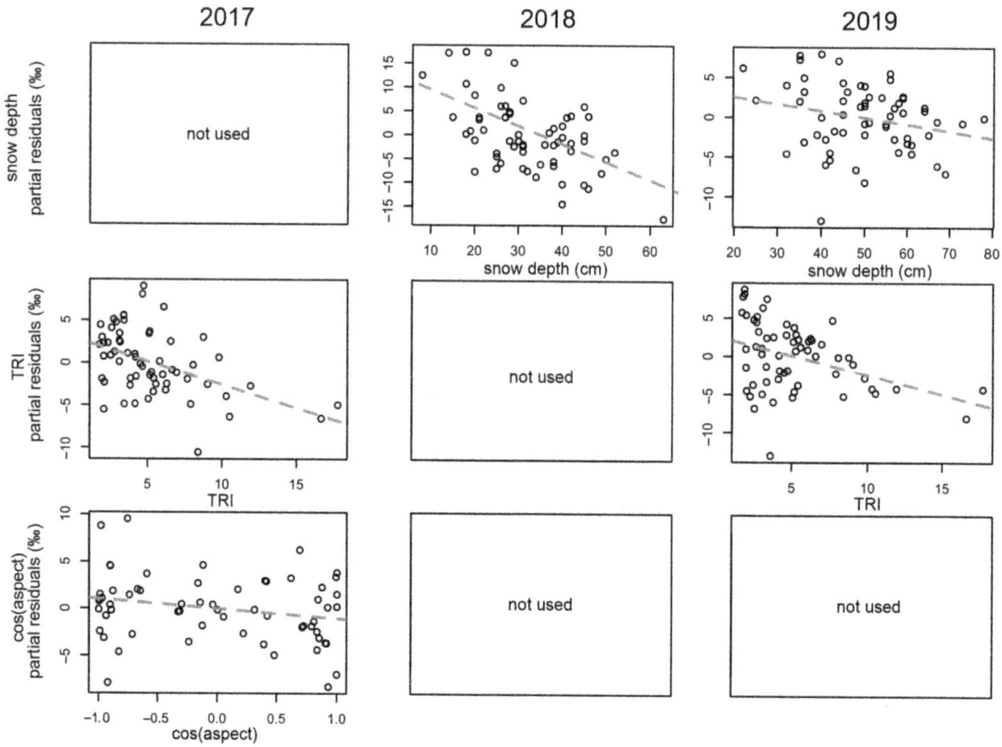

Figure 6. Partial residuals of each predictor for the sampling years selected by the stepwise multiple regression. The dashed lines represent linear relationships.

The TRI effect was most likely related to the wind redistribution of snow from uplands to small valleys. Higher TRI values mean higher ruggedness of the terrain. In the study area, higher TRI values typically belong to river valleys. The coefficients were also negative, indicating a decrease in δ^2H values as we moved toward the valleys. TRI was a significant predictor in 2017 and 2019. These winters had above-average snow depths and quite a high wind intensity, especially in 2017.

SWE, as well as forest ratio, did not show any significant effect on the isotopic composition. SWE was included in the models instead of snow depth, but it was significant only in 2018, and the influence was weaker (adjusted R^2 was about 0.15). A similar SWE value can be formed by increasing both depth and density, which are controlled by distinct processes from the isotopic composition perspective. The forest ratio also did not play a significant role. We attribute this effect to fundamentally different processes occurring in coniferous (canopy interception) and deciduous forests (obstacle of wind transport). However, the sampling points in these areas may have similar forest ratio values.

The d-excess also showed strong correlations with snow depth and SWE, confirming the isotopic composition-depth patterns (Figure 7). Snow depth and d-excess were positively correlated (Table 5) in 2019 (moderate snow) and 2018 (shallow snow). This relationship weakened in the high snow winter season (2017). The relationship was most stable in moderate 2019, while in 2018 several points deviated from the general tendency. We suppose the deviations were related to some local features of snow stratigraphy that attenuate fractionation despite shallow snow. In contrast to H^2, d-excess showed significant correlations with SWE as well (Table 5). However, the strongest relationship was observed in 2019, when the SWE values largely corresponded to the distribution of snow depth. This similarity was largely responsible for the significance of this relationship. In other sampling years, the linear influence of SWE on d-excess was not evident, despite the significance of the relationship (with the presence of outliers and heteroscedasticity).

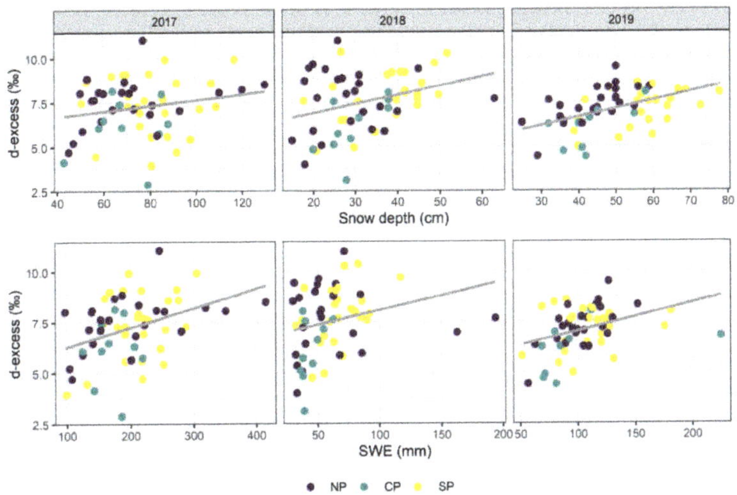

Figure 7. Relationship of the d-excess and snow depth, SWE over the sampling years and basin parts (NP, CP, SP).

Table 5. Adjusted R^2, slopes and intercepts for computed regression equations describing d-excess vs. snow depth and SWE relationships (ns = predictor variable not selected).

Sampling Year	Estimate	Snow Depth	SWE
2017	Slope	ns	0.3803 ***
	Intercept	ns	1.313×10^{-16}
	R^2a	ns	0.13 ***
	Observations	60	60
2018	Slope	0.3124 **	0.2509 *
	Intercept	-4.238×10^{-16}	-3.239×10^{-16}
	R^2a	0.08 **	0.05 *
	Observations	60	60
2019	Slope	0.4891 ***	0.3737***
	Intercept	2.908×10^{-16}	3.214×10^{-16}
	R^2a	0.23 ***	0.13 ***
	Observations	60	60

* $p < 0.1$; ** $p < 0.05$; *** $p < 0.01$.

4. Discussion

The snow isotopic composition showed differences between open and forested areas, mainly in reducing d-excess and the slope of δ^2H vs. $\delta^{18}O$ regressions. The main predictor connected with snow enrichment in heavier isotopes was snow depth. The effect was especially evident in winter seasons with shallow snowpack.

The slope and intercept of δ^2H vs. $\delta^{18}O$ regressions corresponded to the values typical for the continental climate [41]. Additionally, the slope and intercept values agreed well with data previously obtained for the same region [31]. Negative intercept values and their high variability are typical for regions with a cold continental climate due to significant differences in the mechanisms of precipitation formation [41]. Under these conditions, the consistently low slope values and low d-excess values in the forest part most likely indicate the influence of sublimation.

The effect of canopy interception on the isotopic composition was observed not in the direct enrichment in heavier isotopes but through a decrease in d-excess. This pattern slightly differs from the existing studies in which the enrichment in heavier isotopes was correlated with increasing canopy density [26,27]. However, a recent study has shown that canopy density may not have a statistically significant influence on the spatial variability of $\delta^{18}O$ [28]. Forest d-excess may be lower than in open areas (in [27], d-excess was not given but could be calculated from the average $\delta^{2}H$ and $\delta^{18}O$ values) or almost independent of canopy density [26]. We suppose that the lack of direct enrichment occurred due to the lower interception capacity of pine forests and low energy inputs for sublimation. Compared to fir and spruce forests, snow interception, and sublimation losses in pine forests may be up to 10% lower [42], despite the slightly longer deposition of snow on the canopy [43]. The work [27] noted slightly smaller interception effects on isotope ratios at the high-elevation transect, which the authors attributed to lower energy inputs. In the cold continental climate of southern Siberia, this factor may also limit sublimation.

The influence of snow depth on the isotopic composition was the most critical factor of both spatial and temporal variability. Previously, it has been shown that the upper and lower layers of snow are sensible to the changes in the isotopic composition due to the influence of melting, sublimation, and moisture exchange with the soil [6,20,28,44,45]. Vapor flux from the lower snow layers to the upper ones (due to the thermal gradient) usually does not significantly change the isotopic composition since all changes occur within the snowpack. These processes contribute to the homogenization of the isotopic composition between the individual layers during the winter [31,46]. In shallow snow conditions, bulk snowpack isotopic composition is much more sensitive since the upper and lower layers constitute a significant part of the snowpack. In the areas with increased snow accumulation (high snow depth), the fraction of "stable" snow within the snowpack is higher. Such snowpacks tend to be isotopically lighter. In moderate snow winter seasons, the isotopic composition-depth relationship weakens, but it is still recognizable through the d-excess decrease. In high snow winters, the correlation with snow depth almost disappears because a large amount of snow accumulates in the entire basin. The relationship between the isotopic composition and snow depth was earlier observed in alpine conditions [21]. The study [7] previously expected that samples with lower SWE should have been more strongly affected by melt or sublimation and, as a result, become isotopically heavier. However, they did not find a relationship between isotopic composition and SWE in the Canadian prairies. In our work, SWE also did not correlate with the isotopic composition. We argue that snow depth is a more significant factor since both higher density (due to wind exposure or sublimation) and higher snow depth (in a sheltered location) can produce the same SWE value. In terms of isotopic fractionation, these processes may result in different $\delta^{18}O$ and $\delta^{2}H$ ratios, which explains the lack of correlations.

The obtained results suggest that considering spatial variability and inter-annual differences in snow depth is necessary when planning observation strategies involving snow sampling for isotopic composition analysis. Additionally, spatial coverage is important because snowpack isotopic signatures exhibit a high spatial variability over a small scale (<100 m), limiting the usefulness of point samples to estimate an average isotopic composition of snow over a large area [7]. However, our findings also have limitations since we only estimated bulk snowpack isotopic composition, which implies some uncertainty in its evolution. To better understand the mechanisms of isotopic signal transformation, a joint analysis of the isotopic composition of snowpack, primary snowfall, and throughfall during the winter is preferable in further studies.

5. Conclusions

In the West Siberian forest steppe, the expected direct enrichment of forest snowpack in heavier isotopes was not observed. However, we found lower d-excess values and $\delta^{2}H$

vs. $\delta^{18}O$ regressions slopes less than those of GMWL in forested areas compared to open, which may indicate the influence of sublimation. Additionally, isotope ratios between open and forested areas maintained in both shallow and high snow winters. We found out that snow depth is the most critical spatial factor influencing the isotopic composition of snow. As snow depth increased, the $\delta^{18}O$ and $\delta^{2}H$ values of the entire snowpack decreased due to conservation within the snowpack and less influence of sublimation and moisture exchange with the soil. However, this pattern was only evident in winter seasons with below-average snow depth. In above-average snow winter seasons, significant amounts of snow accumulated at most of the sampling sites, which smoothed out the isotopic differences and contributed to the preservation of the $\delta^{18}O$ and $\delta^{2}H$ ratios within the snowpack.

Author Contributions: Conceptualization, D.P., N.M. and D.C.; methodology, D.P.; software, D.P.; validation, D.P., N.M. and D.C.; formal analysis, D.P. and N.M.; investigation, D.P., D.C., R.B., D.Z. and L.L.; resources, N.M. and D.C.; data curation, D.P. and R.B.; writing—original draft preparation, D.P. and N.M.; writing—review and editing, all authors; visualization, D.P. and R.B.; supervision, D.C.; project administration, D.P. and N.M.; funding acquisition, D.C. and N.M. All authors have read and agreed to the published version of the manuscript.

Funding: Data collection was carried out within the State Assignment of IWEP SB RAS FUFZ-2021-0007. Analysis of the topography and land cover influence on the snow isotopic composition was funded by RFBR, project number 19-35-60006. Approbation of the methods for assessing the relationship between atmospheric parameters and landscape heterogeneity was supported by the Russian Science Foundation under grant 21-17-00135 (https://rscf.ru/en/project/21-17-00135/).

Data Availability Statement: The data and R code presented in this study are openly available in Mendeley Data. DOI: 10.17632/2g24d7pgyv.1.

Acknowledgments: The authors would like to thank Tatyana Papina, head of the Chemical Analytical Center at IWEP SB RAS, and researcher Alla Eirikh for performing the laboratory analysis of the snow samples.

Conflicts of Interest: The authors declare no conflict of interest.

References

1. Barnett, T.P.; Adam, J.C.; Lettenmaier, D.P. Potential Impacts of a Warming Climate on Water Availability in Snow-Dominated Regions. *Nature* **2005**, *438*, 303–309. [CrossRef] [PubMed]
2. Strum, M.; Michael, G.; Parr, C. Water and Life from Snow: A Trillion Dollar Science Question. *Water Resour. Res.* **2017**, *53*, 3534–3544. [CrossRef]
3. Blöschl, G. Scaling Issues in Snow Hydrology. *Hydrol. Process.* **1999**, *13*, 2149–2175. [CrossRef]
4. Largeron, C.; Dumont, M.; Morin, S.; Boone, A.; Lafaysse, M.; Metref, S.; Cosme, E.; Jonas, T.; Winstral, A.; Margulis, S.A. Toward Snow Cover Estimation in Mountainous Areas Using Modern Data Assimilation Methods: A Review. *Front. Earth Sci.* **2020**, *8*, 325. [CrossRef]
5. Viallon-Galinier, L.; Hagenmuller, P.; Lafaysse, M. Forcing and Evaluating Detailed Snow Cover Models with Stratigraphy Observations. *Cold Reg. Sci. Technol.* **2020**, *180*, 103163. [CrossRef]
6. Beria, H.; Larsen, J.R.; Ceperley, N.C.; Michelon, A.; Vennemann, T.; Schaefli, B. Understanding Snow Hydrological Processes through the Lens of Stable Water Isotopes. *WIREs Water* **2018**, *5*, e1311. [CrossRef]
7. Pavlovskii, I.; Hayashi, M.; Lennon, M.R. Transformation of Snow Isotopic Signature along Groundwater Recharge Pathways in the Canadian Prairies. *J. Hydrol.* **2018**, *563*, 1147–1160. [CrossRef]
8. N'da, A.B.; Bouchaou, L.; Reichert, B.; Hanich, L.; Ait Brahim, Y.; Chehbouni, A.; Beraaouz, E.H.; Michelot, J.L. Isotopic Signatures for the Assessment of Snow Water Resources in the Moroccan High Atlas Mountains: Contribution to Surface and Groundwater Recharge. *Environ. Earth Sci.* **2016**, *75*, 755. [CrossRef]
9. Rücker, A.; Boss, S.; Kirchner, J.W.; von Freyberg, J. Monitoring Snowpack Outflow Volumes and Their Isotopic Composition to Better Understand Streamflow Generation during Rain-on-Snow Events. *Hydrol. Earth Syst. Sci.* **2019**, *23*, 2983–3005. [CrossRef]
10. Juras, R.; Blöcher, J.R.; Jenicek, M.; Hotovy, O.; Markonis, Y. What Affects the Hydrological Response of Rain-on-Snow Events in Low-Altitude Mountain Ranges in Central Europe? *J. Hydrol.* **2021**, *603*, 127002. [CrossRef]
11. Langs, L.E.; Petrone, R.M.; Pomeroy, J.W. A Δ18O and Δ2H Stable Water Isotope Analysis of Subalpine Forest Water Sources under Seasonal and Hydrological Stress in the Canadian Rocky Mountains. *Hydrol. Process.* **2020**, *34*, 5642–5658. [CrossRef]
12. Sprenger, M.; Leistert, H.; Gimbel, K.; Weiler, M. Illuminating Hydrological Processes at the Soil-Vegetation-Atmosphere Interface with Water Stable Isotopes. *Rev. Geophys.* **2016**, *54*, 674–704. [CrossRef]

13. Jespersen, R.G.; Leffler, A.J.; Oberbauer, S.F.; Welker, J.M. Arctic Plant Ecophysiology and Water Source Utilization in Response to Altered Snow: Isotopic (Δ18O and Δ2H) Evidence for Meltwater Subsidies to Deciduous Shrubs. *Oecologia* **2018**, *187*, 1009–1023. [CrossRef]
14. Thaw, M.; Visser, A.; Bibby, R.; Deinhart, A.; Oerter, E.; Conklin, M. Vegetation Water Sources in California's Sierra Nevada (USA) Are Young and Change over Time, a Multi-Isotope (Δ18O, Δ2H, 3H) Tracer Approach. *Hydrol. Process.* **2021**, *35*, e14249. [CrossRef]
15. Dahlke, H.E.; Lyon, S.W. Early Melt Season Snowpack Isotopic Evolution in the Tarfala Valley, Northern Sweden. *Ann. Glaciol.* **2013**, *54*, 149–156. [CrossRef]
16. Earman, S.; Campbell, A.R.; Phillips, F.M.; Newman, B.D. Isotopic Exchange between Snow and Atmospheric Water Vapor: Estimation of the Snowmelt Component of Groundwater Recharge in the Southwestern United States. *J. Geophys. Res. Atmos.* **2006**, *111*, 1–18. [CrossRef]
17. Taylor, S.; Feng, X.; Kirchner, J.W.; Osterhuber, R.; Klaue, B.; Renshaw, C.E. Isotopic Evolution of a Seasonal Snowpack and Its Melt. *Water Resour. Res.* **2001**, *37*, 759–769. [CrossRef]
18. DeWalle, D.R.; Rango, A. *Principles of Snow Hydrology*; Cambridge University Press: Cambridge, MA, USA, 2008; ISBN 978-0-52-182362-3.
19. Kendall, C.; Mcdonnell, J.J. *Isotope Tracers in Catchment Hydrology*; Elsevier: Amsterdam, The Netherlands, 1998; ISBN 1865843830.
20. Sokratov, S.A.; Golubev, V.N. Snow Isotopic Content Change by Sublimation. *J. Glaciol.* **2009**, *55*, 823–828. [CrossRef]
21. Dietermann, N.; Weiler, M. Spatial Distribution of Stable Water Isotopes in Alpine Snow Cover. *Hydrol. Earth Syst. Sci.* **2013**, *17*, 2657–2668. [CrossRef]
22. Langman, J.B.; Martin, J.; Gaddy, E.; Boll, J.; Behrens, D. Snowpack Aging, Water Isotope Evolution, and Runoff Isotope Signals, Palouse Range, Idaho, USA. *Hydrology* **2022**, *9*, 94. [CrossRef]
23. Moran, T.A.; Marshall, S.J.; Evans, E.C.; Sinclair, K.E. Altitudinal Gradients of Stable Isotopes in Lee-Slope Precipitation in the Canadian Rocky Mountains. *Arct. Antarct. Alp. Res.* **2007**, *39*, 455–467. [CrossRef]
24. Schmieder, J.; Hanzer, F.; Marke, T.; Garvelmann, J.; Warscher, M.; Kunstmann, H.; Strasser, U. The Importance of Snowmelt Spatiotemporal Variability for Isotope-Based Hydrograph Separation in a High-Elevation Catchment. *Hydrol. Earth Syst. Sci.* **2016**, *20*, 5015–5033. [CrossRef]
25. Claassen, H.C.; Downey, J.S. A Model for Deuterium and Oxygen 18 Isotope Changes During Evergreen Interception of Snowfall. *Water Resour. Res.* **1995**, *31*, 601–618. [CrossRef]
26. Koeniger, P.; Hubbart, J.A.; Link, T.; Marshall, J.D. Isotopic Variation of Snow Cover and Streamflow in Response to Changes in Canopy Structure in a Snow-Dominated Mountain Catchment. *Hydrol. Process.* **2008**, *22*, 557–566. [CrossRef]
27. von Freyberg, J.; Bjarnadóttir, T.R.; Allen, S.T. Influences of Forest Canopy on Snowpack Accumulation and Isotope Ratios. *Hydrol. Process.* **2020**, *34*, 679–690. [CrossRef]
28. Carroll, R.W.H.; Deems, J.; Maxwell, R.; Sprenger, M.; Brown, W.; Newman, A.; Beutler, C.; Bill, M.; Hubbard, S.S.; Williams, K.H. Variability in Observed Stable Water Isotopes in Snowpack across a Mountainous Watershed in Colorado. *Hydrol. Process.* **2022**, *36*, e14653. [CrossRef]
29. Gustafson, J.R.; Brooks, P.D.; Molotch, N.P.; Veatch, W.C. Estimating Snow Sublimation Using Natural Chemical and Isotopic Tracers across a Gradient of Solar Radiation. *Water Resour. Res.* **2010**, *46*, 12511. [CrossRef]
30. Kurita, N.; Sugimoto, A.; Fujii, Y.; Fukazawa, T.; Makarov, V.N.; Watanabe, O.; Ichiyanagi, K.; Numaguti, A.; Yoshida, N. Isotopic Composition and Origin of Snow over Siberia. *J. Geophys. Res. Atmos.* **2005**, *110*, 13102. [CrossRef]
31. Papina, T.; Eirikh, A.; Noskova, T. Factors Influencing Changes of the Initial Stable Water Isotopes Composition in the Seasonal Snowpack of the South of Western Siberia, Russia. *Appl. Sci.* **2022**, *12*, 625. [CrossRef]
32. Gan, T.Y. Reducing Vulnerability of Water Resources of Canadian Prairies to Potential Droughts and Possible Climatic Warming. *Water Resour. Manag.* **2000**, *14*, 111–135. [CrossRef]
33. Olson, D.M.; Dinerstein, E.; Wikramanayake, E.D.; Burgess, N.D.; Powell, G.V.N.; Underwood, E.C.; D'amico, J.A.; Itoua, I.; Strand, H.E.; Morrison, J.C.; et al. Terrestrial Ecoregions of the World: A New Map of Life on Earth: A New Global Map of Terrestrial Ecoregions Provides an Innovative Tool for Conserving Biodiversity. *Bioscience* **2001**, *51*, 933–938. [CrossRef]
34. Rudaya, N.; Krivonogov, S.; Słowiński, M.; Cao, X.; Zhilich, S. Postglacial History of the Steppe Altai: Climate, Fire and Plant Diversity. *Quat. Sci. Rev.* **2020**, *249*, 106616. [CrossRef]
35. *Atlas of the Altai Region*; Main Administration of Geodesy and Cartography USSR: Moscow/Barnaul, Russia, 1978.
36. Zanin, G. Geomorphology of the Altai Region. In *Natural Zoning of the Altai Region*; USSR Academy of Sciences: Moscow, Russia, 1958; pp. 62–98.
37. RIHMI–WDC Official Website. Available online: http://www.meteo.ru (accessed on 28 August 2021).
38. Dansgaard, W. Stable Isotopes in Precipitation. *Tellus* **1964**, *16*, 436–468. [CrossRef]
39. Craig, H. Isotopic Variations in Meteoric Waters. *Science* **1961**, *133*, 1702–1703. [CrossRef] [PubMed]
40. Paiva, R.; O'Loughlin, F. Bare-Earth SRTM. Available online: https://data.bris.ac.uk/data/dataset/10tv0p32gizt01nh9edcjzd6wa (accessed on 12 August 2020). [CrossRef]
41. Putman, A.L.; Fiorella, R.P.; Bowen, G.J.; Cai, Z. A Global Perspective on Local Meteoric Water Lines: Meta-Analytic Insight Into Fundamental Controls and Practical Constraints. *Water Resour. Res.* **2019**, *55*, 6896–6910. [CrossRef]

42. Pomeroy, J.W.; Parviainen, J.; Hedstrom, N.; Gray, D.M. Coupled Modelling of Forest Snow Interception and Sublimation. *Hydrol. Process.* **1998**, *12*, 2317–2337. [CrossRef]
43. MacKay, M.D.; Bartlett, P.A. Estimating Canopy Snow Unloading Timescales from Daily Observations of Albedo and Precipitation. *Geophys. Res. Lett.* **2006**, *33*. [CrossRef]
44. Sturm, M.; Benson, C.S. Vapor Transport, Grain Growth and Depth-Hoar Development in the Subarctic Snow. *J. Glaciol.* **1997**, *43*, 42–59. [CrossRef]
45. Ala-Aho, P.; Welker, J.M.; Bailey, H.; Pedersen, S.H.; Kopec, B.; Klein, E.; Mellat, M.; Mustonen, K.R.; Noor, K.; Marttila, H. Arctic Snow Isotope Hydrology: A Comparative Snow-Water Vapor Study. *Atmosphere* **2021**, *12*, 150. [CrossRef]
46. Evans, S.L.; Flores, A.N.; Heilig, A.; Kohn, M.J.; Marshall, H.P.; McNamara, J.P. Isotopic Evidence for Lateral Flow and Diffusive Transport, but Not Sublimation, in a Sloped Seasonal Snowpack, Idaho, USA. *Geophys. Res. Lett.* **2016**, *43*, 3298–3306. [CrossRef]

forests

MDPI

Article

How Potential Evapotranspiration Regulates the Response of Canopy Transpiration to Soil Moisture and Leaf Area Index of the Boreal Larch Forest in China

Zhipeng Xu [1,2], Xiuling Man [1,2], Tijiu Cai [1,2,*] and Youxian Shang [1,2]

1 School of Forestry, Northeast Forestry University, Harbin 150040, China; xuzhipeng34@163.com (Z.X.); xiuling.man@nefu.edu.cn (X.M.); shangyouxian2022@163.com (Y.S.)
2 Key Laboratory of Sustainable Forest Ecosystem Management-Ministry of Education, Northeast Forestry University, Harbin 150040, China
* Correspondence: caitijiu1963@163.com; Tel.: +86-150-4585-3579

Abstract: Transpiration is a critical component of the hydrological cycle in the terrestrial forest ecosystem. However, how potential evapotranspiration regulates the response of canopy transpiration to soil moisture and leaf area index of the boreal larch forest in China has rarely been evaluated. The present study was conducted in the larch (*Larix gmelinii* (Rupr.) Rupr.) forest, which is a typical boreal forest in China. The canopy transpiration was measured using sap flow techniques from May to September in 2021 and simultaneously observing the meteorological variables, leaf area index (LAI) and soil moisture (SWC). The results showed that there were significant differences in canopy transpiration of *Larix gmelinii* among the months. The correlation and regression analysis indicated that canopy transpiration was mainly influenced by potential evapotranspiration (PET), while the effect of soil moisture on canopy transpiration was lowest compared with other environmental factors. Furthermore, our results revealed that the effect of PET on canopy transpiration was not regulated by soil moisture when soil moisture exceeded 0.2 cm^3 cm^{-3}. More importantly, under the condition of sufficient soil moisture, it was demonstrated that the response of canopy transpiration to leaf area index was limited when PET exceeded 9 mm/day. These results provide valuable implications for supporting forest management and water resource utilization in the boreal forest ecosystem under the context of global warming.

Keywords: sap flow; boreal forests; atmospheric evaporative; growth index; soil moisture

Citation: Xu, Z.; Man, X.; Cai, T.; Shang, Y. How Potential Evapotranspiration Regulates the Response of Canopy Transpiration to Soil Moisture and Leaf Area Index of the Boreal Larch Forest in China. *Forests* **2022**, *13*, 571. https://doi.org/10.3390/f13040571

Academic Editors: Yanhui Wang, Karl-Heinz Feger and Lulu Zhang

Received: 5 March 2022
Accepted: 31 March 2022
Published: 4 April 2022

1. Introduction

The terrestrial forest ecosystem provides a vital link to control the water exchange between land surface and atmosphere as well as to sequester the carbon by photosynthesis. In particular, boreal forest biomes cover approximately 11% of the area on the Earth's surface [1], which plays an important role in maintaining the stability of the terrestrial forest ecosystem. Thus, the interest in the functions and role of the boreal forest ecosystem in water resources is increasing, especially for the transpiration across the globe [2–5].

Canopy transpiration is a crucial process of forest hydrological cycling, which couples the soil moisture and atmosphere interactions [6–9]. The results at a global scale showed that canopy transpiration accounts for about 39% of precipitation and more than 60% of evapotranspiration [10,11]. In past decades, due to the frangibility and sensitivity of the boreal forest ecosystem to climatic change [2], canopy transpiration has been increasingly influenced by global climate warming and extreme weather events [12]. With the significant progress of afforestation projects in China, such as the Natural Forest Protection Project, the boreal forest coverage in China has been greatly improved, which has greatly affected the hydrological cycle of the forest ecosystem [5,13]. Thus, determining the canopy transpiration dynamics of boreal forest in a changing environment is greatly important for better understanding the plant survival strategies and the impact of vegetation on related eco-hydrological processes.

Numerous studies analyze the response of canopy transpiration to environmental factors including air temperature, vapor pressure deficit, soil moisture, precipitation, etc. [14–17]. For example, Zhang et al. [18] concluded that increasing vapor pressure deficit promotes canopy transpiration until a certain threshold, while Han et al. [19] found the precipitation has a negative effect on canopy transpiration. However, most often, case studies have only paid attention to the single impact of meteorological factors on canopy transpiration [17,20,21]. We know that environmental factors occur concurrently under natural conditions and that canopy transpiration is affected by the interaction effects of environmental factors. However, the response mechanism of canopy transpiration to the interaction of environmental factors in the boreal forest of China is not clear.

Canopy transpiration is affected by many environmental factors, which can be divided into three aspects including atmospheric evaporative demand, such as potential evapotranspiration [9,22], soil water supply (such as soil moisture) and vegetation phenophase, such as leaf expanding or defoliation [9,23]. In previous work, some studies reported the effects of potential evapotranspiration and soil moisture on transpiration [9,21,22]. For instance, Wan et al. [24] found that soil moisture limits the response of transpiration to potential evapotranspiration. However, numerous studies only consider the atmospheric evaporative demand and water supply and ignore the change in vegetation phenophase during the growing season. Vegetation phenology reflects the dynamic of vegetation growth via leaf area variations and consequently affects the canopy transpiration [23]. Thus, analyzing how potential evapotranspiration mediates the influence of soil water supply and leaf area variation on canopy transpiration is of great significance in the boreal forest regions of China.

The native larch (*Larix gmelinii* (Rupr.) Rupr.) forest represents the typical zonal vegetation and prominent community of boreal forest ecosystems in northeast China, which is the southern margin of the Siberian zone and one of the largest boreal forests underlying wide permafrost in the world [25,26]. In the present study, a consecutive measurement of sap flow from May to September in 2021 in the natural forest of *Larix gmelinii* was performed: (1) to investigate the dynamics of canopy transpiration of *Larix gmelinii* forest in the growing season at different timescales, (2) to clarify the relationships between canopy transpiration and environmental factors, (3) to identify the role of soil moisture in canopy transpiration of *Larix gmelinii* forest, and (4) to reveal how potential evapotranspiration mediates the influence of leaf area variation on canopy transpiration of *Larix gmelinii* forest. These can be exploited to improve forest-water management and ecologically sustainable development, as well as to predict the effect of climate change on vegetation growth in the future.

2. Materials and Methods

2.1. Study Site

The present study was performed in the Mohe Forest Ecosystem Research station (MFERs, 53°27′59″ N, 122°20′06″ E), and the native forest is dominated by the *Larix gmelinii*, which represents the principal community of boreal forest ecosystems in China (Figure 1). The study site has a typical cold temperate continental monsoon climate with mean annual precipitation and temperature being 460.8 mm and −5.5 °C, respectively, from 1959 to 2017. The distribution of intra-annual precipitation was uneven and accounted for approximately 70% of the precipitation that occurred during the growing season. The soil in the study area is the Gleyic Cambisols (https://www.fao.org/soils-portal/soil-survey/soil-maps-and-databases/, accessed on 2 January 2022), and its thickness varies between 20–50 cm [27].

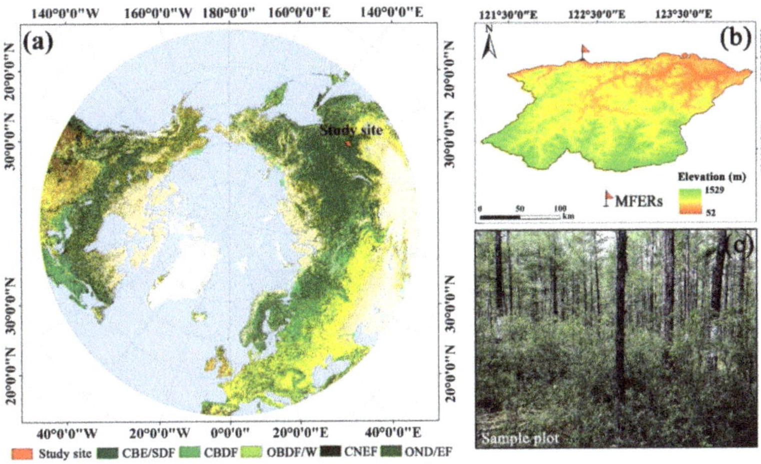

Figure 1. (**a**) The region north of 40° N (Globcover2009: http://due.esrin.esa.int/page_globcover.php, accessed on 2 January 2022). (**b**) The DEM of study site; (**c**) Sample plot. Note that the CBE/SDF, CBDF, OBDF/W, CNEF and OND/EF are closed to open (>15%) broadleaved evergreen or semi-deciduous forest (>5 m), closed (>40%) broadleaved deciduous forest (>5 m), open (15–40%) broadleaved deciduous forest/woodland (>5 m), closed (>40%) needleleaf evergreen forest (>5 m) and open (15–40%) needleleaf deciduous or evergreen forest (>5 m), respectively.

2.2. Forest Structure Characteristics

The field experiment was conducted in the natural forest of *Larix gmelinii* from May to October in 2021. A sample plot with 20 × 20 m was established and the stand density was 1250 trees per hectare. Meanwhile, the forest structure characteristics were measured, which include mean diameter at breast height (DBH), tree height, tree age and total sapwood area (SA) of sample plot, which were 13.1 ± 6.74 cm, 17.35 ± 2.56 m, 75–90 years and 2950.53 cm^2 (Table 1), respectively. The main understory of the experiment plot is mainly dominated by Dahurian rhododendron (*Rhododendron dauricum*), which covered approximately 70% of the area in the study sample plot.

Table 1. Detailed information of forest structure in the *Larix gmelinii* forest.

Forest Structure Characteristics						
Plot area (m^2)	Stand density (tree/ha)	Tree age (year)	Mean DBH (cm)	Total SA (cm^2)	Mean height (m)	Mean LAI (m^2 m^{-2})
400	1250	75–90	13.1 ± 6.74	2950.53	17.35 ± 7.56	1.96 ± 0.36

The leaf area index (LAI, m^2 m^{-2}) is an important indicator of vegetation canopy structure to characterize the growing status and water use of forests [8,9]. In the present study, the LAI of the experiment stand was measured every 5–7 days on sunny days from May to September in 2021 using Plant Canopy Analyzer (LAI-2200, Li-Cor, Lincoln, NE, USA). For each measurement, 25–30 points within the sample plot were selected and we obtained the mean LAI. Subsequently, the functional relationship between LAI and the measurement date (day of the year, DOY) was established. As shown in Figure 2, the coefficient of determination (R^2) reached 0.938 (LAI = $-0.0002 \times DOY^2 + 0.0811 \times DOY - 5.8637$, $p < 0.01$, and consequently the daily LAI value was derived from interpolation using this functional relationship [9].

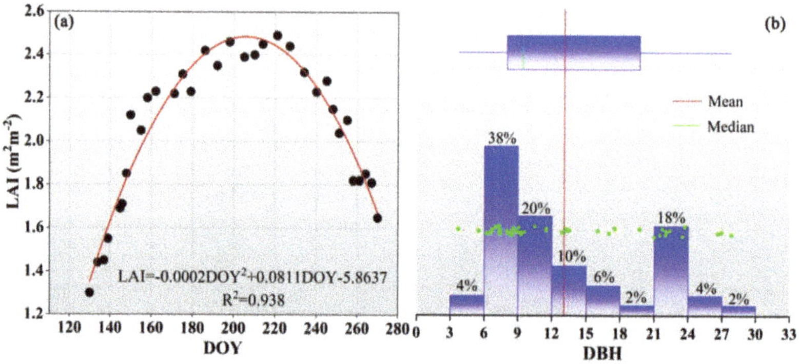

Figure 2. Plot showing the (**a**) LAI variations in the study period; (**b**) frequency distribution of the DBH in study sample plot.

2.3. Sap Flow Measurement, Sapwood Area Determination and Estimation of Canopy Transpiration

In the present study, the nine sample trees were selected for sap flow measurement from 1 May to 30 September in 2021 (Table 2). We adopted the Granier's thermal dissipation probes [28] to monitor the sap flow density (J_s, g cm^{-2} s^{-1}). Each thermal dissipation probe sensor was inserted into the trunk at a height of 1.3 m above the ground and minimizes the effect of radiation on sap flow [28]. The data were saved at 30 min intervals using a CR1000 datalogger (Campbell Scientific, Logan, UT, USA). The sap flow density (J_s, g cm^{-2} s^{-1}) can be expressed as follows:

$$J_s = 0.0119 \times \left(\frac{\Delta T_{max} - \Delta T}{\Delta T}\right)^{1.231} \quad (1)$$

where J_s is sap flow density; ΔT_{max} is the maximum ΔT between sensors at nighttime, for which the sap flow density is close to zero. ΔT is the temperature difference between the two thermal dissipation probes [21,29].

Table 2. Detailed information of sample trees for sap flow measurement.

NO.	DBH (cm)	SA (cm^2)	Tree Height (m)
1	23.7	143.12	19.24
2	23.2	137.89	17.90
3	20.1	107.35	17.33
4	15.1	65.17	15.37
5	12.1	44.28	14.43
6	11.5	40.51	13.50
7	9.2	27.45	11.25
8	9.0	26.41	10.99
9	8.1	21.98	10.38

The 25 trees around the sample plot were selected to estimate the relationship between the DBH and SA using an exponential regression. Firstly, the DBH of each tree was measured using a digital caliper. Then, the sapwood thickness of each tree was determined through distinguishing from the heartwood using the color change method [21,30]. Finally, the relationship between DBH and SA can be expressed as follows:

$$SA = a \times DBH^b \quad (2)$$

where SA and DBH are sapwood area and diameter at breast height, respectively, a and b are parameters.

The canopy transpiration was estimated from sap flow density. Daily canopy transpiration per unit ground area (E_C, mm/d) and canopy transpiration per unit leaf area (E_L, mm/d) can be calculated [21,31–33] as Equations (3) and (4):

$$Ec = J_{sa} \times \frac{SA_{cum}}{A_S} \quad (3)$$

$$E_L = \frac{Ec}{LAI} \quad (4)$$

where J_{sa} is the average sap flow density of all sample trees, SA_{cum} is the cumulative sapwood area of the sample plot and A_s is the sample plot area (Table 1). LAI is the leaf area index (Figure 2a). Meanwhile, the day-time Ec and night-time Ec were determined by the local time of sunrise (at 6:00) and sunset (at 18:00), respectively.

2.4. Canopy Conductance Estimation

Canopy conductance (G_L, mmol m^{-2} s^{-1}) was estimated from E_L using a simplified inverted Penman–Monteith equation [21] can be expressed as follows:

$$G_L = \frac{K_G E_L}{VPD} \quad (5)$$

$$K_G = 115.8 \pm 0.4236 \times Ta \quad (6)$$

where, VPD is the vapor pressure deficit (KPa); Ta is the air temperature (°C); K_G is the conductance coefficient (kPa m^3 kg^{-1}). Furthermore, the data were excluded on rainy days and when VPD < 0.4 kPa to minimize relative errors [17,21,22]

2.5. Meteorological Variables Measurements

In the present study, the meteorological variables were measured by the Gradient meteorological observation system installed on a 36 m tower, including the air temperature (Ta, °C), relative humidity (RH, %), wind speed (Ws, m s^{-1}). Precipitation (P, mm) was measured at the 23 m height of the tower. Photosynthetically active radiation (PAR, mol m^{-2} d^{-1}) and net radiation (Rn, W m^{-2}) was measured by an automatic weather station located in open areas neighboring the study plot. Soil moisture was measured at 5, 10, 20 and 40 cm soil depth using CS650 probes (Campbell Scientific, Logan, UT, USA) in the study plot, respectively. The profile soil moisture (SWC), which refers to the soil moisture covering the depth range between 0–40 cm, was expressed as Equation (5) [34]. All the data were saved at 30 min intervals using a CR3000 datalogger (Campbell Scientific, Logan, UT, USA). Meanwhile, the vapor pressure deficit (VPD, KPa) was calculated using the air temperature and relative humidity [35].

$$SWC = \frac{\theta_1 \times L_1 + \sum_{i=2}^{4} \frac{\theta_{i-1} + \theta_i}{2} \times L_i}{L} \quad (7)$$

where SWC is the mean of SWC at depths of 0–40 cm. θ_i at L_i (i = 1, 2, 3, and 4) represents the soil moisture contents at depths of 5, 10, 20, and 40 cm, and L denotes the observation depth of the soil profile (40 cm).

$$VPD = 0.611 \times (1 - RH) \times \exp\left(\frac{17.502 \times Ta}{Ta + 240.97}\right) \quad (8)$$

where VPD represents the vapor pressure deficit (KPa), and RH and Ta are air temperature (Ta, °C) and relative humidity (RH, %), respectively.

The FAO Penman–Monteith equation was used to estimate the potential evapotranspiration (PET) during the study period [36,37]:

$$\text{PET} = \frac{0.408\,\Delta(\text{Rn} - \text{G}) + \gamma \times \left(\frac{900}{\text{Ta} + 273}\right) \times \text{U}_2 - (\text{e}_\text{s} - \text{e}_\text{a})}{\Delta + \gamma \times (1 + 0.34 \times \text{U}_2)} \quad (9)$$

where Δ is the slope of the vapor press curve ((kPa °C^{-1}), Rn is the net daily radiation (W m^{-2}), G is the soil heat flux into the ground (MJ m^{-2}), γ is the psychrometric constant (kPa °C^{-1}), e_s is the saturation vapor pressure (kPa), e_a is the actual vapor pressure (kPa), U_2 is the mean wind speed (m s^{-1}) at 2 m height, and Ta is the air temperature at a 2 m height. The detailed calculation procedure of each parameter can be found in McMahon et al. [37].

2.6. Statistical Analysis of Data

To analyze the role of soil moisture in the response of canopy transpiration to atmospheric evaporative demand, canopy transpiration per unit leaf area (E_L) was selected to minimize the effect of vegetation growth in the growing season.

The relationship between canopy transpiration and environmental factors was detected by Person correlation coefficient and linear or nonlinear regression. All statistical analyses were performed using IBM SPSS 24.0 statistics software (SPSS Inc., Chicago, IL, USA) and figures were prepared with Origin Pro 2021 software (Origin Lab Inc., Northampton, MA, USA). All statistical analyses were at the 0.05 significance level.

3. Results

3.1. Relationship between SA and DBH

In the present study, 25 trees of different DBH around the sample plot were selected to establish the relationship between SA and DBH using an exponential regression. As shown in Figure 3, the coefficient of determination (R^2) reached 0.96 (SA = 0.5708 × $DBH^{1.7452}$, $p < 0.01$), which demonstrated that the SA was significantly related to DBH. According to the exponential regression, the total SA of the study plot was obtained, which was 2950.53 cm^2.

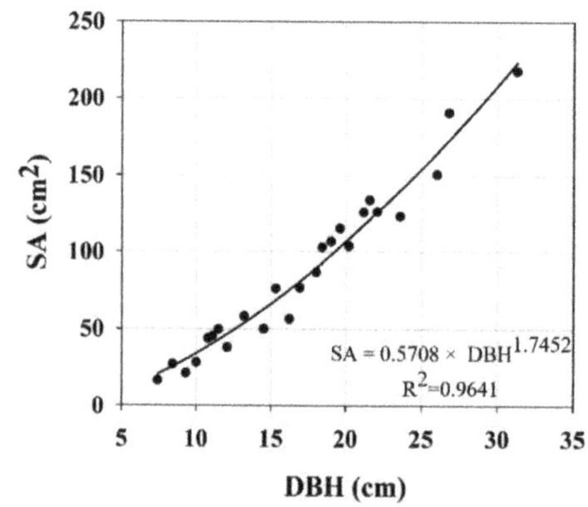

Figure 3. Plot showing the relationship between SA and DBH of the *Larix gmelinii* forest.

3.2. Atmospheric, Soil Moisture and Growth Index

It can be seen from Figure 4 that the daily Ws varied between zero and 0.65 m/s (Figure 4a). The mean daily VPD was 0.39 KPa during the study period (Figure 4b). The daily mean RH and Ta were 77.51% (Figure 4c) and 14.51 °C (Figure 4d), with a range of 34.08% to 99.90% and 2.60 °C to 24.04 °C, respectively. Meanwhile, the mean daily PAR and Rn were 342 μmol m^{-2} s^{-1} and 103.4 W m^{-2} in the growing season ranging from 51.6 to 646.5 μmol m^{-2} s^{-1} (Figure 4e) and 16.6 to 189.6 W m^{-2} (Figure 4h), respectively.

A similar trend was found in PET (Figure 4g), which ranged from 0.3 to 11.9 mm, with mean daily PET being 4.7 mm. As shown in Figure 4f, the SWC in the early stage of the study period was increased due to the soil thawing processes, and subsequently, the change in SWC was significantly related to the precipitation event. The cumulative P (P_{cum}) from May to September was 557.9 mm (Figure 4j), of which the monthly precipitation was unevenly distributed and the highest monthly precipitation was 131.7 mm, occurring in June (Figure 4i).

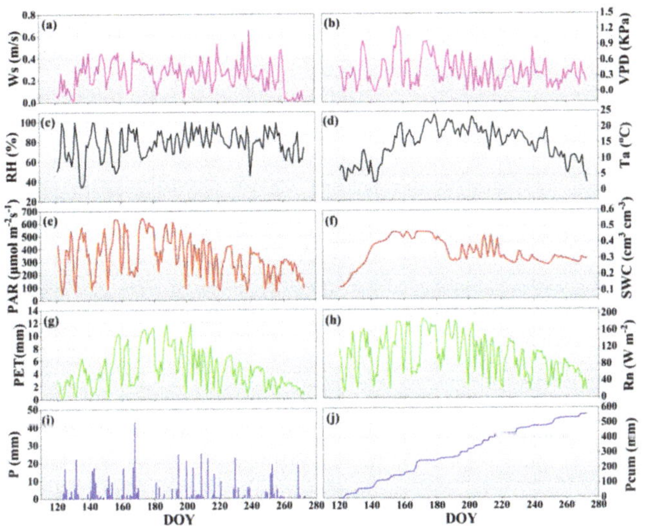

Figure 4. Plots displaying the changes in environmental factors during growing season.

3.3. Seasonal Variation in Canopy Transpiration

3.3.1. Daily Dynamics of Ec and E_L

Figure 5 exhibits the canopy transpiration dynamics that shows that the changing trend in the daily Ec, daily E_L and day-time Ec was similar, which increased gradually from May to July, and then decreased from August to September, while the variations in night-time Ec was relatively stable in the study period as compared with others. The range of daily Ec in the study period was 0.04 to 1.25 mm, with a mean value of 0.60 mm. For the daily E_L, the mean value was 0.29 mm, with a range of 0.02 to 0.58 mm. In addition, the mean values of day-time and night-time Ec were 0.52 and 0.08 mm, with the maximum values reaching 1.12 and 0.19 mm occurring in August and September, respectively, while the minimum values of day-time and night-time Ec both reached 0.01 mm, which was found in September.

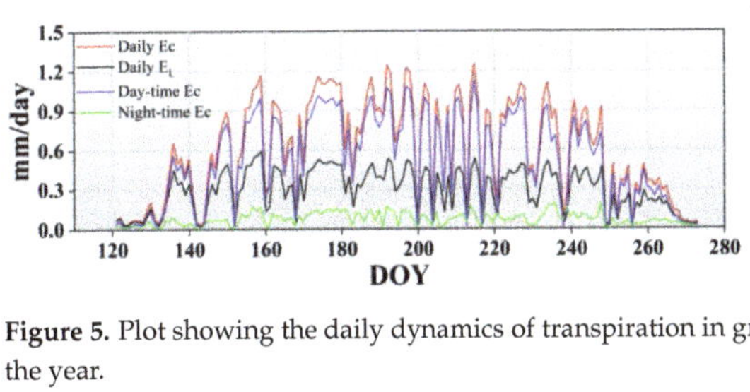

Figure 5. Plot showing the daily dynamics of transpiration in growing season. DOY refers to day of the year.

3.3.2. Monthly Ec Dynamics

As shown in Figure 6a, the mean daily Ec, day-time Ec and night-time Ec in June, July and August were significantly higher ($p < 0.01$) than for May and September. However, no statistically significant ($p > 0.05$) difference was found between May and September. Furthermore, monthly Ec, day-time Ec and night-time Ec show a single peak trend (Figure 6b). The highest monthly Ec and day-time Ec occurred in July, while the highest night-time Ec was found in June. Aside from that, Figure 6c shows that the percentage of monthly Ec, day-time Ec and night-time Ec to P is about 7.37% to 22.49%, 6.30% to 19.60% and 1.06% to 2.90%, respectively. In addition, the total Ec, day-time Ec and night-time Ec were 92.04, 80.20 and 11.84 mm, accounting for 16.50%, 14.38%, and 2.12% of P in the growing season, respectively.

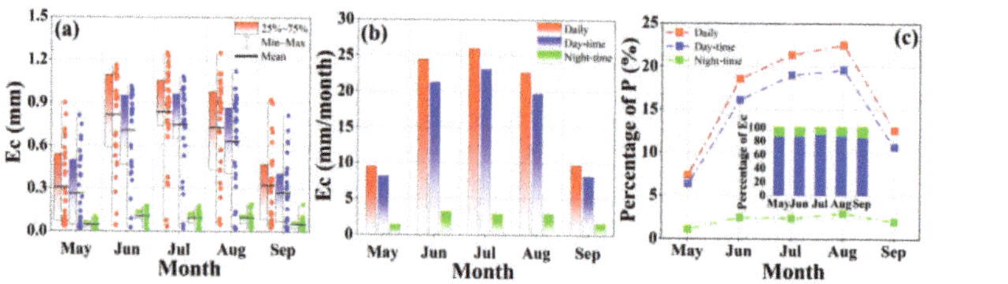

Figure 6. Plot showing the monthly dynamics of Ec in growing season. (**a**) mean daily Ec; (**b**) monthly Ec; (**c**) Percentage of P.

3.4. Relationship between Canopy Transpiration and Environmental Factors

3.4.1. Correlation Analysis

To reveal the relationship between canopy transpiration and environmental factors, Pearson correlation analysis was used. At a monthly scale, the atmospheric evaporative demand indices including VPD, PET, PAR and Rn were significantly positively correlated with Ec in each month (Table 3); however, the RH was negatively related to Ec. For the SWC and LAI, the significant correlation with Ec was only observed in the early and later stages of the growing season. For the whole study period, the correlation between PET and Ec was highest than that for the SWC and LAI (Table 3).

Table 3. Correlation coefficients between environmental factors and canopy transpiration.

Period	VPD	PET	PAR	Ws	Ta	RH	Rn	SWC	LAI
May	0.60 **	0.81 **	0.64 **	0.78 **	0.83 **	−0.44 *	0.66 **	0.66 **	0.72 **
June	0.85 **	0.91 **	0.89 **	0.49 **	0.82 **	−0.81 **	0.86 **	0.25	0.40 *
July	0.88 **	0.83 **	0.91 **	0.71 **	0.33	−0.90 **	0.88 **	−0.14	−0.03
August	0.74 **	0.84 **	0.85 **	0.30	0.40 *	−0.62 **	0.85 **	−0.06	0.02
September	0.46 *	0.94 **	0.75 **	0.54 **	0.74 **	−0.02	0.85 **	−0.31	0.67 **
Whole period	0.66 **	0.86 **	0.58 **	0.74 **	0.78 **	−0.27 **	0.76 **	0.43 **	0.67 **

Note: * and ** means significance level at the 0.05 and 0.01, respectively.

3.4.2. Regression Analysis

To better understand the effects of environmental factors on Ec, the atmospheric evaporative demand indices (PET, VPD, PAR, RH, Ta, Rn and Ws), water supply indicator (SWC) and vegetation growth index (LAI) were selected. For atmospheric evaporative demand indices, Figure 7 shows that the PET impact on Ec was greater than that for the other atmospheric evaporative indices (VPD, PAR, RH, Ta, Rn and Ws), which explained 71% of the variation in Ec. Furthermore, the Ec increased with increased LAI (Figure 7i) and explained 46% of the variation in Ec during the study period. For the water supply indicator (Figure 7h), however, the SWC only explained 19% of the Ec change, which was lower than for the atmospheric evaporative demand indices and vegetation growth index.

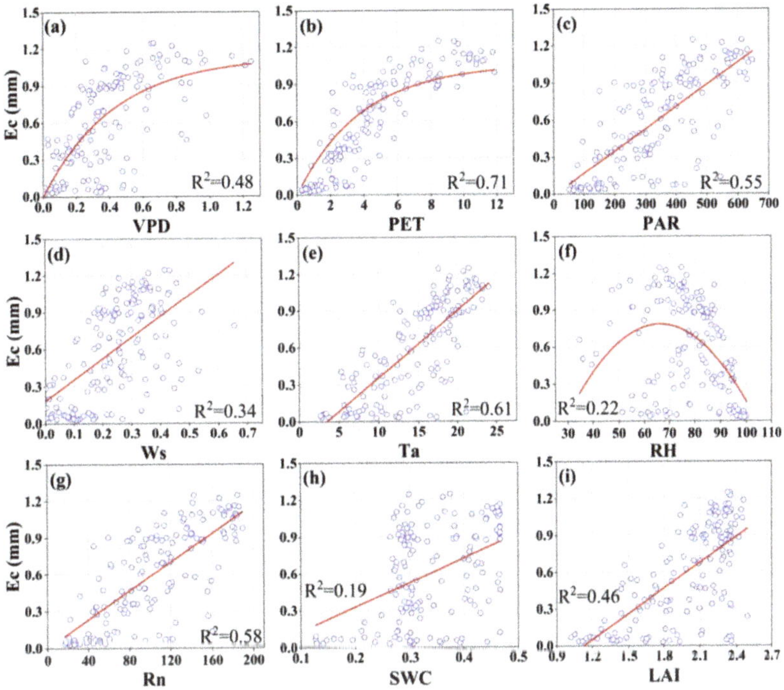

Figure 7. Plot exhibiting the function relationships between environmental factors and Ec in the growing season. (**a**) VPD; (**b**) PET; (**c**) PAR; (**d**) Ws; (**e**) Ta; (**f**) RH; (**g**) Rn; (**h**) SWC; (**i**) LAI.

3.5. Relaotionship between Canopy Conductance and Vapor Pressure Deficit

The relationship between G_L and VPD was estimated using the linear logarithmic function. As shown in Figure 8, the coefficient of determination (R^2) reached 0.54 ($G_L = -55.15 \times \ln(\text{VPD}) + 61.21$, $p < 0.01$), which demonstrated that the G_L significantly decreased with an increasing VPD during the study period. In other words, the VPD explained 54% of the G_L change.

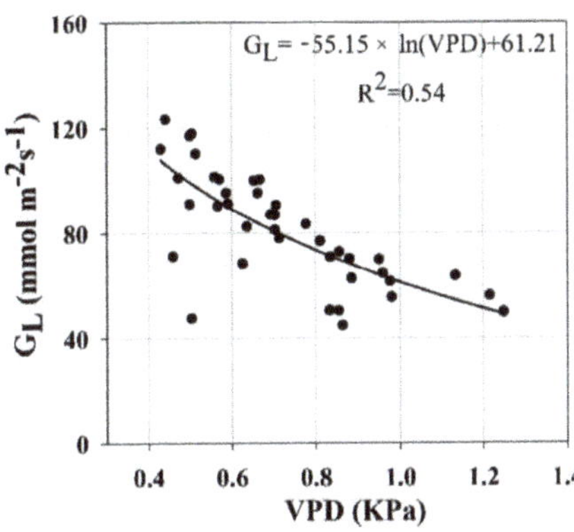

Figure 8. Plot exhibiting the relationship between VPD and canopy conductance. Note that the data were excluded on rainy days and when VPD < 0.4 kPa.

3.6. The Interaction Effects of SWC, LAI and PET on Canopy Transpiration

3.6.1. Responses of Canopy Transpiration to PET under Varying SWC

Through the analysis of Section 3.4, the PET as a compound index can effectively characterize the atmospheric evaporative demand rather than a single climate variable. We used the exponential function to fit the relationship between PET and E_L in varying SWC. The R^2 of each regression was higher than for the 0.51 (Table 4), which reflected that the results were reliable. Meanwhile, Figure 9a exhibits that the response of E_L to PET was similar and mainly restricted by PET when SWC was more than 0.2 cm^3 cm^{-3}. To test this, we also investigated the relationships between the E_L and SWC (more than 0.2 cm^3 cm^{-3}) at varied PET levels (Figure 9b and Table 5). Our results indicated that the E_L was not affected by the SWC at different PET levels when SWC was greater than 0.2 cm^3 cm^{-3} (Figure 9b) and the R^2 of each regression was lower than that the 0.085 (Table 5). These results demonstrated that the PET is the dominant factor controlling the change of canopy transpiration when SWC is more than 0.2 cm^3 cm^{-3}.

Table 4. The response of Ec to PET under varying SWC.

Level	Regression	R^2	Sig.
SWC < 0.2	$E_L = 0.06 \times (1\text{-}e^{-1.57 \times PET})$	0.51	$p < 0.01$
$0.2 \leq$ SWC < 0.3	$E_L = 0.64 \times (1\text{-}e^{-0.16 \times PET})$	0.69	$p < 0.01$
$0.3 \leq$ SWC < 0.4	$E_L = 0.59 \times (1\text{-}e^{-0.16 \times PET})$	0.83	$p < 0.01$
SWC $\geq$ 0.4	$E_L = 0.59 \times (1\text{-}e^{-0.19 \times PET})$	0.81	$p < 0.01$

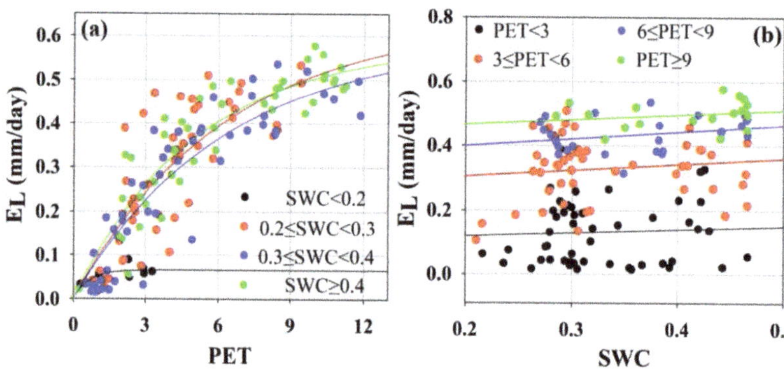

Figure 9. Plot showing the role of SWC in response of canopy transpiration to PET in the study period. (**a**) PET; (**b**) SWC.

Table 5. The response of Ec to PET under varying SWC.

Level	Regression	R^2	Sig.
PET < 3	$E_L = 0.10 \times SWC + 0.10$	0.003	$p > 0.05$
$3 \leq$ PET < 6	$E_L = 0.18 \times SWC + 0.27$	0.019	$p > 0.05$
$6 \leq$ PET < 9	$E_L = 0.22 \times SWC + 0.36$	0.085	$p > 0.05$
PET $\geq$ 9	$E_L = 0.15 \times SWC + 0.44$	0.051	$p > 0.05$

3.6.2. PET Mediates the Response of Canopy Transpiration to LAI

According to the results of Section 3.6.1, the SWC had no significant effects on the Ec when SWC was more than 0.2 cm^3cm^{-3}. Thus, providing an excellent opportunity to assess how the PET regulates the response of Ec to LAI under the elimination of the effect of soil moisture is crucial. Figure 10a and Table 6 show that the significant linear relationship between Ec and LAI was observed ($p < 0.05$) when PET < 9 mm/day, for which Ec increases when LAI increases. However, when PET > 9 mm/day, there was no statistically significant relationship between Ec and LAI ($p > 0.05$). Moreover, as shown in Figure 10b and Table 7, when PET > 9 mm/day, the Ec did not increase, but showed a slightly decreasing trend with the increase in PET ($p > 0.05$). These results demonstrated that the PET limits the effect of LAI on Ec when PET > 9mm/day.

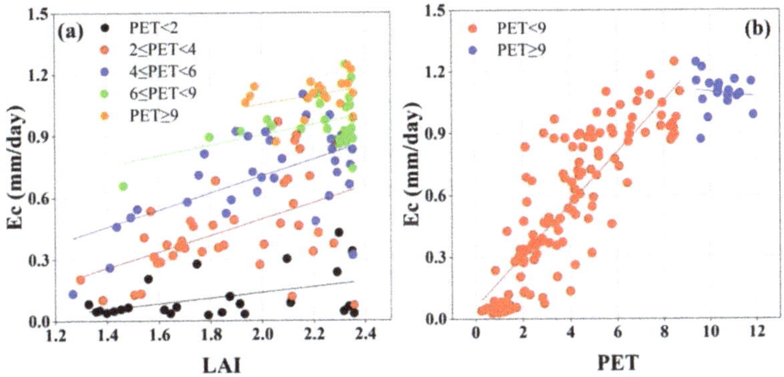

Figure 10. Plot showing the role of PET in response of canopy transpiration to LAI in study period. (**a**) LAI; (**b**) PET.

Table 6. The response of Ec to LAI at varied PET levels.

Level	Regression	R^2	Sig.
PET < 2	$E_C = 0.14 \times LAI - 0.13$	0.19	$p < 0.05$
2 ≤ PET < 4	$E_C = 0.40 \times LAI - 0.30$	0.23	$p < 0.01$
4 ≤ PET < 6	$E_C = 0.41 \times LAI - 0.12$	0.34	$p < 0.01$
6 ≤ PET < 9	$E_C = 0.25 \times LAI + 0.40$	0.18	$p < 0.05$
PET ≥ 9	$E_C = 0.21 \times LAI + 0.64$	0.11	$p > 0.05$

Table 7. The response of Ec to PET.

Level	Regression	R^2	Sig.
PET < 9	$E_C = 0.13 \times PET + 0.05$	0.71	$p < 0.01$
PET ≥ 9	$E_C = -0.01 \times PET + 1.23$	0.01	$p > 0.05$

4. Discussion

Canopy transpiration is one of the crucial methods of water consumption in the forest ecosystem, which affects the forest hydrological processes [19,38]. Knowledge of canopy transpiration is greatly important for the forest–water relationship under the background of carbon neutrality. In the present study, the daily mean Ec was 0.60 mm/day accounting for 16.50% of P during the measurement period, which was in a reasonable range compared with previous studies [9,17,19,21]. For instance, Wang et al. [9] reported that the mean daily canopy transpiration of Larch was 0.70 mm/day, ranging from 0.02 to 1.55 mm in the growing season of 2018. However, we also found that the canopy transpiration of the present study was lower than the same tree species in the previous study [26]. The difference in Ec may be caused by forest structure and climate condition of the measurement period. In the present study, the stand density was 1250 trees ha^{-1}, which was significantly lower than that for the 2350 trees ha^{-1} of Liu et al. [26]. Wang et al. [4] demonstrated that the thinning significantly reduced daily transpiration at a stand scale. Moreover, the precipitation in the study period was greatly higher than the mean annual precipitation from 1957 to 2017. Wullschleger and Hanson [39] indicated that excessive precipitation limited the level of transpiration. These reasons largely account for the lower daily canopy transpiration in the present study.

During the growing season, there were significant monthly dynamics in canopy transpiration. The higher canopy transpirations were found in June, July and August, which is consistent with other studies [19,38,40]. However, previous studies also suggested that the canopy transpiration of Mongolian pine trees in May was higher than the other months attributed to the high-growth period [41]. It is maybe closely related to the monthly dynamic of vegetation growth and climate conditions. It can be seen from Figure 4 that the LAI and Ta in June, July and August were significantly higher than that of May and September in the present study, which leads to the higher canopy transpiration. Additionally, although the VPD and PAR in August were similar to May and September, the LAI and Ta were higher and subsequently promote canopy transpiration.

In addition, the results of correlation and regression analysis showed a significant relationship between Ec and environmental factors (Table 3 and Figure 7), which was consistent with the previous studies [42–45]. For instance, the relationship between Ec and PET has exhibited the exponential threshold function (Figure 7) and indicated that Ec increases rapidly with the increasing of PET and then tended to be saturation when PET continued to increase. Wieser et al. [42] suggested that the canopy stomatal conductance decreased with the increase in VPD. As shown in Figure 8, our study also demonstrated that the canopy conductance (G_L) significantly decreased with the increase in VPD. Another possible reason is that the stomata need to be closed through the plant self-protection mechanism to maintain canopy transpiration to avoid excessive water consumption at higher atmospheric evaporative demand [42,46,47]. Furthermore, in the present study, PET as a compound index is the major driver of transpiration rather than a single environmental variable. Similarly, Song et al. [21] reported that the effects of PET on transpiration in three forest management areas of NE China were stronger than those of the other environmental variables. It is revealed that the compound index characterizing the atmospheric evaporative demand can better explain the change in transpiration. However, we found that the effect of soil moisture on the Ec of the *Larix gmelinii* forest was lowest compared with other environment variables. Similar results were observed in the semiarid region and boreal forest [3,21,48,49]. On the contrary, some studies revealed that canopy transpiration is significantly influenced by soil moisture [50]. These different results may be closely attributed to the different tree species, root distribution and climate conditions. Chu et al. [51] reported that the root of *Larix gmelinii* forest was mainly distributed in topsoil [52] combined with the excessive precipitation in the study period, which led to sufficient soil moisture in topsoil to maintain water consumption of canopy transpiration and consequently the soil moisture was not the main limiting environmental factor of canopy transpiration of *Larix gmelinii* forest.

Generally, canopy transpiration is influenced by the interaction effects of multiple environmental variables under natural conditions [53]. In the present study, when soil moisture was more than 0.2 cm^3cm^{-3} (accounting for 94.77% (145 days) of the whole study period (153 days)), we found that the effect of PET on canopy transpiration was not limited by soil moisture. It is indicated that the water supply can meet the water consumption of canopy transpiration of *Larix gmelinii* in the current soil moisture conditions. In addition, when PET was higher than the threshold value (PET > 9 mm/day), the response of canopy transpiration to LAI will be limited. For example, Song et al. [45] reported that the soil moisture limits the response of transpiration to atmospheric water demand in seasonal drought, while the effect was not significant in the well-water soil moisture conditions of subtropical coniferous. Some similar results were also observed in Clausnitzer et al. [33] and Zha et al. [54]. Numerous studies have indicated that the limitations of soil moisture on the response of canopy transpiration to atmospheric evaporative demand will occur in drought conditions [45,54,55]. When the soil moisture supply is insufficient or in high atmospheric evaporation demand, the leaf stomata and leaf water content of plants was closed and declined to prevent excessive water consumption [56–58], which is consistent with the results in Figure 8. On the other hand, it is well known that the most important water source of the forest ecosystem is from precipitation input. However, for the above-mentioned result that total canopy transpiration only accounts for 16.50% of P during the measurement period, this indicates that the water input though excessive precipitation can effectively increase soil moisture to maintain canopy transpiration in the present study. Moreover, the age of *Larix gmelinii* in the study sample plot ranged from 75 to 90 years, which means it is a relatively mature stand [59]. Bretfeld et al. [60] reported that transpiration is more likely to be limited by soil moisture in young forests than that of old forests due to plant root volume difference. Altogether, *Larix gmelinii* forest adjust its water consumption through canopy transpiration by responding to changes in the external environment.

5. Conclusions

The present study was performed to evaluate the effects of soil moisture, potential evapotranspiration and leaf area index on the canopy transpiration of a boreal larch (*Larix gmelinii*) forest in China. Results showed that the average canopy transpiration was 0.60 mm/day during the study period and exhibited significant monthly change. The dominant driving force of canopy transpiration is potential evapotranspiration, while the effect of soil moisture on canopy transpiration was weakest compared with other environmental factors. Meanwhile, canopy transpiration was not limited by soil moisture when soil moisture exceeded 0.2 cm^3cm^{-3}. Furthermore, under the condition of sufficient soil moisture, the response of canopy transpiration to the changes in leaf area index can be limited by potential evapotranspiration when potential evapotranspiration exceeded 9 mm/day. Thus, we conclude that the ecological benefits such as carbon sequestration and water resource consumption of *Larix gmelinii* forest could be largely decided by future climate warming.

Author Contributions: Conceptualization, Z.X. and X.M.; methodology, Z.X.; software, Z.X.; validation, X.M., Y.S. and Z.X.; formal analysis, X.M.; investigation, Z.X. and Y.S.; resources, X.M. and T.C.; data curation, Z.X. and Y.S.; writing—original draft preparation, Z.X.; writing—review and editing, X.M. and T.C.; funding acquisition, T.C. All authors have read and agreed to the published version of the manuscript.

Funding: This research was funded by the Fundamental Research Funds for the Central Universities, grant number 2572020AW14, and the National Natural Science Foundation of China, grant number 31971451.

Institutional Review Board Statement: Not applicable.

Informed Consent Statement: Not applicable.

Data Availability Statement: Not applicable.

Acknowledgments: We acknowledge the Mohe Forest Ecological Research Station for providing the meteorological data.

Conflicts of Interest: The authors declare no conflict of interest.

References

1. Bonan, G.B.; Shugart, H.H. Environmental Factors and Ecological Processes in Boreal Forests. *Annu. Rev. Ecol. Syst.* **1989**, *20*, 1–28. [CrossRef]
2. Peng, C.; Ma, Z.; Lei, X.; Zhu, Q.; Chen, H.; Wang, W.; Liu, S.; Li, W.; Fang, X.; Zhou, X. A drought-induced pervasive increase in tree mortality across Canada's boreal forests. *Nat. Clim. Chang.* **2011**, *1*, 467. [CrossRef]
3. Wang, H.L.; Tetzlaff, D.; Dick, J.J.; Soulsby, C. Assessing the environmental controls on Scots pine transpiration and the implications for water partitioning in a boreal headwater catchment. *Agric. For. Meteorol.* **2017**, *240–241*, 58–66. [CrossRef]
4. Wang, Y.; Wei, X.; del Campo, A.D.; Winkler, R.; Wu, J.; Li, Q.; Liu, W. Juvenile thinning can effectively mitigate the effects of drought on tree growth and water consumption in a young Pinus contorta stand in the interior of British Columbia, Canada. *For. Ecol. Manag.* **2019**, *454*, 117667. [CrossRef]
5. Xu, Z.; Man, X.; Duan, L.; Cai, T. Assessing the relative contribution of increased forest cover to decreasing river runoff in two boreal forested watersheds of Northeastern China. *Ecohydrol. Hydrobiol.* **2021**, *22*, 113–125. [CrossRef]
6. Ungar, E.D.; Rotenberg, E.; Raz-Yaseef, N.; Cohen, S.; Yakir, D.; Schiller, G. Transpiration and annual water balance of Aleppo pine in a semiarid region: Implications for forest management. *For. Ecol. Manag.* **2013**, *298*, 39–51. [CrossRef]
7. Kumagai, T.; Tateishi, M.; Miyazawa, Y.; Kobayashi, M.; Yoshifuji, N.; Komatsu, H.; Shimizu, T. Estimation of annual forest evapotranspiration from a coniferous plantation watershed in Japan (1): Water use components in Japanese cedar stands. *J. Hydrol.* **2014**, *508*, 66–76. [CrossRef]

8. Liu, Z.; Wang, Y.; Tian, A.; Webb, A.A.; Yu, P.; Xiong, W.; Xu, L.; Wang, Y. Modeling the Response of Daily Evapotranspiration and its Components of a Larch Plantation to the Variation of Weather, Soil Moisture, and Canopy Leaf Area Index. *J. Geophys. Res. Atmos.* **2018**, *123*, 7354–7374. [CrossRef]
9. Wang, L.; Liu, Z.; Guo, J.; Wang, Y.; Ma, J.; Yu, S.; Yu, P.; Xu, L. Estimate canopy transpiration in larch plantations via the interactions among reference evapotranspiration, leaf area index, and soil moisture. *For. Ecol. Manag.* **2021**, *481*, 118749. [CrossRef]
10. Jasechko, S.; Sharp, Z.D.; Gibson, J.J.; Birks, S.J.; Yi, Y.; Fawcett, P.J. Terrestrial water fluxes dominated by transpiration. *Nature* **2013**, *496*, 347–350. [CrossRef]
11. Good, S.P.; Noone, D.; Bowen, G. Hydrologic connectivity constrains partitioning of global terrestrial water fluxes. *Science* **2015**, *349*, 175–177. [CrossRef] [PubMed]
12. Sheffield, J.; Andreadis, K.M.; Wood, E.; Lettenmaier, D.P. Global and Continental Drought in the Second Half of the Twentieth Century: Severity–Area–Duration Analysis and Temporal Variability of Large-Scale Events. *J. Clim.* **2009**, *22*, 1962–1981. [CrossRef]
13. Zhu, Y.; Zhang, J.; Zhang, Y.; Qin, S.; Shao, Y.; Gao, Y. Responses of vegetation to climatic variations in the desert region of northern China. *Catena* **2019**, *175*, 27–36. [CrossRef]
14. Cinnirella, S.; Magnani, F.; Saracino, A.; Borghetti, M. Response of a mature Pinus laricio plantation to a three-year restriction of water supply: Structural and functional acclimation to drought. *Tree Physiol.* **2002**, *22*, 21–30. [CrossRef] [PubMed]
15. Pfautsch, S.; Bleby, T.M.; Rennenberg, H.; Adams, M.A. Sap flow measurements reveal influence of temperature and stand structure on water use of Eucalyptus regnans forests. *For. Ecol. Manag.* **2010**, *259*, 1190–1199. [CrossRef]
16. Tie, Q.; Hu, H.C.; Tian, F.Q.; Guan, H.D.; Lin, H. Environmental and physiological controls on sap flow in a subhumid mountainous catchment in North China. *Agric. For. Meteorol.* **2017**, *240–241*, 46–57. [CrossRef]
17. Deng, J.; Yao, J.; Zheng, X.; Gao, G. Transpiration and canopy stomatal conductance dynamics of Mongolian pine plantations in semiarid deserts, Northern China. *Agric. Water Manag.* **2021**, *249*, 106806. [CrossRef]
18. Zhang, H.; Wei, W.; Chen, L.; Wang, L. Effects of terracing on soil water and canopy transpiration of Pinus tabulaeformis in the Loess Plateau of China. *Ecol. Eng.* **2017**, *102*, 557–564. [CrossRef]
19. Han, C.; Chen, N.; Zhang, C.K.; Liu, Y.J.; Khan, S.; Lu, K.L.; Li, Y.G.; Dong, X.X.; Zhao, C.M. Sap flow and responses to me-teorological about the Larix principis-rupprechtii plantation in Gansu Xinlong mountain, northwestern China. *For. Ecol. Manag.* **2019**, *451*, 117519. [CrossRef]
20. Cooper, A.E.; Kirchner, J.W.; Wolf, S.; Lombardozzi, D.L.; Sullivan, B.W.; Tyler, S.W.; Harpold, A.A. Snowmelt causes different limitations on transpiration in a Sierra Nevada conifer forest. *Agric. For. Meteorol.* **2020**, *291*, 108089. [CrossRef]
21. Song, L.; Zhu, J.; Zheng, X.; Wang, K.; Lü, L.; Zhang, X.; Hao, G. Transpiration and canopy conductance dynamics of Pinus sylvestris var. mongolica in its natural range and in an introduced region in the sandy plains of Northern China. *Agric. For. Meteorol.* **2020**, *281*, 107830. [CrossRef]
22. Jiao, L.; Lu, N.; Fang, W.; Li, Z.; Wang, J.; Jin, Z. Determining the independent impact of soil water on forest transpiration: A case study of a black locust plantation in the Loess Plateau, China. *J. Hydrol.* **2019**, *572*, 671–681. [CrossRef]
23. Hayat, M.; Zha, T.; Jia, X.; Iqbal, S.; Qian, D.; Bourque, C.P.-A.; Khan, A.; Tian, Y.; Bai, Y.; Liu, P.; et al. A multiple-temporal scale analysis of biophysical control of sap flow in Salix psammophila growing in a semiarid shrubland ecosystem of northwest China. *Agric. For. Meteorol.* **2020**, *288–289*, 107985. [CrossRef]
24. Wan, Y.; Yu, P.; Wang, Y.; Wang, B.; Yu, Y.; Wang, X.; Liu, Z.; Liu, X.; Wang, S.; Xiong, W. The Variation in Water Consumption by Transpiration of Qinghai Spruce among Canopy Layers in the Qilian Mountains, Northwestern China. *Forests* **2020**, *11*, 845. [CrossRef]
25. Yang, G.; Di, X.-Y.; Guo, Q.-X.; Shu, Z.; Zeng, T.; Yu, H.-Z.; Wang, C. The impact of climate change on forest fire danger rating in China's boreal forest. *J. For. Res.* **2011**, *22*, 249–257. [CrossRef]
26. Liu, J.; Cheng, F.; Munger, W.; Jiang, P.; Whitby, T.; Chen, S.; Ji, W.; Man, X. Precipitation extremes influence patterns and partitioning of evapotranspiration and transpiration in a deciduous boreal larch forest. *Agric. For. Meteorol.* **2020**, *287*, 107936. [CrossRef]
27. Duan, B.; Man, X.; Cai, T.; Xiao, R.; Ge, Z. Increasing soil organic carbon and nitrogen stocks along with secondary forest succession in permafrost region of the Daxing'an mountains, northeast China. *Glob. Ecol. Conserv.* **2020**, *24*, e01258. [CrossRef]
28. Granier, A. Evaluation of transpiration in a Douglas-fir stand by means of sap flow measurements. *Tree Physiol.* **1987**, *3*, 309–320. [CrossRef]
29. Peters, R.L.; Fonti, P.; Frank, D.; Poyatos, R.; Pappas, C.; Kahmen, A.; Carraro, V.; Prendin, A.L.; Schneider, L.; Baltzer, J.L.; et al. Quantification of uncertainties in conifer sap flow measured with the thermal dissipation method. *New Phytol.* **2018**, *219*, 1283–1299. [CrossRef]
30. Otieno, D.; Li, Y.; Ou, Y.; Cheng, J.; Liu, S.; Tang, X.; Zhang, Q.; Jung, E.-Y.; Zhang, D.; Tenhunen, J. Stand characteristics and water use at two elevations in a sub-tropical evergreen forest in southern China. *Agric. For. Meteorol.* **2014**, *194*, 155–166. [CrossRef]
31. Kostner, B.; Tenhunen, J.D.; Alsheimer, M.; Wedler, M.; Scharfenberg, H.-J.; Zimmermann, R.; Falge, E.; Joss, U. Controls on Evapotranspiration in a Spruce Forest Catchment of the Fichtelgebirge. *Ecol. Stud.* **2001**, *146*, 377–415. [CrossRef]

32. Tang, J.; Bolstad, P.V.; Ewers, B.E.; Desai, A.R.; Davis, K.J.; Carey, E.V. Sap flux-upscaled canopy transpiration, stomatal conductance, and water use efficiency in an old growth forest in the Great Lakes region of the United States. *J. Geophys. Res. Biogeosci.* **2006**, *111*. [CrossRef]
33. Clausnitzer, F.; Köstner, B.; Schwärzel, K.; Bernhofer, C. Relationships between canopy transpiration, atmospheric conditions and soil water availability-analyses of long-term sap-flow measurements in an old Norway spruce forest at the Ore Mountains/Germany. *Agric. For. Meteorol.* **2011**, *151*, 1023–1034. [CrossRef]
34. Gao, X.; Zhao, X.; Brocca, L.; Pan, D.; Wu, P. Testing of observation operators designed to estimate profile soil moisture from surface measurements. *Hydrol. Process.* **2019**, *33*, 575–584. [CrossRef]
35. Howell, T.A.; Dusek, D.A. Comparison of Vapor-Pressure-Deficit Calculation Methods—Southern High Plains. *J. Irrig. Drain. Eng.* **1995**, *121*, 191–198. [CrossRef]
36. Allen, R.G.; Pereira, L.S.; Raes, D.; Smith, M. Crop evapotranspiration-Guidelines for computing crop water requirements-FAO Irrigation and drainage paper 56. FAO: Rome, Italy, 1998; Volume 300, p. D05109.
37. Mcmahon, T.A.; Peel, M.C.; Lowe, L.; Srikanthan, R.; Mcvicar, T. Estimating actual, potential, reference crop and pan evap-oration using standard meteorological data: A pragmatic synthesis. *Hydrol. Earth Syst. Sci.* **2013**, *17*, 1331–1363. [CrossRef]
38. Ouyang, L.; Wu, J.; Zhao, P.; Li, Y.Q.; Zhu, L.W.; Ni, G.Y. Consumption of precipitation by evapotranspiration indicates po-tential drought for broadleaved and coniferous plantations in hilly lands of South China. *Agric. Water Manag.* **2021**, *252*, 106927. [CrossRef]
39. Wullschleger, S.; Hanson, P.J. Sensitivity of canopy transpiration to altered precipitation in an upland oak forest: Evidence from a long-term field manipulation study. *Glob. Chang. Biol.* **2005**, *12*, 97–109. [CrossRef]
40. Bai, Y.; Li, X.; Zhou, S.; Yang, X.; Yu, K.; Wang, M.; Liu, S.; Wang, P.; Wu, X.; Wang, X.; et al. Quantifying plant transpiration and canopy conductance using eddy flux data: An underlying water use efficiency method. *Agric. For. Meteorol.* **2019**, *271*, 375–384. [CrossRef]
41. Song, L.; Zhu, J.; Li, M.; Yu, Z. Water utilization of Pinus sylvestris var. mongolica in a sparse wood grassland in the semiarid sandy region of Northeast China. *Trees* **2014**, *28*, 971–982. [CrossRef]
42. Wieser, G.; Leo, M.; Oberhuber, W. Transpiration and canopy conductance in an inner alpine Scots pine (*Pinus sylvestris* L.) forest. *Flora-Morphol. Distrib. Funct. Ecol. Plants* **2014**, *209*, 491–498. [CrossRef] [PubMed]
43. Ghimire, C.P.; Lubczynski, M.W.; Bruijnzeel, L.A.; Chavarro-Rincón, D. Transpiration and canopy conductance of two con-trasting forest types in the Lesser Himalaya of Central Nepal. *Agric. For. Meteorol.* **2014**, *197*, 76–90. [CrossRef]
44. Urban, J.; Rubtsov, A.V.; Urban, A.V.; Shashkin, A.V.; Benkova, V.E. Canopy transpiration of a *Larix sibirica* and *Pinus sylvestris* forest in Central Siberia. *Agric. For. Meteorol.* **2019**, *271*, 64–72. [CrossRef]
45. Song, X.; Lyu, S.; Wen, X. Limitation of soil moisture on the response of transpiration to vapor pressure deficit in a subtropical coniferous plantation subjected to seasonal drought. *J. Hydrol.* **2020**, *591*, 125301. [CrossRef]
46. Hogg, E.H.; Hurdle, P.A. Sap flow in trembling aspen: Implications for stomatal responses to vapor pressure deficit. *Tree Physiol.* **1997**, *17*, 501–509. [CrossRef]
47. Ge, Y.; Ghang, J.; Liu, K.; Qin, G.Q. A Physio-ecological Study on the Transpiration of *Mosla hangchowensis* Matsuda. *Chin. J. Plant Ecol.* **1999**, *23*, 320–326.
48. Brito, P.; Lorenzo, J.R.; Gonzalez-Rodriguez, A.M.; Morales, D.; Wieser, G.; Jimenez, M.S. Canopy transpiration of a semiarid Pinus canariensis forest at a treeline ecotone in two hydrologically contrasting years. *Agric. For. Meteorol.* **2015**, *201*, 120–127. [CrossRef]
49. Jiao, L.; Lu, N.; Sun, G.; Ward, E.; Fu, B. Biophysical controls on canopy transpiration in a black locust (Robinia pseudoacacia) plantation on the semi-arid Loess Plateau, China. *Ecohydrology* **2016**, *9*, 1068–1081. [CrossRef]
50. Lagergren, F.; Lindroth, A. Transpiration response to soil moisture in pine and spruce trees in Sweden. *Agric. For. Meteorol.* **2002**, *112*, 67–85. [CrossRef]
51. Chu, X.; Di, X.Y.; Zhang, J.L.; Cai, H.Y.; Yan, B.Z. Distribution and seasonal dynamics of fine root biomass of two types of forests in Great Xing'an Mountains. *J. Northeast For. Univ.* **2011**, *39*, 36–39.
52. Xu, Z.; Man, X.; Duan, L.; Cai, T. Improved subsurface soil moisture prediction from surface soil moisture through the integration of the (de)coupling effect. *J. Hydrol.* **2022**, *608*, 127634. [CrossRef]
53. Fernandes, T.; del Campo, A.; García-Bartual, R.; González-Sanchis, M. Coupling daily transpiration modelling with forest management in a semiarid pine plantation. *iForest-Biogeosci. For.* **2016**, *9*, 38–48. [CrossRef]
54. Zha, T.; Qian, D.; Jia, X.; Bai, Y.; Tian, Y.; Bourque, C.P.-A.; Ma, J.; Feng, W.; Wu, B.; Peltola, H. Soil moisture control of sap-flow response to biophysical factors in a desert-shrub species, Artemisia ordosica. *Biogeosciences* **2017**, *14*, 4533–4544. [CrossRef]
55. Yuan, W.; Zheng, Y.; Piao, S.; Ciais, P.; Lombardozzi, D.; Wang, Y.; Ryu, Y.; Chen, G.; Dong, W.; Hu, Z.; et al. Increased atmospheric vapor pressure deficit reduces global vegetation growth. *Sci. Adv.* **2019**, *5*, eaax1396. [CrossRef] [PubMed]
56. McAdam, S.A.M.; Brodribb, T.J. Stomatal innovation and the rise of seed plants. *Ecol. Lett.* **2012**, *15*, 1–8. [CrossRef]
57. Moshelion, M.; Halperin, O.; Wallach, R.; Oren, R.; Way, D.A. Role of aquaporins in determining transpiration and photo-synthesis in water-stressed plants: Crop water use efficiency, growth and yield. *Plant Cell Environ.* **2014**, *38*, 1785–1793.
58. Konings, A.G.; Williams, P.; Gentine, P. Sensitivity of grassland productivity to aridity controlled by stomatal and xylem regulation. *Nat. Geosci.* **2017**, *10*, 284–288. [CrossRef]

59. Xiao, R.H.; Man, X.L.; Duan, B.X. Carbon and Nitrogen Stocks in Three Types of *Larix gmelinii* Forests in Daxing'an Mountains, Northeast China. *Forests* **2020**, *11*, 305. [CrossRef]
60. Bretfeld, M.; Ewers, B.E.; Hall, J.S. Plant water use responses along secondary forest succession during the 2015–2016 El Niño drought in Panama. *New Phytol.* **2018**, *219*, 885–899. [CrossRef]

forests

MDPI

Article

Canopy Transpiration and Stomatal Conductance Dynamics of *Ulmus pumila* L. and *Caragana korshinskii* Kom. Plantations on the Bashang Plateau, China

Yu Zhang [1,2,3], Wei Li [4,5,6,*], Haiming Yan [4,5,6], Baoni Xie [4,5,6], Jianxia Zhao [7], Nan Wang [4] and Xiaomeng Wang [4]

1 School of Geographical Sciences, Hebei Normal University, Shijiazhuang 050024, China; yuzhang89@mail.bnu.edu.cn
2 Hebei Key Laboratory of Environmental Change and Ecological Construction, Shijiazhuang 050024, China
3 Hebei Technology Innovation Center for Remote Sensing Identification of Environmental Change, Shijiazhuang 050024, China
4 School of Land Science and Space Planning, Hebei GEO University, Shijiazhuang 050031, China; haiming.yan@hgu.edu.cn (H.Y.); xbn-feya@nwafu.edu.cn (B.X.); wn305099@163.com (N.W.); wxm975000@163.com (X.W.)
5 International Science and Technology Cooperation Base of Hebei Province, Hebei International Joint Research Center for Remote Sensing of Agricultural Drought Monitoring, Hebei GEO University, Shijiazhuang 050031, China
6 Hebei Province Collaborative Innovation Center for Sustainable Utilization of Water Resources and Optimization of Industrial Structure, Hebei GEO University, Shijiazhuang 050031, China
7 College of Geography and Land Engineering, Yuxi Normal University, Yuxi 653100, China; zjx@yxnu.edu.cn
* Correspondence: weil87land@hgu.edu.cn

Abstract: Constructing protective forests to control water and soil erosion is an effective measure to address land degradation in the Bashang Plateau of North China, but forest dieback has occurred frequently due to severe water deficits in recent decades. However, transpiration dynamics and their biophysical control factors under various soil water contents for different forest functional types are still unknown. Here, canopy transpiration and stomatal conductance of a 38-year-old *Ulmus pumila* L. and a 20-year-old *Caragana korshinskii* Kom. were quantified using the sap flow method, while simultaneously monitoring the meteorological and soil water content. The results showed that canopy transpiration averaged 0.55 ± 0.34 mm d^{-1} and 0.66 ± 0.32 mm d^{-1} for *U. pumila*, and was 0.74 ± 0.26 mm d^{-1} and 0.77 ± 0.24 mm d^{-1} for *C. korshinskii* in 2020 and 2021, respectively. The sensitivity of canopy transpiration to vapor pressure deficit (*VPD*) decreased as soil water stress increased for both species, indicating that the transpiration process is significantly affected by soil drought. Additionally, canopy stomatal conductance averaged 1.03 ± 0.91 mm s^{-1} and 1.34 ± 1.22 mm s^{-1} for *U. pumila*, and was 1.46 ± 0.90 mm s^{-1} and 1.51 ± 1.06 mm s^{-1} for *C. korshinskii* in 2020 and 2021, respectively. The low values of the decoupling coefficient (Ω) showed that canopy and atmosphere were well coupled for both species. Stomatal sensitivity to *VPD* decreased with decreasing soil water content, indicating that both *U. pumila* and *C. korshinskii* maintained a water-saving strategy under the stressed water conditions. Our results enable better understanding of transpiration dynamics and water-use strategies of different forest functional types in the Bashang Plateau, which will provide important insights for planted forests management and ecosystem stability under future climate changes.

Keywords: Bashang Plateau; transpiration; canopy stomatal conductance; soil water; afforestation

Citation: Zhang, Y.; Li, W.; Yan, H.; Xie, B.; Zhao, J.; Wang, N.; Wang, X. Canopy Transpiration and Stomatal Conductance Dynamics of *Ulmus pumila* L. and *Caragana korshinskii* Kom. Plantations on the Bashang Plateau, China. *Forests* **2022**, *13*, 1081. https://doi.org/10.3390/f13071081

Academic Editors: Yanhui Wang, Karl-Heinz Feger and Lulu Zhang

Received: 17 June 2022
Accepted: 8 July 2022
Published: 9 July 2022

1. Introduction

Afforestation is an important ecological measure for restoring fragile and degraded land. China has the largest afforested area in the world, with the area of planted forests accounting for about 23% of the global plantation area [1]. The area of planted forests in China now stands at about 69.33 × 10^6 ha. Planted forests can effectively contribute to improving the functions of water and soil conservation, sand-fixing, carbon storage, and adjusting the microclimate of weak ecological environment systems [2,3]. However, under the influence of global climate change-induced drought and human activities, these planted forests, especially in arid and semiarid regions, have suffered severe canopy dieback and mortality approximately 30–35 years after planting [4,5]. Immense afforestation may consume too much water and cause a severe water deficit, and, compared with native plantation species, introduced plantation species cause a greater reduction in soil water availability, negatively impacting ecosystem services and functions [6,7]. Transpiration and related water-use strategies are fundamental to understanding the physiological processes of plantations and play a vital role in their survival and growth, especially in semiarid and arid regions, where water availability is greatly affected by the increased frequency and intensity of droughts [8,9]. However, previous studies have shown that the effects of excessive water loss caused by transpiration and its environmental driving factors differ significantly, resulting in different growth performances among different functional types [10–12]. Therefore, it is imperative to understand the transpiration processes and dynamics of planted tree and shrub species and how these processes change from their normal functions.

Transpiration, the plant process in which water is consumed from the soil and released into the atmosphere, is an important parameter in understanding forest hydrological processes and further determining the forest water balance [13,14]. The accurate quantification of forest transpiration is essential for tree physiology and ecohydrology [15,16]. Based on their advantages of simple operation, relatively low cost, fewer limitations in environmental conditions, and increasingly robust features, sap flow measurements have been considered the most practical method to estimate tree transpiration at both the individual and stand scales [17,18]. Environmental factors, including solar radiation, air temperature, wind speed, vapor pressure deficit (*VPD*), and soil water content, have significant impacts on the transpiration process. These abiotic factors can be divided into two aspects: evaporative demand (potential evapotranspiration, PET) and water supply [19]. In semiarid and arid regions, soil water availability is a critical factor for plantation growth and vegetation productivity due to its influence on plant transpiration [20]. Previous studies showed that soil water stress in plantations occurs when relative extractable water content (REW) is below a threshold of 0.4 [21,22]. A linear or nonlinear relationship, including exponential and polynomial functions, was found between transpiration and evaporative demand when REW > 0.4, whereas, when REW < 0.4, increased soil water stress influenced transpiration, which was expressed following exponential and polynomial functions for evaporative demand for different planted forests [23]. Moreover, transpiration is regulated via canopy stomatal conductance (G_c) in response to variations in *VPD* and soil water content. High *VPD* causes a decrease in canopy stomatal conductance. In order to prevent irreversible damage to tree hydraulic traits, transpiration then gradually increases nonlinearly with increasing *VPD*, and is maintained or decreases when the canopy stomata begin to close. However, different specific nonlinear functions, such as exponential functions [24], logarithmic functions, and polynomial functions [25], have been reported for different plantation types in different artificial ecosystems. Additionally, reduced soil water availability may decrease hydraulic conductance from soil to leaves, causing canopy stomatal closure to avoid embolism and hydraulic failure, and thereby reducing tree transpiration. Denham et al. quantified the collaborative influence of soil water content and *VPD* along a hydroclimatological gradient, and indicated that a more sensitive relationship between canopy stomatal conductance and soil water content was found in dry than in wet sites [26]. Most climate models predict that soil droughts will be more frequent and more severe in semiarid and arid regions [27]. However, few studies have explored the impact of environmental factors on transpiration and G_c under different soil water availability conditions, and influenced by climate change-induced drought. Moreover, it is also unclear whether these relationships exhibit interspecific differences among different forest functional types.

The Bashang Plateau is one of the most vulnerable and sensitive portions of northern China, and is a major source of dust and sand storms that affect Beijing and its surrounding areas. Land degradation and desertification are the most serious environmental problems in this area. A series of afforestation projects have been implemented by China's government during the past four decades, e.g., the Grain-for-Green Program, the Three-North Shelterbelt Program, and the Beijing-Tianjin Sand Source Restoration Project, to address land degradation and improve the regional ecological environment [28,29]. Many fast-growing and stress-tolerant species, such as *Populus simonii* Carr, *Pinus sylvestris* var. *mongolica Litv*, *Ulmus pumila* L, and *Caragana korshinskii* Kom were planted to improve vegetation cover and control wind and sand erosion. The degree of desertification on the Bashang Plateau has improved significantly during the past four decades. However, these plantations have frequently suffered canopy dieback and mortality in recent years. The main reason for this is related to water deficiency, which affected plantations' transpiration dynamics and water-use strategies. *U. pumila* is a major native tree species and *C. korshinskii* is an introduced shrub species. Nevertheless, the transpiration processes of the two functional types are still not fully understood, and the biophysical control factors of transpiration under various soil water availability conditions are still unknown. The objectives of this study were to explore: (1) canopy transpiration and stomatal conductance processes during the growing season for *U. pumila* (native species) and *C. korshinskii* (introduced species); and (2) the biophysical mechanisms of canopy transpiration control under different soil water content conditions. The results can provide a deep understanding of the water-use dynamics of planted tree and shrub species, and thus provide scientific guidance for afforestation species selection and optimal allocation.

2. Materials and Methods

2.1. Site Description

Our study was conducted in the Kangbao pasture region, Kangbao County, Zhangjiakou City, Hebei Province, China (Figure 1). This area is located in the temperate climate zone. The average annual air temperature is about 2.3 °C, and the average annual precipitation is 330 mm, with over 85% falling between May and September. The mean annual potential evapotranspiration is 880 mm and the mean wind speed is 3.15 m s^{-1}. The major soil type is chestnut soil. The main growing season runs from May to September. The dominant tree species here are *U. pumila*, *P. simonii*, and *Pinus sylvestris* var. mongolica Litv, and the dominant shrub species are *C. korshinskii*, *Hippophae rhamnoides* Linn, and *Armeniaca sibirica* (L.) Lam.

This experiment was conducted in stands of *U. pumila* (114°47′ E, 42°06′ N, altitude 1325 m) and *C. korshinskii* (114°48′ E, 42°07′ N, altitude 1305 m) in 2020 and 2021. The two sites were approximately 3.0 km apart. A plot having an area of 20 by 20 m was set up in each of the *U. pumila* and *C. korshinskii* stands. The *U. pumila* were about 38 years old and the *C. korshinskii* were about 20 years old. The corresponding maximum leaf area indexes (LAI) were 0.31 and 0.43 m^2 m^{-2}, respectively. The mean diameter at breast height (DBH, 1.3 m) of *U. pumila* was 16.60 cm and the mean stem basal diameter (SBD, 10–15 cm above the ground) of *C. korshinskii* was 2.38 cm. The physical characteristics of the soil differed between the *U. pumila* and *C. korshinskii* stands. The soil particle size distribution was 39.64% sand (0.05–2 mm), 50.85% silt (0.002–0.05 mm), and 9.69% clay (<0.002 mm) in the *U. pumila* stand, and 52.89% sand, 39.25% silt, and 7.86% clay in the *C. korshinskii* stand. The soil organic matter (SOM) in the *U. pumila* stand at the 0–100 cm depth was 1.34%, and that of the *C. korshinskii* stand was 0.85% (Table 1).

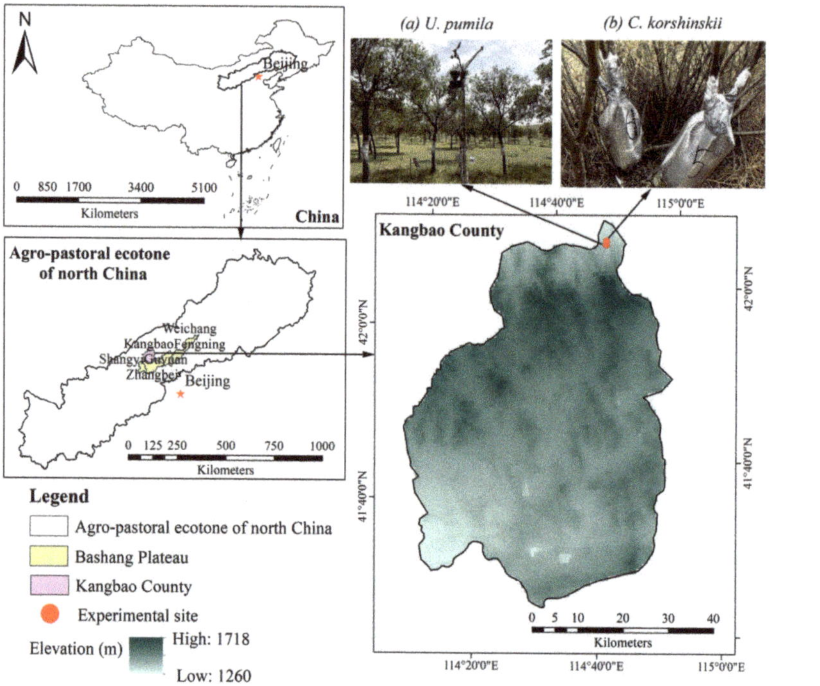

Figure 1. Geographical research sites in this study, namely, the (**a**) *U. pumila* site and (**b**) *C. korshinskii* site on the Bashang Plateau.

Table 1. Detailed information about vegetation characteristics and soil properties of *U. pumila* and *C. korshinskii* stands.

Species	Vegetation Characteristics				Soil Properties (%)			
	Age	DBH/SBD (cm)	SA (cm^2)	LAI (m^2 m^{-2})	Soil Particle-Size Distributions			SOM
					Sand	Silt	Clay	
U. pumila	38	16.60	120.63	0.31	39.46	50.85	9.69	1.34
C. korshinskii	20	2.38	2.65	0.43	52.89	39.25	7.86	0.85

Note: DBH = diameter at breast height; SBD = stem basal diameter; SA = sapwood area; LAI = leaf area index; SOM = soil organic matter.

2.2. Measurement of Sap Flux Density and Estimation of Canopy Transpiration

Measurements of sap flow were carried out using Granier-type thermal dissipation probes (TDPs) between 1 May and 30 September in 2020 and 2021 [30]. The sensors consisted of a pair of probes having a 20 mm length and 2 mm diameter, each of which had a cooper constant thermocouple. Based on the distributions of DBH and SBD, six sampled trees in the *U. pumila* stand and six sampled stems in the *C. korshinskii* stand were selected for sap flux measurements. The probes were inserted into the north-facing side of the truck (stem) at a height of 1.3 m (sampled *U. pumila*) and 15 cm (sampled *C. korshinskii*) above ground level to minimize exposure to the sun. In order to protect the probes against physical damage, solar radiation, temperature fluctuations, and rain, they were mounted in waterproof silicone and covered with an aluminum cover. The temperature differences between heated and reference probes were recorded on a data logger at 10 min intervals

(CR1000, Campbell Scientific Inc., Logan, UT, USA). The sap flux density (F_d) was estimated using the standard calibration for the TDP method:

$$F_d = 0.714 \times \left(\frac{\Delta T_{max} - \Delta T}{\Delta T}\right)^{1.231} \quad (1)$$

where F_d (mL cm^{-2} min^{-1}) is the sap flux density, ΔT (°C) is the temperature difference between the two probes at any given time, and $\Delta Tmax$ (°C) is the maximum temperature difference between sensors when the sap flux density is close to zero. Due to the fact that we only used one pair of probes per sample tree, azimuthal and radial variations in sap flux within trees were ignored In calculating canopy transpiration per area of ground (E_c, mm h^{-1}); we multiplied the mean F_d for each sample tree (F_d-avg) by the total sapwood area of the study plot per unit area of ground:

$$E_c = F_{d-avg} \times \frac{A_{si}}{A_g} \times 600 \quad (2)$$

where F_d-avg is the average sap flux density of sampled trees (or shrubs), A_g is the ground surface area of the studied plots (400 m^2), and A_{si} is the total sapwood area in the studied plot. In order to measure the sapwood area of *U. pumila*, the staining method was applied to 19 trees both inside and outside the plot. Exponential growth function relationships between sapwood and DBH ($A_s = 27.10 \times e^{0.09DBH}, n = 19, R^2 = 0.83$) were established. Furthermore, we used the stem basal cross-sectional area as the scalar to estimate the E_c and stand transpiration for the *C. korshinskii* stand. The total sapwood area in the plot was 6016.04 cm^2 and 6877.70 cm^2 for the *U. pumila* and *C. korshinskii* stands, respectively.

2.3. Meteorological Data and Soil Water Measurement

Precipitation (P, mm), air temperature (T_a, °C), relative humidity (RH, %), wind speed (u_2, m s^{-1}), and photosynthetically active radiation (PAR, µmol m^{-2} s^{-1}) were measured continuously with an Onset HOBO U21 automatic weather station (Onset Computer Corp., Bourne, MA, USA) located in the open areas neighboring the *U. pumila* plantation stand. Air temperature and relative humidity were measured at a height of 1.5 m, and precipitation, wind speed, and photosynthetically active radiation were measured at a height of 2.0 m (Figure 1). These variables were recorded every 10 min as averages using data loggers of the CR1000 series (Campbell Scientific, Logan, UT, USA). VPD (kPa) was derived from air temperature and the relative humidity:

$$VPD = 0.611 \times \exp\left(\frac{17.502T_a}{T_a + 240.97}\right) \times (1 - RH) \quad (3)$$

The soil volumetric water content (θ, %) was continuously monitored at a single location in *U. pumila* and *C. korshinskii* stands with an EC–5TE sensor (Decagon, Inc. Pullman, WA, USA). Both EC–5TE sensors were installed at depths of 0–10 cm, 10–20 cm, 20–40 cm, 40–60 cm, and 60–100 cm below the ground surface. Measurements were taken every 10 s, and the 30 min averages were recorded by CR1000 data loggers (Campbell Scientific, Logan, UT, USA). The soil water content values used in the paper were averaged for the depths of 0–100 cm for each plot. Relative extractable soil water (REW) was calculated using averaged θ across 0–100 cm:

$$REW = \frac{\theta - \theta_{min}}{\theta_{max} - \theta_{min}} \quad (4)$$

where θ_{min} and θ_{max} are the minimum and maximum daily average soil water content, respectively, during May–September in 2020 and 2021.

To account for climate impacts on stand canopy transpiration, daily potential evapotranspiration (PET, mm d^{-1}) was calculated by the Penman–Monteith equation based on FAO–56 [31]:

$$PET = \frac{0.408\Delta(R_n - G) + \gamma(900/(T_a + 273))u_2 VPD}{\Delta + \gamma(1 + 0.34u_2)} \tag{5}$$

where R_n is the net radiation (MJ m^{-2} d^{-1}), G is the soil heat flux density ($G \approx 0$ for daily scale), γ is the psychrometric constant (kPa °C^{-1}), and Δ is the slope of the vapor pressure curve (kPa °C^{-1}).

2.4. Canopy Conductance and Decoupling Coefficient

G_c (mm s^{-1}) was calculated by multiplying the conductance coefficient (K_G) by E_c and dividing the result by VPD [32]:

$$G_c = \frac{K_G \times E_c}{VPD} \tag{6}$$

where K_G is the conductance coefficient as a function of air temperature (115.8 + 0.4236 T_a, kPa m^3 kg^{-1}), which accounts for the temperature effect on the psychometric constant, the latent heat of vaporization, and the specific heat and density of air.

We estimated the decoupling coefficient ($0 < \Omega < 1$) as a means of quantifying the degree of coupling between the canopy and atmosphere. Canopies are aerodynamically well coupled to the atmosphere when Ω approaches 0, and canopy transpiration is determined primarily by stomatal opening. As Ω approaches 1, stomatal control on canopy transpiration becomes weaker, and canopy transpiration is more influenced by solar radiation. The Ω was estimated as follow [33]:

$$\Omega = \frac{1}{1 + [\gamma/(\Delta + \gamma)](g_a/G_c)} \tag{7}$$

where g_a is the aerodynamic conductance (m s^{-1}), calculated as follows:

$$g_a = \frac{1 + 0.54U_z}{[\ln (z - d)/z_0]^2} \tag{8}$$

where U_z is the wind speed above the canopy (m s^{-1}), which is derived from the measured wind speed at a 2.0 m height (u_2). z is usually equal to the canopy height (m), z_0 is the roughness height (usually 0.1 * H, where H is the canopy height) and d is the displacement height (0.75 * H).

2.5. Data Analysis

In the present study, significant differences in daily E_c, G_c, and Ω between the two growing seasons and the two planted forests were tested using two-way repeated measures ANOVA. The differences in the micrometeorological variables during the two growing seasons of 2020 and 2021 were testing using a paired samples t-test. Then, the relationships between PAR, VPD, and E_c were mainly analyzed using an exponential threshold function after partitioning the data into two classes (stressed water conditions: REW < 0.40 and non-stressed water conditions: REW > 0.40), as follows:

$$E_c = a\left(1 - e^{-bx}\right) \tag{9}$$

where a and b are the fitting parameters, and x is the corresponding meteorological variable.

The response of G_c to VPD was quantified using a linear logarithmic function [34]:

$$G_c = -m \ln VPD + G_{ref} \tag{10}$$

where m is the canopy conductance's sensitivity to VPD and G_{ref} is the reference canopy conductance when VPD = 1 kPa. An upper boundary line was derived based on the data of

E_c at least one standard deviation greater than the mean E_c of each *VPD* (0.5 kPa) interval, from which the parameters of m and G_{ref} were determined.

All statistical analyses were performed using the SPSS 21 software program (SPSS Inc., Chicago, IL, USA) and all the figures were created using Sigmaplot 11.0 software (Hearne Scientific Software Plc, Melbourne, Australia).

3. Results

3.1. Environmental Variables

During the growing season (May–September), the total precipitation was 256 and 283 mm in 2020 and 2021, respectively (Figure 2), accounting for 88.90% and 98.31% of the long-term average precipitation (288 mm, 1980–2020). The photosynthetically active radiation was 402.47 ± 124.65 μmol m^{-2} s^{-1} and 442.56 ± 126.41 μmol m^{-2} s^{-1}, whereas *PET* averaged 3.65 ± 1.36 mm d^{-1} and 3.72 ± 1.30 mm d^{-1} during the growing season in 2020 and 2021, respectively (Figure 2a). During both growing seasons, daily air temperatures displayed marked seasonal variations, with a range of 5.1 to 22.6 °C and 1.5 to 23.1 °C, and mean values of 15.14 ± 3.88 °C and 15.02 ± 4.28 °C, in 2020 and 2021, respectively (Figure 2b). The average *VPD* was 0.63 ± 0.35 kPa and 0.61 ± 0.32 kPa, whereas u_2 was 3.11 ± 1.56 m s^{-1} and 3.31 ± 1.77 m s^{-1} in 2020 and 2021, respectively (Figure 2c). The mean soil water content at the 0–100 cm soil layer was 8.03% for *U. pumila* and 5.09% for *C. korshinskii* in 2020, and 10.42% for *U. pumila* and 5.33% for *C. korshinskii* in 2021. The soil water content in the 0–40 cm layer increased quickly after precipitation events, whereas that of the 40–100 cm layer was stable during the growing season. In the vertical profile, the soil water content in the 40–100 cm layer was lower than that in other profiles in *U. pumila*, whereas the lowest soil water content occurred at 20–40 cm in *C. korshinskii* (Figure 3).

3.2. Canopy Transpiration per Unit Ground Area

During the measurement period, E_c for the *U. pumila* stand ranged from 0.08 to 1.77 mm d^{-1}, with a mean value of 0.55 ± 0.34 mm d^{-1} in 2020; whereas, in 2021, E_c ranged from 0.08 to 2.57 mm d^{-1}, with a mean value of 0.66 ± 0.32 mm d^{-1}. For *C. korshinskii*, E_c ranged from 0.10 to 1.41 mm d^{-1} and from 0.21 to 1.30 mm d^{-1}, with mean values of 0.74 ± 0.26 mm d^{-1} and 0.77 ± 0.24 mm d^{-1} in 2020 and 2021, respectively (Figure 4a,c). The accumulated E_c over the entire growing season was 83.72 and 113.65 mm for *U. pumila* in 2020 and 2021, respectively, accounting for 32.70% and 40.14% of the precipitation over the same period. The accumulated E_c was 101.29 and 117.77 mm for the *C. korshinskii* stand in 2020 and 2021, respectively, accounting for 39.57% and 41.60% of the precipitation over the same period (Figure 4b,d).

3.3. Response of Canopy Transpiration to Environmental Variables

Statistical analysis indicated that the positive effect of environmental variables on E_c had a ranking of $\theta > T_a > PAR > u_2 > VPD > PET$ in 2020 and $PAR > u_2 > \theta > T_a > P > PET$ in 2021 for *U. pumila*. The positive effect of environmental variables on E_c had a ranking of $PAR > \theta > u_2 > T_a > VPD > P$ in 2020 and $P > \theta > T_a > VPD > PAR > u_2 > PET$ in 2021 for *C. korshinskii* (Table 2). This suggests that *PAR*, θ, and *VPD* were major determinants of transpiration for both *U. pumila* and *C. korshinskii*.

The E_c responded to REW following a quadratic polynomial function for *U. pumila* both in 2020 and 2021 (Figure 5a,b). E_c reached the maximum values when REW = 0.4–0.5 in 2020 and REW = 0.4 in 2021. The E_c responded to REW following a saturated exponential function for *C. korshinskii* both in 2020 and 2021 (Figure 5c,d). E_c rapidly increased with rising REW when REW < 0.4, E_c then increased slowly when REW > 0.4, and, finally, E_c tended to be saturated at 1.31 and 1.14 mm when REW was close to 1 in 2020 and 2021, respectively. Additionally, the response of E_c to *PAR* and *VPD* was affected by different soil water conditions. The E_c increased with an increasing *PAR* and *VPD* for *U. pumila* under REW < 0.40, while it tended to be saturated when *VPD* reached 2.0 kPa (Figure 6a,b). *PAR* and *VPD* explained 28% and 22% of the variation in E_c for *U. pumila* under REW < 0.40, respectively, whereas E_c maintained a high value and stability when REW > 0.4. The E_c increased significantly with an increasing *PAR* and *VPD* for *C. korshinskii* under different soil water conditions (Figure 6c,d). The E_c exhibited an exponentially saturating response to *VPD* and tended to level off at 1.5 kPa under REW < 0.4, and there was no obvious threshold value of *Ec* under REW > 0.4. *VPD* explained 16% and 81% of the variation in E_c for *C. korshinskii* under REW < 0.40 and REW > 0.4, respectively, whereas *PAR* explained 23% and 52% of the variation in E_c for *C. korshinskii* under REW < 0.40 and REW > 0.4, respectively.

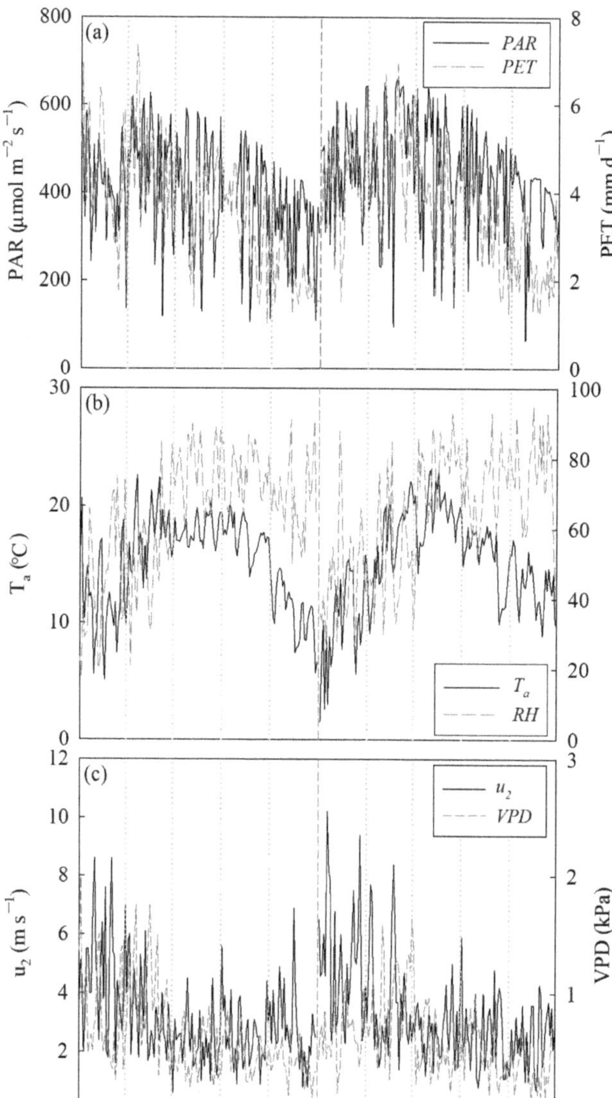

Figure 2. Daily variations in (**a**) photosynthetically active radiation (*PAR*) and daily potential evapotranspiration (*PET*), (**b**) temperature (T_a) and relative humidity (*RH*), (**c**) wind speed (u_2) and vapor pressure deficit (*VPD*).

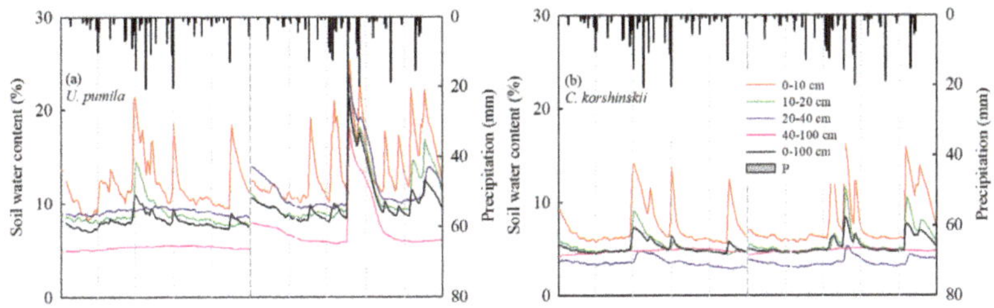

Figure 3. Daily precipitation and volumetric soil water content (0–100 cm) from May 1 to September 30 in (**a**) *U. pumila* and (**b**) *C. korshinskii*.

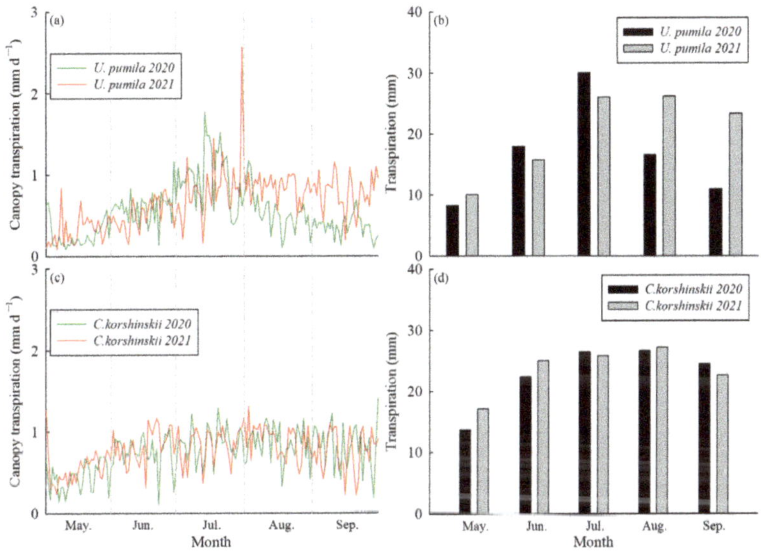

Figure 4. Variations in daily canopy transpiration of per ground area (E_c, **a**,**c**), and monthly transpiration (**b**,**d**) for *U. pumila* and *C. korshinskii*.

Table 2. Correlation coefficients between canopy transpiration (E_c) and environmental factors for *U. pumila* and *C. korshinskii* in 2020 and 2021.

		PAR	T_a	*RH*	*VPD*	u_2	*p*	*PET*	θ
U. pumila	*Ec* (2020)	0.522 **	0.615 **	-	0.366 **	−0.404 *	-	0.277 **	0.691 **
	Ec (2021)	0.513 **	0.369 **	-	-	−0.508 **	−0.253 **	0.176 *	0.502 **
C. korshinskii	*Ec* (2020)	0.475 **	0.362 **	-	0.354 **	−0.462 **	−0.262 **	-	0.474 **
	Ec (2021)	0.349 **	0.402 **	-	0.389 **	−0.330 **	−0.497 **	0.185 *	0.444 **

Note: *PAR* = photosynthetically active radiation; T_a = air temperature; *RH* = relative humidity; *VPD* = vapor pressure deficit; u_2 = wind speed; *P* = precipitation; *PET* = daily potential evapotranspiration; θ = soil water content. ** indicates $p < 0.01$, * indicates $p < 0.05$.

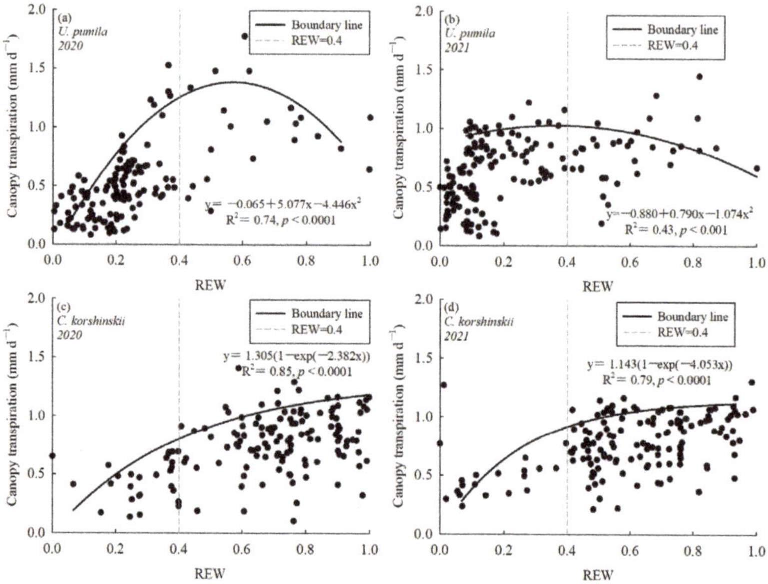

Figure 5. Response of canopy transpiration per ground area to the relative extractable soil water and boundary line analysis for *U. pumila* and *C. korshinskii* in 2020 (**a**,**c**) and 2021 (**b**,**d**).

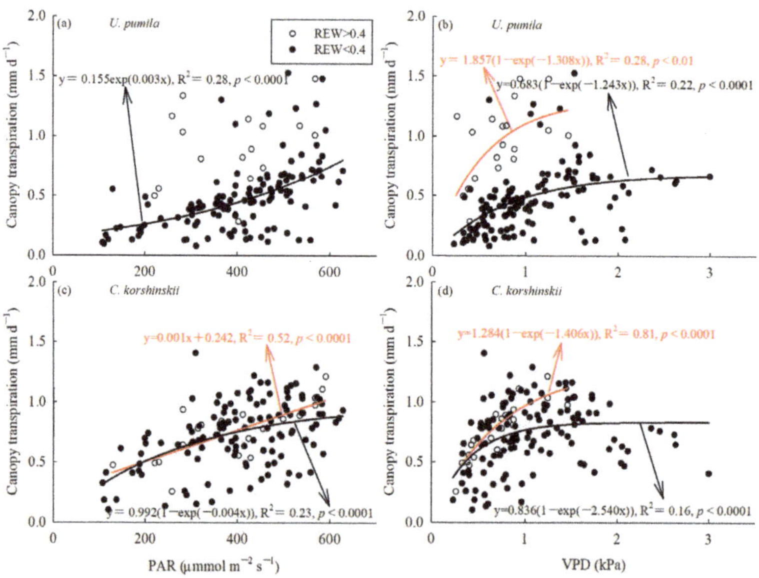

Figure 6. Relationships between canopy transpiration per ground area and daily environmental variables for *U. pumila* and *C. korshinskii* under stressed (REW < 0.40) and non-stressed (REW > 0.4) conditions across the measurement periods of 2020: (**a**,**c**) photosynthetically active radiation; (**b**,**d**) vapor pressure deficit.

3.4. Response of Canopy Stomatal Conductance to VPD

The G_c showed similar seasonal variations for *U. pumila* and *C. korshinskii* during 2020 and 2021. The maximum monthly G_c of *U. pumila* was 1.98 ± 1.07 mm s^{-1} and

2.02 ± 1.42 mm s^{-1} in July during 2020 and 2021, respectively. However, in both years, the G_c of *C. korshinskii* was significantly higher in July–September than in May–June ($F = 15.517$, $p < 0.001$ in 2020 and $F = 11.477$, $p < 0.001$ in 2021) (Figure 7). The G_c significantly decreased with an increase in *VPD* for both afforestation species (Figure 8). *VPD* explained 15% and 41% of total variation in G_c for *U. pumila* in 2020 and 2021, respectively (Figure 8a,b), and 45% and 51% of total variation in G_c for *C. korshinskii* in 2020 and 2021, respectively (Figure 8c,d). In addition, *m* had a significantly higher effect in 2021 than in 2020 for *U. pumila* (3.01 > 2.50) and *C. korshinskii* (2.83 > 1.96) after applying boundary line analysis, whereas no significant difference in the reference canopy stomatal conductance (G_{ref}) was observed for the two planted species. Moreover, the higher G_c of *U. pumila* and *C. korshinskii* when REW > 0.4 than when REW < 0.4 (*U. pumila*: 2.46 > 0.78, $F = 26.383$, $p < 0.001$ in 2020; *C. korshinskii*: 2.02 > 1.36, $F = 14.359$, $p < 0.001$ in 2021) indicates the significant effect of soil water content on canopy stomatal conductance. *m* was also higher under REW > 0.4 conditions than under REW < 0.4 conditions for both of the planted species (*U. pumila*: 3.30 > 0.87 in 2020 and 3.59 > 2.98 in 2021; *C. korshinskii*: 2.26 > 1.97 in 2020 and 3.70 > 2.08 in 2021) (Figure 8e–h).

Throughout the measurement period, the daily Ω of *U. pumila* ranged from 0.0001 to 0.0112, with mean values of 0.0015 ± 0.0014 and 0.0020 ± 0.0019 in 2020 and 2021, respectively (Figure 7). The daily Ω of *C. korshinskii* ranged from 0.0003 to 0.0145, with mean values of 0.0024 ± 0.0016 and 0.0024 ± 0.0018 in 2020 and 2021, respectively. No differences were found for the value of Ω between 2020 and 2021 for both *U. pumila* ($F = 5.896$, $p = 0.016$) and *C. korshinskii* ($F = 0.021$, $p = 0.884$).

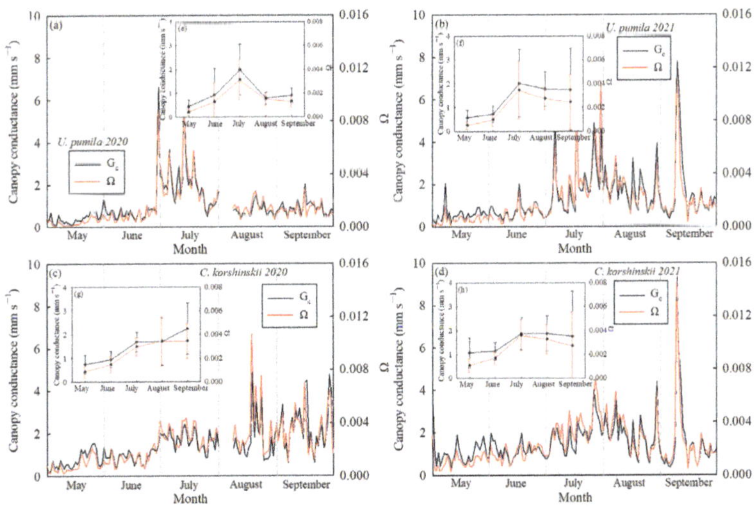

Figure 7. Seasonal variations in daily mean canopy conductance per ground area (G_c) and mean decoupling coefficient (Ω) for *U. pumila* and *C. korshinskii* during the growing season of 2020 (**a**,**c**) and 2021 (**b**,**d**); monthly G_c and Ω for *U. pumila* and *C. korshinskii* of 2020 (**e**,**g**) and 2021 (**f**,**h**).

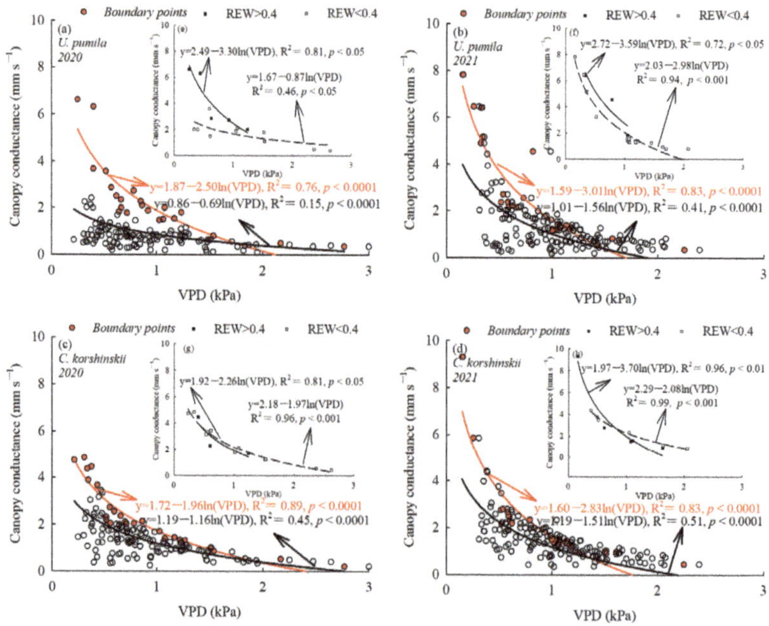

Figure 8. Relationships between canopy conductance per ground area and daily vapor pressure deficit in the measurement periods (black line) and boundary line analysis (red line) for *U. pumila* and *C. korshinskii* in 2020 (**a**,**c**) and 2021 (**b**,**d**); boundary line analysis under stressed (REW < 0.4) and non-stressed (REW > 0.4) conditions for *U. pumila* and *C. korshinskii* in 2020 (**e**,**g**) and 2021 (**f**,**h**).

4. Discussion

4.1. Canopy Transpiration of U. pumila and C. korshinskii

The E_c of the *U. pumila* stand in this study was 0.55 and 0.66 mm d^{-1}, which is similar to the values reported for Mongolian pine planted in the southeastern Keerqin Sandy Land [35] and half-mature Mongolian pine planted in the Rare Psammophytes Protection Botanical Base [36], but significantly lower than those of Chinese pine (0.7 mm d^{-1}), Mongolian pine (1.1 mm d^{-1}), and *Populus* × *xiaozhuanica* plantations (1.2–1.5 mm d^{-1}) planted in the Keerqin Sandy Land [16]; *Populus simonii* (175.2 kg d^{-1}) planted on the Loess Plateau [37]; and *Populus euphratica* Oliv. planted in the lower reaches of the Heihe River Basin [38]. The E_c of the *C. korshinskii* stand in this study was 0.74 and 0.77 mm d^{-1} in the two seasons, respectively. These values are lower than those for the same species planted in the Loess Plateau under different precipitation regions [37,39]. Compared with other major planted shrubs, a similar range of transpiration was reported for a *Vitex negundo* L. plantation (107.21 mm cumulative in the growing season), but a higher transpiration of *Hippophae rhamnoides* L. was found in the Loess Plateau (122.33–180.28 mm cumulative in the growing season) [11]. This may be related to the higher evaporative demand in drier and colder regions and the lower soil water availability in the Bashang Plateau. The mean annual precipitation in our study site was 100–150 mm lower than that for planted forests in the Keerqin Sandy Land and the Loess Plateau. In addition, a relatively lower soil water content may also increase the hydraulic resistance of the soil–root system, preventing water movement between soil and plant leaves, which ultimately decreases plantations' transpiration rate [40]. Finally, short heights and small sapwood areas, resulting from severe canopy dieback and planted forest degradation, were found for the *U. pumila* stand in this study (Table 2), and canopy transpiration was thus exhibited at a lower level.

E_c responded to REW as a saturated exponential function for *C. korshinskii* in this study (Figure 5c,d), which was in agreement with results reported for broadleaved and coniferous trees under different climates [41], a larch plantation in the semiarid Northwest China [42], a black locust plantation [43], and *C. korshinskii* [44] on the Loess Plateau. However, the REW threshold in this study is similar to or relatively higher than those in the studies reported above. Li et al. indicated that the REW threshold is generally higher under dry conditions than under wet conditions [23]. Averaged soil water content of the *C. korshinskii* stand was 5.09% in 2020 and 5.33% in 2021, which was lower than those in the plantations reported above (Figure 3b). E_c presented a polynomial pattern in REW for *U. pumila* in this study (Figure 5a,b), which was consistent with that in *Haloxylon ammodendron* and *Calligonum mongolicum* in Northwest China [45]. The reason that E_c was lower during the higher REW period is linked to the much lower VPD and T_a when compared to the dry period (Figures 2 and 6b). The daily E_c of both the *U. pumila* and *C. korshinskii* stands increased with increasing PAR and VPD, and E_c tended to be saturated at high VPD values. These results are similar to those reported for poplar trees [16], Mongolian pine [46], and other planted species [12]. In both plantations, however, the VPD threshold decreased as soil water content decreased, ranging from non-stressed to water stressed. An obvious VPD threshold was observed at approximately 2.0 kPa for the *U. pumila* stand and 1.5 kPa for the *C. korshinskii* stand when REW < 0.4. Liu et al. indicated that the utilization of surface soil water gradually increased with the increase in the degree of degradation of the agroforestry shelterbelts [47]. We then concluded that the *U. pumila* and *C. korshinskii* stands primarily obtained water from the shallow and middle soil layers because of the root distribution pattern and the degree of canopy dieback. As soil water decreased, the increasing whole-tree hydraulic resistance triggered canopy stomatal closure, and hence limited the response of transpiration to major environmental variables. In addition, the Ω values (Figure 6) were also lower than those reported for mature Mongolian pine (0.041), half-mature Mongolian pine (0.15), and young Mongolian pine (0.18) [36], and *Schima superba* (0.22) [48]. The smaller LAI of the *U. pumila* (0.31) and *C. korshinskii* (0.43) plantations in this study, compared to that of the Mongolian pine (0.96–1.54) in the Horqin desert and the Mu Us desert, indicates that decreasing the canopy resistance resulted in strong canopy–atmosphere coupling. Therefore, under high VPD conditions, canopy stomata must be closed to prevent water potential from dropping below the threshold and thus inhibiting tree transpiration. Moreover, a relatively lower VPD threshold under water stress was observed for both *C. korshinskii* (1.5 kPa) and *U. pumila* (2.0 kPa), indicating that the transpiration process of the two species was affected significantly by soil drought [49]. Planted species usually have a shallower fine root distribution and lower stem and leaf hydraulic conductivity [38,50]. As concluded above, the transpiration process and water-use strategy of *U. pumila* and *C. korshinskii* were revealed in the Bashang Plateau, and further research on the water-use pattern of typical planted forests should be undertaken in the future. However, in the present study, we did not investigate seasonal variations in LAI and its effects on canopy transpiration during the growing season for *U. pumila* and *C. korshinskii*. Additional studies on canopy transpiration and stomatal conductance combined with different plantation ages and densities would provide further insights into the ecohydrological processes and afforestation management on the Bashang Plateau.

4.2. Canopy Stomatal Conductance of U. pumila and C. korshinskii

The G_c values were significantly higher under non-stressed conditions than under stressed conditions for both plantations, indicating the strong limitations of soil water on canopy stomatal conductance. In addition, the significantly nonlinear negative relationship between G_c and VPD in both the *U. pumila* and *C. korshinskii* plantations suggested that stronger stomatal regulation of canopy transpiration responded to higher VPD. Stomatal closure thus played an important role in maintaining water status to avoid a catastrophic loss of xylem function [51]. This drought adaptation mechanism can be characterized as a water-saving strategy, consistent with those reported for *P. sylvestris* plantations [36,52,53].

We also found that the parameters of m and G_{ref} for the *U. pumila* plantation were higher than those for the *C. korshinskii* plantation (Figure 8). This was contrary to the findings of Song et al. [46], who observed that, in comparison with Mongolian pine (an introduced species), Chinese pine (a native species) had relatively higher values of m and G_{ref}. We further suggest that m and G_{ref} for both the *U. pumila* and *C. korshinskii* plantations were higher under REW > 0.4 than under REW < 0.4 (Figure 8e–h), indicating that the sensitivity of canopy stomatal conductance to *VPD* decreased with soil drought; this was consistent with the findings of Novick et al. [54], Jiao et al. [19] on the Loess Plateau, and Song et al. [35] in the Horqin desert. Soil drought can negatively affect the hydraulic conductance of soil-to-plant leaf transporting paths and decrease m values, thus limiting the transpiration rate. Grossiord et al. indicated that reduced soil water availability negated the benefits of stomatal and hydraulic adjustments and resulted in reduced transpiration in juniper [55]. Similar results were found for grapevines, both in pot grown and field-grown experiments [56]. Therefore, further decreasing hydraulic conductance to prevent xylem cavitation as soil water stress increases results in a less sensitive response of G_c to *VPD*. This may be one of the important reasons why planted forests are vulnerable to dieback and degradation. Further research on the hydraulic traits of typical planted forests should be undertaken in the future.

5. Conclusions

In the present study, canopy transpiration and stomatal conductance were estimated for an *U. pumila* and a *C. korshinskii* stand. During the growing season, canopy transpiration was 83.72 and 113.65 mm in 2020 and 2021, respectively, for the *U. pumila* stand, and 101.29 and 117.77 mm in 2020 and 2021, respectively, for the *C. korshinskii* stand. Transpiration of *U. pumila* and *C. korshinskii* were better correlated to soil water content, photosynthetically active radiation, and *VPD*. Both planted species showed reduced *VPD* sensitivity of canopy transpiration as soil water decreased, indicating the transpiration process was affected significantly by soil drought. The *VPD* threshold was 1.50 and 2.0 kPa for *U. pumila* and *C. korshinskii*, respectively. Furthermore, the transpiration of both planted species was mainly regulated by stomatal opening due to low values of the decoupling coefficients (0.0015 and 0.0020 for *U. pumila*; 0.0024 for *C. korshinskii*). Both *U. pumila* and *C. korshinskii* gradually displayed a reduced canopy stomatal conductance with increasing *VPD*, but decreasing stomatal sensitivity to *VPD* in response to soil drought was also observed. Based on the results above, we conclude that both *U. pumila* and *C. korshinskii* plantations exhibited water-saving strategies under a cold and arid environment on the Bashang Plateau. These findings provide a deeper understanding of transpiration dynamics of different forest functional types on the Bashang Plateau, which may be applied for afforestation management in semiarid regions.

Author Contributions: Conceptualization, Y.Z.; methodology, Y.Z. and W.L.; investigation, J.Z. and N.W.; data curation, H.Y. and B.X.; Supervision, X.W.; Writing—original draft, Y.Z.; Writing—review and editing, Y.Z. and W.L. All authors have read and agreed to the published version of the manuscript.

Funding: This research was funded by the National Natural Science Foundation of China (No. 42101019, 42001027, 51909052), Natural Science Foundation of Hebei Province (D2019205274, D2019403115, D2021403023), Science and Technology Project of Hebei Education Department (QN2019152, BJK2022022), Key Laboratory of Agricultural Water Resources & Hebei Key Laboratory of Agricultural Water-Saving, Center for Agricultural Resources Research, Institute of Genetics and Developmental Biology (KFKT201903), Science Foundation of Hebei Normal University (L2018B22), Special soft science research project of Hebei Provincial Science and Technology Plan (21557401D).

Informed Consent Statement: Not applicable.

Acknowledgments: We thank the reviewers and editors for their work.

Conflicts of Interest: The authors declare no conflict of interest.

References

1. Peng, S.; Piao, S.; Zeng, Z.; Ciais, P.; Zhou, L.; Li, L.Z.; Myneni, R.B.; Yin, Y.; Zeng, H. Afforestation in China cools local land surface temperature. *Proc. Natl. Acad. Sci. USA* **2014**, *111*, 2915–2919. [CrossRef] [PubMed]
2. Zhang, M.; Wei, X. Deforestation, forestation, and water supply. *Science* **2021**, *371*, 990–991. [CrossRef] [PubMed]
3. Zhang, B.; Tian, L.; Yang, Y.; He, X. Revegetation does not decrease water yield in the Loess Plateau of China. *Geophys. Res. Lett.* **2022**, *49*, e2022GL098025. [CrossRef]
4. Zhu, J.; Li, F.; Xu, M.; Kang, H.; Wu, X. The role of ectomycorrhizal fungi in alleviating pine decline in semiarid sandy soil of northern China: An experimental approach. *Ann. For. Sci.* **2008**, *65*, 1. [CrossRef]
5. Song, L.; Zhu, J.; Zheng, X.; Wang, K.; Lü, L.; Zhang, X.; Hao, G. Transpiration and canopy conductance dynamics of *Pinus sylvestris* var. *mongolica* in its natural range and in an introduced region in the sandy plains of Northern China. *Agric. For. Meteorol.* **2020**, *281*, 107830. [CrossRef]
6. Caldeira, M.C.; Lecomte, X.; David, T.S.; Pinto, J.G.; Bugalho, M.N.; Werner, C. Synergy of extreme drought and shrub invasion reduce ecosystem functioning and resilience in water-limited climates. *Sci. Rep.* **2015**, *5*, 15110. [CrossRef] [PubMed]
7. Wu, H.; Li, X.; Jiang, Z.; Chen, H.; Zhang, C.; Xiao, X. Contrasting water use pattern of introduced and native plants in an alpine desert ecosystem, Northeast Qinghai–Tibet Plateau, China. *Sci. Total Environ.* **2016**, *542*, 182–191. [CrossRef] [PubMed]
8. Anderegg, W.R.; Anderegg, L.D.; Kerr, K.L.; Trugman, A.T. Widespread drought-induced tree mortality at dry range edges indicates that climate stress exceeds species' compensating mechanisms. *Glob. Chang. Biol.* **2019**, *25*, 3793–3802. [CrossRef] [PubMed]
9. Ji, Y.; Zhou, G.; Li, Z.; Wang, S.; Zhou, H.; Song, X. Triggers of widespread dieback and mortality of poplar (*Populus* spp.) plantations across northern China. *J. Arid Environ.* **2020**, *174*, 104076. [CrossRef]
10. Yu, S.; Guo, J.; Liu, Z.; Wang, Y.; Ma, J.; Li, J.; Liu, F. Assessing the impact of soil moisture on canopy transpiration using a modified Jarvis-Stewart model. *Water* **2021**, *13*, 2720. [CrossRef]
11. Fang, W.; Lu, N.; Liu, J.; Jiao, L.; Zhang, Y.; Wang, M.; Fu, B. Canopy transpiration and stand water balance between two contrasting hydrological years in three typical shrub communities on the semiarid Loess Plateau of China. *Ecohydrology* **2019**, *12*, e2064. [CrossRef]
12. Urban, J.; Rubtsov, A.V.; Urban, A.V.; Shashkin, A.V.; Benkova, V.E. Canopy transpiration of a *Larix sibirica* and *Pinus sylvestris* forest in Central Siberia. *Agric. For. Meteorol.* **2019**, *271*, 64–72. [CrossRef]
13. Munoz-Villers, L.E.; Holwerda, F.; Alvarado-Barrientos, M.S.; Geissert, D.R.; Dawson, T.E. Reduced dry season transpiration is coupled with shallow soil water use in tropical montane forest trees. *Oecologia* **2018**, *188*, 303–317. [CrossRef]
14. He, Q.; Yan, M.; Miyazawa, Y.; Chen, Q.; Cheng, R.; Otsuki, K.; Yamanaka, N.; Du, S. Sap flow changes and climatic responses over multiple-year treatment of rainfall exclusion in a sub-humid black locust plantation. *For. Ecol. Manag.* **2020**, *457*, 117730. [CrossRef]
15. Zhang, Z.; Zhao, P.; Zhao, X.; Zhou, J.; Zhao, P.; Zeng, X.; Hu, Y.; Ouyang, L. The tree height-related spatial variances of tree sap flux density and its scale-up to stand transpiration in a subtropical evergreen broadleaf forest. *Ecohydrology* **2018**, *11*, e1979. [CrossRef]
16. Song, L.; Zhu, J.; Zhang, T.; Wang, K.; Wang, G.; Liu, J. Higher canopy transpiration rates induced dieback in poplar (*Populus* × *xiaozhuanica*) plantations in a semiarid sandy region of Northeast China. *Agric. Water Manag.* **2021**, *243*, 106414. [CrossRef]
17. Bosch, D.D.; Marshall, L.K.; Teskey, R. Forest transpiration from sap flux density measurements in a Southeastern Coastal Plain riparian buffer system. *Agric. For. Meteorol.* **2014**, *187*, 72–82. [CrossRef]
18. Chen, Z.; Zhang, Z.; Sun, G.; Chen, L.; Xu, H.; Chen, S. Biophysical controls on nocturnal sap flow in plantation forests in a semi-arid region of northern China. *Agric. For. Meteorol.* **2020**, *284*, 107904. [CrossRef]
19. Jiao, L.; Lu, N.; Fang, W.; Li, Z.; Wang, J.; Jin, Z. Determining the independent impact of soil water on forest transpiration: A case study of a black locust plantation in the Loess Plateau, China. *J. Hydrol.* **2019**, *572*, 671–681. [CrossRef]
20. Dymond, S.F.; Bradford, J.B.; Bolstad, P.V.; Kolka, R.K.; Sebestyen, S.D.; DeSutter, T.M. Topographic, edaphic, and vegetative controls on plant-available water. *Ecohydrology* **2017**, *10*, e1897. [CrossRef]
21. Granier, A.; Bréda, N.; Biron, P.; Villette, S. A lumped water balance model to evaluate duration and intensity of drought constraints in forest stands. *Ecol. Model.* **1999**, *116*, 269–283. [CrossRef]
22. MacKay, S.L.; Arain, M.A.; Khomik, M.; Brodeur, J.J.; Schumacher, J.; Hartmann, H.; Peichl, M. The impact of induced drought on transpiration and growth in a temperate pine plantation forest. *Hydrol. Process.* **2012**, *26*, 1779–1791. [CrossRef]
23. Li, Z.; Yu, P.; Wang, Y.; Webb, A.A.; He, C.; Wang, Y.; Yang, L. A model coupling the effects of soil moisture and potential evaporation on the tree transpiration of a semi-arid larch plantation. *Ecohydrology* **2017**, *10*, e1764. [CrossRef]
24. Chang, X.; Zhao, W.; Liu, H.; Wei, X.; Liu, B.; He, Z. Qinghai spruce (*Picea crassifolia*) forest transpiration and canopy conductance in the upper Heihe River Basin of arid northwestern China. *Agric. For. Meteorol.* **2014**, *198*, 209–220. [CrossRef]
25. Ouyang, L.; Zhao, P.; Rao, X.; Zhu, L.; Ni, G. Interpreting the water use strategies of plantation tree species by canopy stomatal conductance and its sensitivity to vapor pressure deficit in South China. *For. Ecol. Manag.* **2022**, *505*, 119940. [CrossRef]
26. Denham, S.O.; Oishi, A.C.; Miniat, C.F.; Wood, J.D.; Yi, K.; Benson, M.C.; Novick, K.A. Eastern US deciduous tree species respond dissimilarly to declining soil moisture but similarly to rising evaporative demand. *Tree Physiol.* **2021**, *41*, 944–959. [CrossRef] [PubMed]
27. Grossiord, C.; Sevanto, S.; Dawson, T.E.; Adams, H.D.; Collins, A.D.; Dickman, L.T.; Newman, B.D.; Stockton, E.A.; McDowell, N.G. Warming combined with more extreme precipitation regimes modifies the water sources used by trees. *New Phytol.* **2017**, *213*, 584–596. [CrossRef] [PubMed]

28. Wang, H.; Sun, B.; Yu, X.; Xin, Z.; Jia, G. The driver-pattern-effect connection of vegetation dynamics in the transition area between semi-arid and semi-humid northern China. *Catena* **2020**, *194*, 104713. [CrossRef]
29. Chu, X.; Zhan, J.; Wang, C.; Hameeda, S.; Wang, X. Households' Willingness to Accept Improved Ecosystem Services and Influencing Factors: Application of Contingent Valuation Method in Bashang Plateau, Hebei Province, China. *J. Environ. Manag.* **2020**, *255*, 109925. [CrossRef]
30. Granier, A. Evaluation of transpiration in a Douglas-fir stand by means of sap flow measurements. *Tree Physiol.* **1987**, *3*, 309–320. [CrossRef]
31. Allen, R.G.; Pereira, L.S.; Raes, D.; Smith, M. Crop evapotranspiration-Guidelines for computing crop water requirements-FAO Irrigation and drainage paper 56. *Fao Rome* **1998**, *300*, D05109.
32. Naithani, K.J.; Ewers, B.E.; Pendall, E. Sap flux-scaled transpiration and stomatal conductance response to soil and atmospheric drought in a semi-arid sagebrush ecosystem. *J. Hydrol.* **2012**, *464*, 176–185. [CrossRef]
33. Kumagai, T.O.; Saitoh, T.M.; Sato, Y.; Morooka, T.; Manfroi, O.J.; Kuraji, K.; Suzuki, M. Transpiration, canopy conductance and the decoupling coefficient of a lowland mixed dipterocarp forest in Sarawak, Borneo: Dry spell effects. *J. Hydrol.* **2004**, *287*, 237–251. [CrossRef]
34. Oren, R.; Sperry, J.; Katul, G.; Pataki, D.; Ewers, B.; Phillips, N.; Schäfer, K. Survey and synthesis of intra-and interspecific variation in stomatal sensitivity to vapour pressure deficit. *Plant Cell Environ.* **1999**, *22*, 1515–1526. [CrossRef]
35. Song, L.; Zhu, J.; Li, X.; Wang, K.; Wang, G.; Sun, H. Transpiration of *Pinus sylvestris* var. *mongolica* trees at different positions of sand dunes in a semiarid sandy region of Northeast China. *Trees* **2022**, *36*, 749–762. [CrossRef]
36. Deng, J.; Yao, J.; Zheng, X.; Gao, G. Transpiration and canopy stomatal conductance dynamics of Mongolian pine plantations in semiarid deserts, Northern China. *Agric. Water Manag.* **2021**, *249*, 106806. [CrossRef]
37. Wang, S.; Fan, J.; Ge, J.; Wang, Q.; Fu, W. Discrepancy in tree transpiration of *Salix matsudana*, *Populus simonii* under distinct soil, topography conditions in an ecological rehabilitation area on the Northern Loess Plateau. *For. Ecol. Manag.* **2019**, *432*, 675–685. [CrossRef]
38. Li, W.; Si, J.; Yu, T.; Li, X. Response of *Populus euphratica* Oliv. sap flow to environmental variables for a desert riparian forest in the Heihe River Basin, Northwest China. *J. Arid Land* **2016**, *8*, 591–603. [CrossRef]
39. Jian, S.; Zhao, C.; Fang, S.; Yu, K. Evaluation of water use of *Caragana korshinskii* and *Hippophae rhamnoides* in the Chinese Loess Plateau. *Can. J. For. Res.* **2015**, *45*, 15–25. [CrossRef]
40. Ghimire, C.P.; Bruijnzeel, L.A.; Lubczynski, M.W.; Zwartendijk, B.W.; Odongo, V.O.; Ravelona, M.; Van Meerveld, H. Transpiration and stomatal conductance in a young secondary tropical montane forest: Contrasts between native trees and invasive understorey shrubs. *Tree Physiol.* **2018**, *38*, 1053–1070. [CrossRef]
41. Granier, A.; Loustau, D.; Bréda, N. A generic model of forest canopy conductance dependent on climate, soil water availability and leaf area index. *Ann. For. Sci.* **2000**, *57*, 755–765. [CrossRef]
42. Wang, Y.; Wang, Y.; Li, Z.; Yu, P.; Han, X. Interannual variation of transpiration and its modeling of a larch plantation in semiarid Northwest China. *Forests* **2020**, *11*, 1303. [CrossRef]
43. Wu, Y.; Huang, M.; Warrington, D.N. Black locust transpiration responses to soil water availability as affected by meteorological factors and soil texture. *Pedosphere* **2015**, *25*, 57–71. [CrossRef]
44. She, D.; Xia, Y.; Shao, M.; Peng, S.; Yu, S. Transpiration and canopy conductance of *Caragana korshinskii* trees in response to soil moisture in sand land of China. *Agrofor. Syst.* **2013**, *87*, 667–678. [CrossRef]
45. Ji, X.; Zhao, W.; Kang, E.; Jin, B.; Xu, S. Transpiration from three dominant shrub species in a desert-oasis ecotone of arid regions of Northwestern China. *Hydrol. Process.* **2016**, *30*, 4841–4854. [CrossRef]
46. Song, L.; Zhu, J.; Zheng, X.; Wang, K.; Zhang, J.; Hao, G.; Wang, G.; Liu, J. Comparison of canopy transpiration between *Pinus sylvestris* var. *mongolica* and *Pinus tabuliformis* plantations in a semiarid sandy region of Northeast China. *Agric. For. Meteorol.* **2022**, *314*, 108784. [CrossRef]
47. Liu, Z.; Jia, G.; Yu, X. Variation of water uptake in degradation agroforestry shelterbelts on the North China Plain. *Agric. Ecosyst. Environ.* **2020**, *287*, 106697. [CrossRef]
48. Zhang, Z.Z.; Zhao, P.; McCarthy, H.R.; Zhao, X.H.; Niu, J.F.; Zhu, L.W.; Ni, G.Y.; Ouyang, L.; Huang, Y.Q. Influence of the decoupling degree on the estimation of canopy stomatal conductance for two broadleaf tree species. *Agric. For. Meteorol.* **2016**, *221*, 230–241. [CrossRef]
49. Du, S.; Wang, Y.; Kume, T.; Zhang, J.; Otsuki, K.; Yamanaka, N.; Liu, G. Sapflow characteristics and climatic responses in three forest species in the semiarid Loess Plateau region of China. *Agric. For. Meteorol.* **2011**, *151*, 1–10. [CrossRef]
50. Liu, Y.; Wang, A.; An, Y.; Lian, P.; Wu, D.; Zhu, J.; Meinzer, F.C.; Hao, G. Hydraulics play an important role in causing low growth rate and dieback of aging *Pinus sylvestris* var. *mongolica* trees in plantations of Northeast China. *Plant Cell Environ.* **2018**, *41*, 1500–1511. [CrossRef]
51. Gao, J.; Zhao, P.; Shen, W.; Niu, J.; Zhu, L.; Ni, G. Biophysical limits to responses of water flux to vapor pressure deficit in seven tree species with contrasting land use regimes. *Agric. For. Meteorol.* **2015**, *200*, 258–269. [CrossRef]
52. Leo, M.; Oberhuber, W.; Schuster, R.; Grams, T.E.; Matyssek, R.; Wieser, G. Evaluating the effect of plant water availability on inner alpine coniferous trees based on sap flow measurements. *Eur. J. For. Res.* **2014**, *133*, 691–698. [CrossRef]
53. Song, L.; Zhu, J.; Li, M.; Zhang, J.; Zheng, X.; Wang, K. Canopy transpiration of *Pinus sylvestris* var. *mongolica* in a sparse wood grassland in the semiarid sandy region of Northeast China. *Agric. For. Meteorol.* **2018**, *250*, 192–201. [CrossRef]

54. Novick, K.A.; Ficklin, D.L.; Stoy, P.C.; Williams, C.A.; Bohrer, G.; Oishi, A.C.; Papuga, S.A.; Blanken, P.D.; Noormets, A.; Sulman, B.N. The increasing importance of atmospheric demand for ecosystem water and carbon fluxes. *Nat. Clim. Chang.* **2016**, *6*, 1023–1027. [CrossRef]
55. Grossiord, C.; Sevanto, S.; Borrego, I.; Chan, A.M.; Collins, A.D.; Dickman, L.T.; Hudson, P.J.; McBranch, N.; Michaletz, S.T.; Pockman, W.T. Tree water dynamics in a drying and warming world. *Plant Cell Environ.* **2017**, *40*, 1861–1873. [CrossRef] [PubMed]
56. Romero, P.; Botía, P.; Keller, M. Hydraulics and gas exchange recover more rapidly from severe drought stress in small pot-grown grapevines than in field-grown plants. *J. Plant Physiol.* **2017**, *216*, 58–73. [CrossRef] [PubMed]

forests

MDPI

Article

Differences in Transpiration Characteristics among Eucalyptus Plantations of Three Species on the Leizhou Peninsula, Southern China

Zhichao Wang [1,2], Siru Liu [3], Yuxing Xu [1,2], Wankuan Zhu [1,2] and Apeng Du [1,2,*]

1 Research Institute of Fast-Growing Trees (RIFT), Chinese Academy of Forestry (CAF), Zhanjiang 524022, China
2 Guangdong Zhanjiang Eucalyptus Plantation Ecosystem Research Station, Zhanjiang 524022, China
3 South Subtropical Crops Research Institute, CATAS, Zhanjiang 524091, China
* Correspondence: cercdap@caf.ac.cn; Tel.: +86-759-3382163; Fax: +86-759-3380674

Abstract: How much transpiration water consumption varies between eucalyptus species is unknown, making the suitability of a particular eucalyptus species for large-scale planting in a given area, or whether interspecific differences need to be taken into account for eucalyptus water consumption estimates, uncertain. Here, *Eucalyptus camaldulensis* Dehnh. (Ec), *Eucalyptus pellita* F. v. Muell. (Ep), the most resistant species, and *Eucalyptus urophylla* S.T. Blake × *Eucalyptus grandis* Hill ex Maiden (Eug), the most widely planted species, were monitored for sap flow. Their stand transpiration was also estimated and its relationship to various influencing factors analyzed for the same stand age and site, and predictive models for daily transpiration (T) developed. The results showed that the T of all eucalyptus species was jointly influenced by meteorological factors, soil water content (SWC), and leaf area index (LAI), with great variation in the T response to each influencing factor among species. Accordingly, we developed species-specific transpiration prediction models that could adequately explain the changed T of each species (R^2-values: 0.863–0.911). There were significant differences in the stand daily mean sap flow density (J_C) and transpiration among the three species. Although Ec had a significantly lower J_C than Ep, it was significantly higher than Eug on all timescales, where the mean annual J_C of Ep (0.11 cm min^{-1}) was 1.4 and 2.6 times that of Ec (0.08 cm min^{-1}) and Eug (0.042 cm min^{-1}), respectively. Transpiration of Eug was significantly less than Ep, but significantly greater than Ec on all timescales, where the annual transpiration of Ep (743.41 mm) was 2.4 and 1.5 times that of Ec (311.52 mm) and Eug (493.58 mm), respectively. These results suggest that interspecific differences cannot be ignored when estimating transpiration rates in Chinese eucalyptus plantations, whose amount of water use should be considered when choosing the most optimal species to plant regionally.

Keywords: eucalyptus species; influencing factors; sap flow density; transpiration model; water consumption

Citation: Wang, Z.; Liu, S.; Xu, Y.; Zhu, W.; Du, A. Differences in Transpiration Characteristics among Eucalyptus Plantations of Three Species on the Leizhou Peninsula, Southern China. *Forests* **2022**, *13*, 1544. https://doi.org/10.3390/f13101544

Academic Editors: Yanhui Wang, Karl-Heinz Feger and Lulu Zhang

Received: 15 July 2022
Accepted: 19 September 2022
Published: 21 September 2022

1. Introduction

Eucalyptus is among the world's most important tree species cultivated for timber, and over the last 20 years, its plantations in China have expanded to 5.6 million hectares due to its rapid growth performance, high wood yield and strong economic value [1]. However, there is widespread debate and concern in the public sphere and industry regarding the large-scale planting of eucalyptus because of its impact on the regional water cycle and groundwater [2–4]. Previous studies had proposed that the higher water consumption by eucalyptus would result in lower water yield in the watershed and the depletion of groundwater resources [5–8], while others argued that the water used by eucalypt plantations does not severely deplete the local water supply [9,10]. Such ongoing disputes have severely restricted the sustainable development of the eucalyptus industry in some parts of southern China. Therefore, it is imperative that the water use dynamics of eucalyptus be studied using rigorous science.

Transpiration by trees is the most important hydrological process in forest ecosystems and a major cause of water loss from forest ecosystems [11,12]. Therefore, accurate estimation of the transpiration of eucalyptus plantations is crucial for resolving the current debate about their impact on regional water resources [11,13]. Many studies done in China have estimated transpiration for eucalypts, but most of those focused on the main cultivar, *Eucalyptus urophylla* S.T. Blake × *Eucalyptus grandis* Hill ex Maiden (Eug) [3,14] or *Eucalyptus urophylla* S.T. Blake [15], leaving other eucalypt species understudied. With the expansion of eucalyptus plantations in China, the problems of instability and weak disease/pest resistance of monocultures have become increasingly prominent, hence, the diversification of eucalyptus species and popularization of highly resistant species has begun in China [16]. Nevertheless, the differences in transpiration between Eucalyptus species remain unclear. Only by clarifying these differences can we increase the accuracy of estimating regional water consumption by eucalyptus and accordingly assess the suitability of various species for planting.

Tree species identity largely determines the hydraulic structural properties, growth characteristics, leaf longevity, leaf area index, and sapwood area of a stand [17]. Specifically, a tree's hydraulic structure influences water transport, beginning with root uptake and ending with water dissipation at its crown foliage [18]. The sapwood area determines the potential area of water conduction in the trunk [19], while leaf area index directly determines the transpiration capacity of a tree stand's canopy [12,20]. Therefore, transpiration by forests tends to considerably vary depending on the tree species.

Because the estimation of transpiration and water consumption in forests on a large spatial scale often involves numerous complex tree species, it is meaningful to explore the differences in transpiration water consumption characteristics among different tree species to increase the accuracy of forest transpiration estimation and refine water use estimation models. In addition, the transpiration of trees is closely related to many environmental factors such as atmospheric temperature, relative humidity, wind speed, vapor pressure deficit, solar radiation, and soil moisture [21–23], as well as canopy structural characteristics such as leaf area index (LAI), as demonstrated by many studies [20,24–27]. Differences among tree species in the biological characteristics can also lead to large differences in the magnitude of their transpiration response to changes in various factors that influence it [28]. In the context of global climate change, it is also essential to clarify the relationship between transpiration and influencing factors across tree species and to predict future changes in their water use via modeling.

Eucalyptus camaldulensis Dehnh (Ec), a perennial single-stemmed tree which can reach heights of 20 to 30 m, has a vigorously growing, deep, and extensive root system which reaches depths of at least 10–30 m with lateral or fan roots and sinker roots extending from the lateral roots, as well as vertical tap roots [29], which makes it highly drought- and wind-tolerant [30]. *Eucalyptus pellita* F. v. Muell. (Ep) is a well-formed medium-size-to-tall tree species that can reach 40 m or more in height and 1 m in diameter, with the advantages of rapid growth and high resistance to pests and diseases [31,32]. Because of their robust resistance to various factors, both eucalyptus species are increasingly favored for commercial planting in China, and the plantation area dedicated to them is gradually expanding. However, the transpiration and water consumption characteristics of Ec and Ep are not yet known in China, which generates considerable uncertainty regarding their suitability for planting on a large scale and their effect on water security. Therefore, here, we studied the sap flow characteristics of Ec and Ep of the same stand age, under the same site conditions, to estimate their stand-level transpiration as well as analyzing the relationship between their stand transpiration and various influencing factors; for this, Eug, the species with the largest planted area, served as the control. Our study had three objectives: (1) to define the differences in transpiration and sap flow characteristics among the three eucalyptus species; (2) to determine how climatic variables differ in their effects on transpiration by the three species and to develop transpiration prediction models for each; and (3) to test whether it is necessary for regional eucalyptus transpiration estimates to take into account differences in transpiration characteristics between species when choosing what to plant. The results are expected to provide a timely reference for later assessment of the impact that changes to the planting structure of eucalyptus will have on regional water resources, the selection of tree species for regional eucalyptus planting plans, and the refinement of models for estimating how much water is consumed by eucalyptus transpiration.

2. Materials and Methods

2.1. Experimental Site and Plantation

This study was conducted at the Eucalyptus Plantation Ecosystem Research Station (latitude 21°16′ N, longitude 110°05′ E), located in Zhanjiang City, Guangdong Province, China, lying at an elevation of 80–220 m a.s.l. (Figure 1), and which has a typical maritime monsoon climate. The average annual temperature and precipitation at this site is, respectively, 23.1 °C and 1319.5 mm (23.6 °C and 1462.3 mm for the monitoring year) (Figure 1). The extreme minimum temperature was 1.4 °C, which occurred in January, while the extreme maximum temperature was 38.1 °C, which occurred in July. In addition, precipitation is highly concentrated between May and October (the rainy season), which accounts for 77%–85% of the annual precipitation [33]. The soil here is classified as Rhodi-Udic Ferralosols according to the World Reference Base for Soil Resources [34], having been developed from weathered deposits of basalt, and it is acidic (average pH of 4.9 at a depth of 0–80 cm).

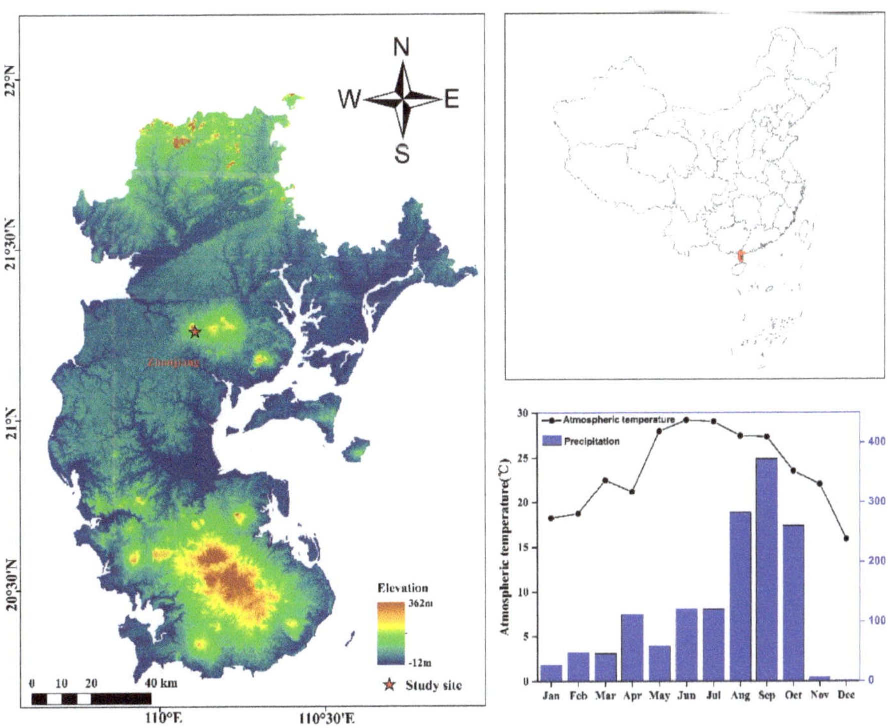

Figure 1. Location of the study area, topography, and mean monthly temperature and precipitation during the monitoring period.

The selected plantations of three eucalyptus species (Ep, Ec, and Eug) were all planted in 2008. Given the small distance (<1 km) between the three plantations, there is little difference in the meteorological conditions among them. Three experimental plots with an area of 20 m × 20 m were established in the center of each plantation for sap flow measurements. The trees in the plots were all numbered, and the height and the diameter at breast height (DBH) of each were measured. All three plantations had a flat topography. The details of the three experimental plantations of eucalyptus species are presented in Table 1.

Table 1. Summary of the stand characteristics of the three experimental plantations of eucalyptus species.

Index	Ep	Ec	Eug
Age (year)	12	12	12
Mean DBH (cm)	22.93 ± 1.78 a	19.01 ± 1.65 b	23.52 ± 1.19 a
Mean height (m)	21.67 ± 1.99 b	19.77 ± 0.93 b	27.1 ± 0.32 a
Stand LAI	2.43 ± 0.13 a	1.87 ± 0.05 c	2.08 ± 0.05 b
Stem density (plants/ha)	950	1050	1175
Mean sapwood area (cm^2/plant)	134.31 ± 18.15 b	70.78 ± 7.85 c	189.88 ± 12.02 a
Stand sapwood area (m^2/ha)	12.34	7.18	21.49

Note: Some values in this table are expressed as the mean ± standard error. Ep, Ec, and Eug represent the eucalyptus species of *E. pellita*, *E. camaldulensis*, and *E. urophylla* × *E. grandis*, respectively. Different lowercase letters indicate significant differences among eucalyptus species ($p < 0.05$). LAI: leaf area index.

2.2. Measurement of Environmental Factor Parameters

Meteorological parameters: The meteorological parameters were continuously measured in an open area near the experimental plantations. The air temperature (Ta, °C) and relative humidity (RH, %) were quantified using a thermo recorder (HMP155A, Vaisala, Helsinki, Finland); the solar radiation (Rs, W m^{-2}) was measured by a photon sensor (LI-200R, LICOR, Lincoln, NE, USA); the precipitation (*P*, mm) was recorded with a tilting rain gauge (TE525MM, Campbell, Logan, UT, USA); and the wind speed (WS, m s^{-1}) was measured with an anemometer (ATMOS 22, Decagon, Pullman, WA, USA). All measurements were recorded every minute, and the 10-min average value for each variable was stored in a data logger (CR3000, Campbell, USA). The vapor pressure deficit (VPD, kPa) was calculated from Ta and RH according to the following equation [35]:

$$\mathrm{VPD} = 0.611 \times e^{\frac{17.502 \times \mathrm{Ta}}{(\mathrm{Ta}+240.97)}} \times (1 - \mathrm{RH}) \quad (1)$$

Soil water content (SWC): The SWC was measured using six soil moisture sensors (CS616, Campbell, USA) buried at soil depths of 10, 20, 40, 60, 80, and 100 cm near the sample trees in each stand. The measured SWC data included the soil volumetric water content (cm^3 cm^{-3}) and was recorded every 30 min by a data logger (CR1000, Campbell) that was synchronized with the monitoring of the trees' sap flow rate, as described below.

2.3. Leaf Area Index (LAI) Measurement

The LAI was measured using a digital plant canopy analysis system (HemiView; Delta-T, Cambridge, UK). On a single day per week, more than 20 points were randomly selected in each experimental plantation to obtain an image of the upper canopy. Images were obtained using a camera with a fish-eye lens at sunset on a sunny day to minimize LAI measurement errors caused by strong light. Then, HemiView software, supported by the digital plant canopy analysis system, was used to analyze the forest canopy images and estimate the LAI at each location. Finally, the LAI for the three plantations was estimated for each day using an interpolation method.

2.4. Sap Flow Measurements

Based on the investigation of each tree in the experimental plots, 18 healthy trees representing the DBH size classes in each test plantation were chosen for the sap flow density measurements using thermal dissipation probes (SF-G, Ecomatik, Munich, Bavaria,

Germany), carried out from January to December 2020. Each set of probes consisted of two sensors (S_0: the heated probe, S_1: the reference probe) that were 23 mm in length and 2 mm in diameter. Each thermal dissipation probe was inserted into the active xylem at breast height (1.3 m) on the north side of the trunk and covered with radiation-proof aluminum foil to not only avoid incurring direct solar radiation and physical damage, but also to reduce the effects of ambient temperature fluctuation and precipitation. The S_0-S_1 connection provided the temperature differences (ΔT), these recorded every 30 s and averaged over 30 min by the data logger (CR1000, Campbell). The recorded ΔT was then converted into sap flow density (J_s) based on the calibration equation of Granier [36]:

$$J_s = 0.714 \times \left(\frac{\Delta T_{max} - \Delta T}{\Delta T} \right)^{1.231} \quad (2)$$

where J_s denotes the sap flow density (cm min^{-1}), ΔT_{max} denotes the value of ΔT when the sap flow is nil or close to zero. However, because the sap flow density cannot reach zero every day, we recorded the maximum daily ΔT over 7–10 days (ΔT_{max}) to avoid underestimation [37].

2.5. Sapwood Area Determination

Under the assumption that the stem cross-sections were circular, individual values of sapwood area (SA, cm^2) were calculated by measuring DBH, sapwood thickness, and bark thickness. To avoid damaging the sampled trees used for sap flow measurements, 18 non-monitored trees with different DBH classes outside the experimental plots were selected in each eucalyptus species plantation and felled, in order to obtain the cross-sectional at breast height. We identified the boundary between sapwood and heartwood by their visible wood color difference. Following Vertessy et al. [38], we used an allometric equation to relate SA to DBH:

$$SA = a \times DBH^b, \quad (3)$$

where SA denotes the sapwood area (cm^2), and a and b indicate the estimated equation parameters.

2.6. Estimation Methods for Individual Tree and Stand-Level Transpiration

Assuming a consistent sap flow rate throughout the sapwood, the individual tree transpiration E_i (g day^{-1}) was estimated as the product of sap flow density and the sapwood area, according to this published empirical calibration equation [36]:

$$E_i = J_d \times SA \times 60 \times 24, \quad (4)$$

where E_i denotes the individual tree transpiration (g day^{-1}), and J_d denotes the mean daily sap flow density of a monitored tree (cm min^{-1}).

The daily transpiration (T, mm day^{-1}) was calculated from individual trees to the stand level using the following equation [12]:

$$T = J_C \times \frac{\sum SA}{S \times 1000} \times 60 \times 24 \quad (5)$$

where T denotes the stand-level daily transpiration (mm day^{-1}), $\sum SA$ (cm^2) denotes the sum of the sapwood area of all trees in a given experimental plot, S (m^2) denotes the ground area of an experimental plot, and J_C (cm min^{-1}) denotes the stand daily mean sap flow density (the sapwood area-weighted average of J_d for each DBH class). J_C was calculated as:

$$J_C = \frac{\sum_i^n J_{di} \times SA_i}{\sum SA} \quad (6)$$

where J_{di} is the average J_d of the ith *DBH* class, SA_i is the total sapwood area of the sample trees of the ith *DBH* class in a given experimental plot, and $\sum SA$ (cm^2) denotes the sum of the sapwood area of all trees in an experimental plot.

2.7. Data Analysis and Statistics

One-way analysis of variance (ANOVA) followed by Tukey's honestly significant difference (HSD) test was used to examine the differences in J_C and T in different months or seasons among eucalyptus species. Regression analysis was used to examine the response of T to different environmental parameters and LAI, and the correspondence between the measured and predicted T values. Using regression, a power function curve was fitted to quantify the relationship between SA and DBH. Stepwise regression analysis was performed with 5% and 10% confidence levels as the threshold values for selection and rejection, respectively, to develop a multivariate linear model of T and f(X) (the relationship between the influencing factor and T is expressed by f(X)). The R v3.6.1 software platform (R Development Core Team 2019) was used to perform all statistical analyses and to draw the figures.

3. Results

3.1. Stands Structure and the Relationship between Sapwood Area and DBH

Despite the similarity in age and stand conditions of the three eucalyptus species, there were differences in the structural characteristics (Table 1). The DBH of Ep varied from 15.5 to 29.7 cm, whose mean value of 22.93 cm was similar to that of Eug (23.52 cm) ($p > 0.05$), but both were significantly greater than that of Ec (19.1 cm) ($p < 0.05$). The LAI of Eug stands (2.08) was significantly higher than that of Ec (1.87), but significantly lower than that of Ep (2.43). On average, Eug (27.1 m) was significantly taller than the other two eucalyptus species, whereas Ep and Ec were similar in height (Table 1).

The SA of each eucalyptus species was described well by a power function of DBH (Figure 2). Accordingly, the mean SA of Ep trees was 134.31 cm^2, which was significantly greater than that of Ec (70.78 cm^2), but significantly less than that of Eug (189.88 cm^2) (Table 1). The total SA of the three plantations was 12.34 m^2 ha^{-1} (Ep), 7.18 m^2 ha^{-1} (Ec), and 21.49 m^2 ha^{-1} (Eug) (Table 1).

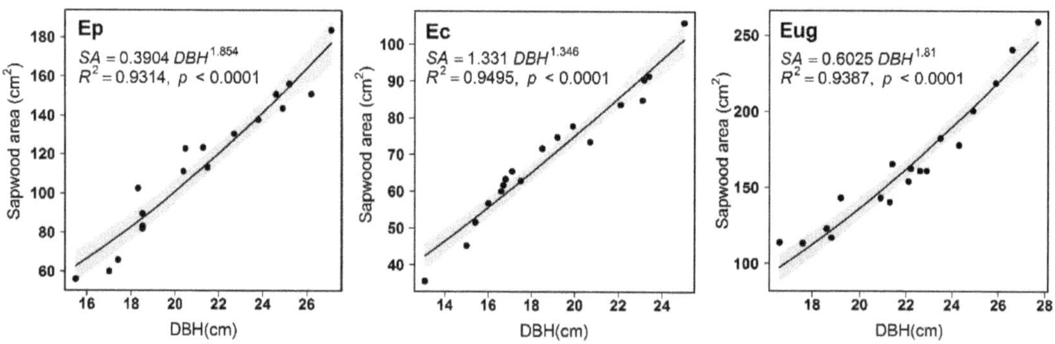

Figure 2. Variation in sapwood area as a function of individual trees' DBH of three eucalyptus species. Ep, Ec, and Eug represent the eucalyptus species of *E. pellita, E. camaldulensis and E. urophylla* × *E. grandis*, respectively. The shaded bands denote the 95% confidence intervals.

3.2. Characteristics of Sap Flow Density of the Three Eucalyptus Species

There are many similarities in the sap flow density characteristics of the three eucalyptus species. In both dry and wet seasons, all species showed a typical single-peaked curve during sunny days. The times when the sap flow started to increase, reached its peak, and began to decrease remained consistent for the three species (Figure 3). The Jc of the three species followed the same seasonal pattern: undergoing a gradual increase from January to May, with high values maintained in May, June, and July, and then declining from August to December (Figure 4B). The mean *Jc* was significantly higher in the wet season than in the dry season for all three eucalyptus species. In addition, the CV (coefficient of variation) of *Jc* for three eucalyptus species was similar throughout the year, at approximately 40%.

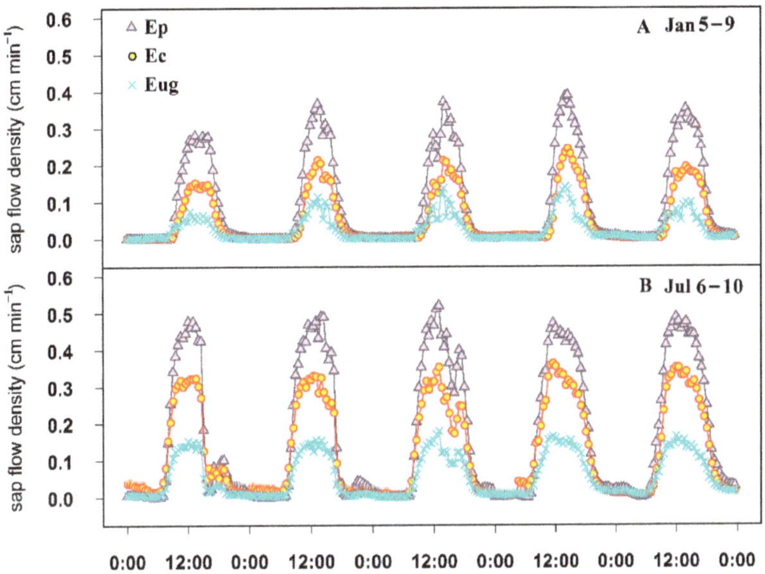

Figure 3. Diurnal variation of sap flow density in the three eucalyptus species plantations on sunny days during typical months for the (**A**) dry and (**B**) wet season. Ep, Ec, and Eug represent the eucalyptus species of *E. pellita*, *E. camaldulensis*, and *E. urophylla* × *E. grandis*, respectively.

However, many differences in sap flow density were evident between the three eucalyptus species. The daily peaks differed considerably on sunny days, with the highest daily peak found for Ep, followed by Ec, and least for Eug (Figure 3). The *Jc* over the entire year ranged from 0.0027 cm min^{-1} to 0.1937 cm min^{-1} for Ep, 0.0118 cm min^{-1} to 0.1394 cm min^{-1} for Ec, and 0.0049 cm min^{-1} to 0.0866 cm min^{-1} for Eug (Figure 4A). The annual mean *Jc* of Ec (0.080 cm min^{-1}) was significantly lower than that of Ep (0.11 cm min^{-1}), but significantly greater than that of Eug (0.042 cm min^{-1}) (Figure 4E). The same hierarchical pattern of mean *Jc* values for the three eucalyptus species was likewise observed in the dry (Figure 4C) and wet seasons (Figure 4D).

3.3. The Water Use Characteristics of the Three Species of Eucalyptus Stands

The total transpiration during the monitoring period was 743.41 mm for Ep, 311.52 mm for Ec, and 493.58 mm for Eug; these, respectively, amounting to 50.8%, 21.3%, and 33.8% of the annual precipitation. Throughout the year, the mean T for Eug reached 1.35 mm day^{-1}, which was significantly greater than Ec (0.85 mm day^{-1}) but significantly less than Ep (2.03 mm day^{-1}). This pattern also applied to the mean monthly transpiration and the mean T for all months, as well as for the dry and wet seasons except for April, August, and October, when no significant differences were found between the mean T of Eug and Ec (Table 2).

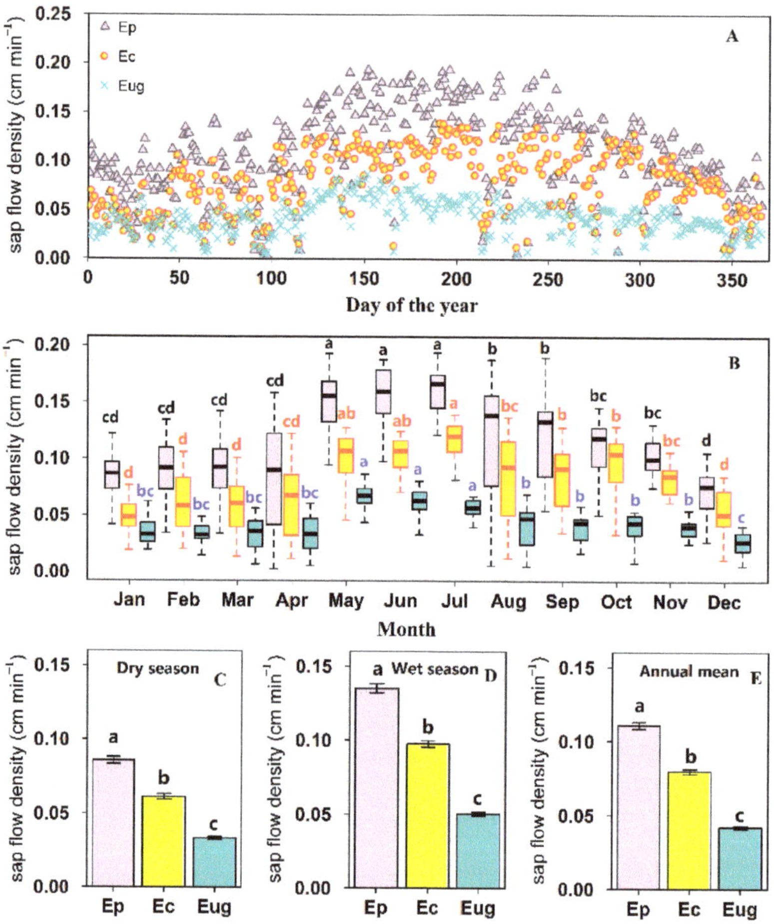

Figure 4. Panels (**A**,**B**) show the (**A**) variation of J_C and the (**B**) monthly mean for J_C in three eucalyptus species' plantations during the monitoring period. Panels (**C**–**E**) indicate the mean J_C of each species during the (**C**) dry season, (**D**) wet season, and (**E**) throughout the year. Error bars indicate the standard error. Different lowercase letters indicate significant differences ($p < 0.05$). Ep, Ec, and Eug represent the eucalyptus species of *E. pellita*, *E. camaldulensis*, and *E. urophylla* × *E. grandis*, respectively.

The T and monthly cumulative T for the three eucalyptus species' plantations all followed the same seasonal pattern of gradual increases from January to May, with large values in May, June, and July, before decreasing again from August to December (Table 2). The total transpiration during the wet season (from May to October) was 456.52 mm for Ep, 192.00 mm for Ec, and 298.20 mm for Eug, which collectively accounted for approximately 60% of their total annual transpiration and 37.5%, 15.8%, and 24.5% of simultaneous precipitation, respectively. However, the total T of each species during the dry season accounted for 117.7% (Ep), 49.1% (Ec), and 80.1% (Eug) of the simultaneous precipitation.

3.4. Factors Influencing Transpiration by the Three Eucalyptus Species

To elucidate the relationship between T and meteorological parameters, the effects of Rs, VPD, Ta, RH, and WS on T of the three eucalyptus species were analyzed (Figure 5). This analysis revealed that the T of all species significantly increased as a convex parabolic function of Rs (R^2 = 0.862 for Ep, 0.808 for Ec, and 0.764 for Eug, all $p < 0.001$) (Figure 5A) and VPD (R^2 = 0.674 for Ep, R^2 = 0.751 for Ec, and 0.639 for Eug, all $p < 0.001$) (Figure 5B), but significantly increased as a concave parabolic function of Ta (R^2 = 0.546 for Ep, 0.392 for Ec, and 0.461 for Eug, all $p < 0.001$) (Figure 5C). Additionally, the T of each species was significantly related to RH in a convex parabolic function that first increased and then decreased at a threshold value of 70% (R^2 = 0.392 for Ep, 0.407 for Ec, and 0.325 for Eug, all $p < 0.001$) (Figure 5D). The T showed a significant negative linear relationship with WS for each species (R^2 = 0.138 for Ep, 0.079 for Ec, and 0.089 for Eug, all $p < 0.001$) (Figure 5E). Among the five meteorological parameters, Rs best explained the variation in T of each eucalyptus species, which indicates its dominant control over transpiration in eucalyptus plantations in the study area.

Table 2. Summary of stand-level canopy transpiration values for the three eucalyptus species plantations at different time scales.

Species / Time Scale	Ep		Ec		Eug		
	Monthly Water Use (mm)	Daily Mean Water Use (mm/d)	Monthly Water Use (mm)	Daily Mean Water Use (mm/d)	Monthly Water Use (mm)	Daily Mean Water Use (mm/d)	Precipitation (mm)
Jan	47.85	$1.54 \pm 0.07^{cd/A}$	16.09	$0.52 \pm 0.03^{d/C}$	35.73	$1.15 \pm 0.07^{bc/B}$	27.6
Feb	48.97	$1.69 \pm 0.08^{cd/A}$	18.84	$0.65 \pm 0.05^{d/C}$	30.14	$1.04 \pm 0.06^{bc/B}$	48.9
Mar	49.37	$1.59 \pm 0.11^{cd/A}$	18.90	$0.61 \pm 0.05^{d/C}$	32.81	$1.06 \pm 0.08^{bc/B}$	46.6
Apr	45.63	$1.52 \pm 0.16^{cd/A}$	21.13	$0.70 \pm 0.06^{cd/B}$	31.41	$1.05 \pm 0.10^{bc/B}$	112.3
May	85.00	$2.74 \pm 0.09^{a/A}$	33.59	$1.08 \pm 0.04^{ab/C}$	66.24	$2.14 \pm 0.07^{a/B}$	58.8
Jun	84.76	$2.83 \pm 0.11^{a/A}$	32.42	$1.08 \pm 0.04^{ab/C}$	58.18	$1.94 \pm 0.09^{a/B}$	121.6
Jul	91.92	$2.97 \pm 0.06^{a/A}$	38.55	$1.24 \pm 0.03^{a/C}$	56.29	$1.82 \pm 0.04^{a/B}$	121.3
Aug	68.08	$2.20 \pm 0.18^{b/A}$	28.66	$0.92 \pm 0.07^{bc/B}$	40.57	$1.31 \pm 0.10^{b/B}$	282.8
Sep	66.67	$2.22 \pm 0.11^{b/A}$	27.85	$0.93 \pm 0.05^{b/C}$	38.43	$1.28 \pm 0.07^{b/B}$	373.3
Oct	60.09	$1.94 \pm 0.12^{bc/A}$	30.93	$1.00 \pm 0.06^{b/B}$	38.49	$1.24 \pm 0.08^{b/B}$	260.7
Nov	55.26	$1.84 \pm 0.06^{bc/A}$	26.78	$0.89 \pm 0.03^{bc/C}$	38.23	$1.27 \pm 0.05^{b/B}$	6.9
Dec	39.81	$1.28 \pm 0.08^{d/A}$	17.78	$0.57 \pm 0.04^{d/C}$	27.06	$0.87 \pm 0.05^{c/B}$	1.5
Monthly mean	61.95^{A}	—	25.96^{C}	—	41.13^{B}	—	121.86
Annual total	743.41	2.03 ± 0.04^{A}	311.52	0.85 ± 0.02^{C}	493.58	1.35 ± 0.03^{B}	1462.3
Dry season total	286.89	1.58 ± 0.04^{A}	119.52	0.66 ± 0.02^{C}	195.38	1.07 ± 0.03^{B}	243.8
Wet season total	456.52	2.48 ± 0.05^{A}	192.00	1.04 ± 0.02^{C}	298.20	1.62 ± 0.04^{B}	1218.5

Note: Some values in this table are the mean ± standard error. Ep, Ec, and Eug represent the eucalyptus species of *E. pellita*, *E. camaldulensis*, and *E. urophylla* × *E. grandis*, respectively. Different lowercase letters indicate significant differences between different months for each eucalyptus species ($p < 0.05$). Different capital letters indicate significant differences among eucalyptus species within the same month or season or year ($p < 0.05$).

The relationships between T and SWC are also presented in Figure 5F, which shows that the T of the three eucalyptus species plantations was highly dispersed for any given SWC due to the strong influence of other factors. However, when the SWC fell below a certain threshold value (19.3% for Ep, 17.5% for Ec, and 18.6% for Eug), the upper boundary lines indicate a significant non-linear increase of T with rising SWC, which can be well characterized by a saturated exponential equation (R^2 = 0.963 for Ep, 0.951 for Ec, and 0.970 for Eug, $p < 0.001$). Once the respective thresholds were exceeded, there was no longer a clear relationship between SWC and T.

The T and LAI data for the three eucalyptus species are plotted in Figure 6. Evidently, for any given LAI, the T of the three plantations showed a high degree of dispersion due to the strong influence of other factors. However, all species showed a clear Gompertz model curve for the upper boundary of T associated with LAI (R^2 = 0.941 for Ep, 0.961 for Ec, and 0.948 for Eug, all $p < 0.001$), which represents the relationship between T and LAI when no other factors are limiting (Figure 6). These curves show that there was an initial non-linear increase in the T of all three species with an increasing LAI that finally stabilized after reaching their respective potential maximum; this suggested an LAI threshold exists for each eucalyptus species, beyond which LAI will no longer act as a limiting factor for their T.

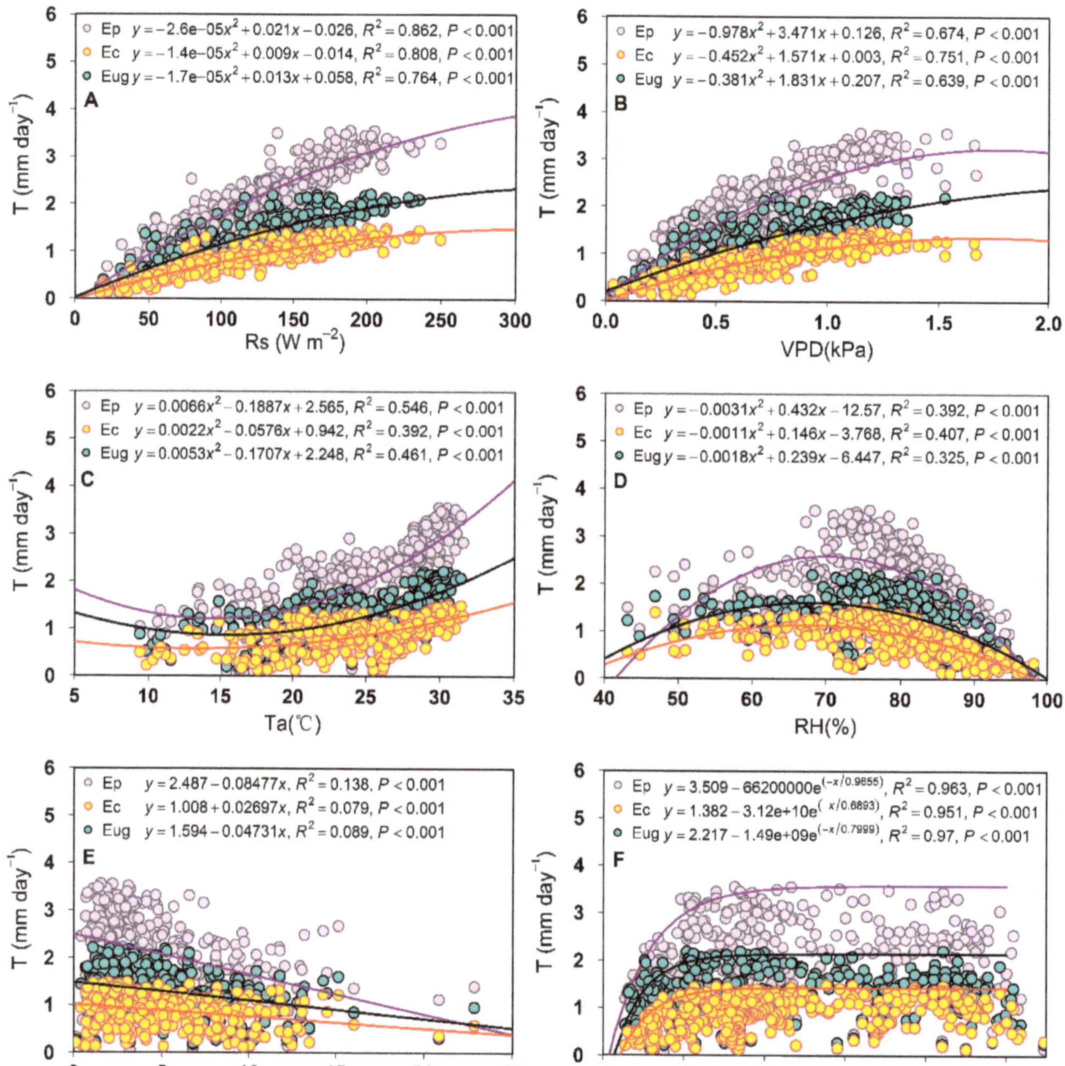

Figure 5. The relationships between the daily stand transpiration (T, mm day^{-1}) and six environmental factors: (**A**) solar radiation (Rs, W m^{-2}), (**B**) vapor pressure deficit (VPD, kPa), (**C**) air temperature (Ta, °C), (**D**) air relative humidity (RH, %), (**E**) wind speed (WS, m s^{-1}), and (**F**) soil moisture content (SWC, %) during the experimental periods. Ep, Ec, and Eug represent the eucalyptus species of *E. pellita*, *E. camaldulensis*, and *E. urophylla* × *E. grandis*, respectively.

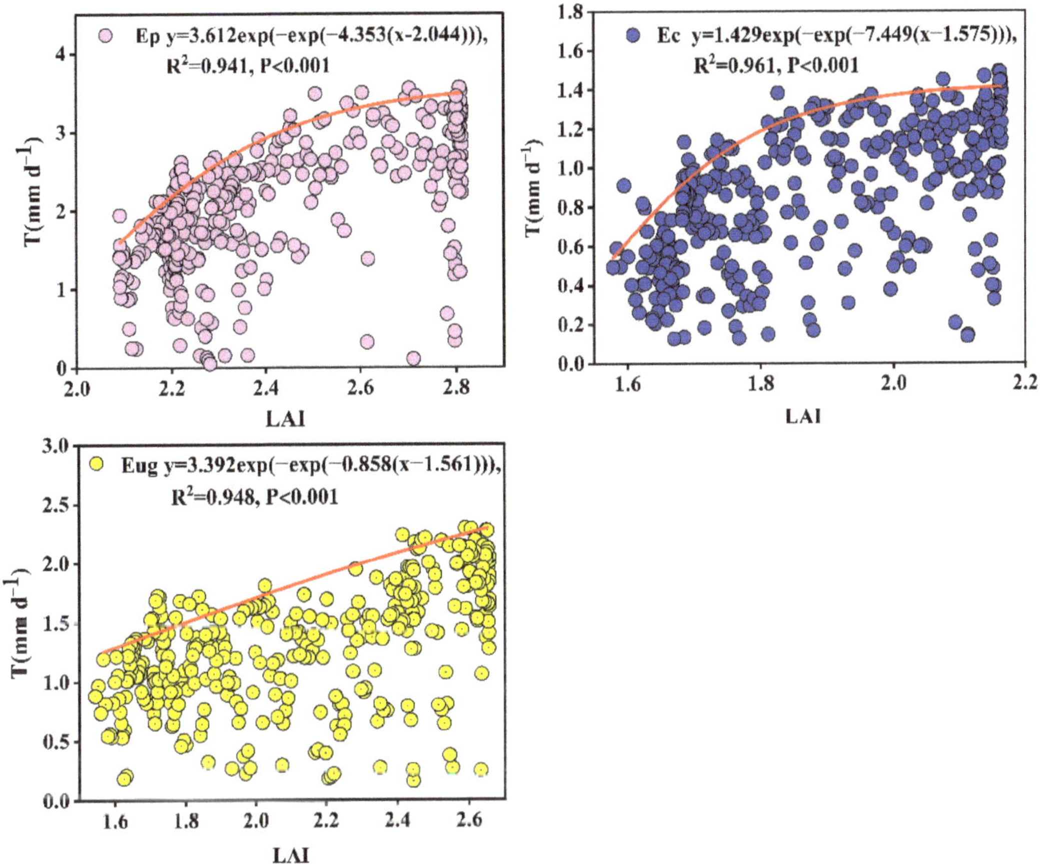

Figure 6. Response of the daily stand transpiration (T, mm day^{-1}) to the variation in canopy LAI for the three eucalyptus species (the red solid line is the upper boundary line). Ep, Ec, and Eug represent the eucalyptus species of *E. pellita*, *E. camaldulensis*, and *E. urophylla* × *E. grandis*, respectively.

3.5. Model to Estimate the Daily T of Each Eucalyptus Species

Our results showed that, for all three species, their T was influenced by the combination of Rs, VPD, Ta, RH, WS, SWC, and LAI. We established a relationship equation with T for each influence factor separately. Considering the interaction between many influencing factors and to further reveal their combined effect on T of the three stands, a stepwise regression model based on 10 months of data (one month of data from each of the dry and wet seasons was retained for model validation) was used to derive a multivariate relationship between T and f (Rs), f(VPD), f(Ta), f(RH), f(WS), f(SWC), and f(LAI). A final integrated model for predicting T for each eucalyptus species was constructed (Table 3).

The summed T for the omitted two months (set aside for validation) calculated using the predictive model was 82.13 mm for Eug, 124.30 mm for Ep, and 48.04 mm for Ec, which was 106.78%, 95.15%, and 96.50% of their actual measured values, respectively. The Nash and Sutcliffe coefficient was 0.897 for Eug, 0.938 for Ep, and 0.912 for Ec. Further, the slopes of the linear fit between the measured and predicted sets of values are 0.901 for Eug, 0.939 for Ep, and 0.915 for Ec, respectively, with corresponding R^2 values of 0.917, 0.951, and 0.945, respectively (Figure 7), indicating excellent agreement between these two sets of values. These results thus provide ample evidence that respective predictive models built for the T of the three eucalyptus species are accurate.

Table 3. The integrated model for predicting daily stand transpiration (T) of three eucalyptus species.

Eucalyptus Species	Predictive Model of T	R^2	n	Sig.
Ep	$T = -1.9 \times 10^{-5}Rs^2 + 0.0153Rs + 2.05 \times 10^{-3}Ta^2 - 0.0586Ta - 8.57 \times 10^{-4}RH^2 + 0.119 RH + 1.03\times 10^7 exp(-SWC/0.9655) + 0.161exp(-exp(-4.35(LAI - 2.044))) - 3.515$	0.911	305	0.000
Ec	$T = -9.19 \times 10^{-6}Rs^2 + 5.91 \times 10^{-3}Rs - 0.224VPD^2 + 0.78VPD - 6.22 \times 10^{-4}Ta^2 + 0.016 Ta + 1.97\times 10^9 exp(-SWC/0.6893) + 0.514 exp(-exp(-7.45(LAI - 1.575))) - 0.568$	0.872	305	0.000
Eug	$T = -1.01 \times 10^{-5}Rs^2 + 7.76 \times 10^{-3}Rs - 0.202VPD^2 + 0.97VPD + 5.85 \times 10^8 exp(-SWC/0.7999) + 0.822exp(-exp(-0.858(LAI - 1.561))) - 0.541$	0.863	305	0.000

Note: Ep, Ec, and Eug represent the eucalyptus species of *E. pellita*, *E. camaldulensis*, and *E. urophylla* × *E. grandis*, respectively. T: daily stand transpiration (mm day^{-1}), Rs: solar radiation (W m^{-2}), Ta: air temperature (°C), RH: relative humidity (%), VPD: vapor pressure deficit (kPa), SWC: soil water content (%), LAI: leaf area index, n: data for the multiple linear regression model, Sig.: significance of the data.

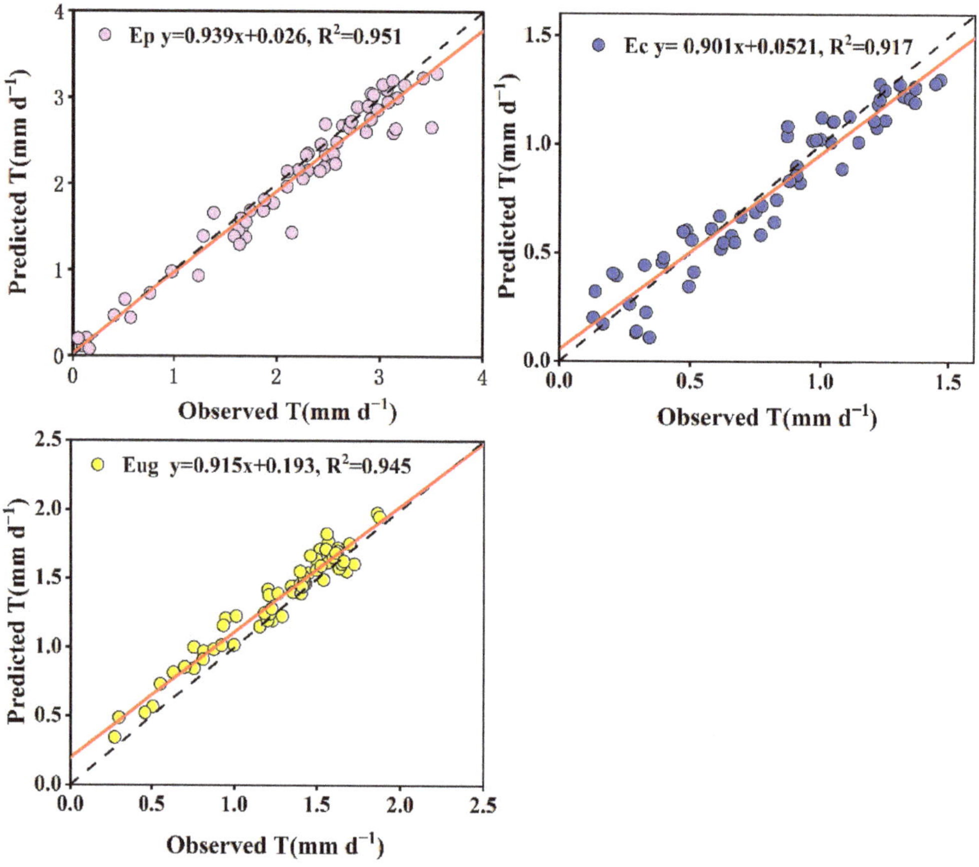

Figure 7. The predicted T (mm d^{-1}) plotted against the corresponding observed T (mm d^{-1}) for the three eucalyptus species. Note: The dotted line denotes the 1:1 relationship, and the solid line indicates the actual relationship between the predicted and observed values. Ep, Ec, and Eug represent the eucalyptus species of *E. pellita*, *E. camaldulensis*, and *E. urophylla* × *E. grandis*, respectively.

4. Discussion

4.1. Sap Flow Density and Water Use of the Three Eucalyptus Species

The sap flow is a key indicator of a tree's water use characteristics and the water transport mechanism [39,40]. Previous studies have reported that sap flow density is not only controlled by environmental factors [25,41,42], but is also closely related to the trees' physiological structure as well as stand characteristics [11,43]. Here, the sap flow density of three eucalyptus species' plantations showed a similar daily and seasonal variation pattern due to the same environmental conditions (Figures 3 and 4). Nevertheless, pronounced differences in sap flow density among the three eucalyptus species were also evident.

The peak daily sap flow and the mean sap flow density of Ec on all time scales were significantly greater than Eug, but significantly less than Ep (Figures 3 and 4). These differences were mainly caused by differing biological characteristics and stand structure among the three eucalyptus species. Applying a previous empirical formula [36], it was clear that sap flow density is proportional to canopy transpiration and inversely proportional to the area of trunk water transport. Being affected by LAI, the T of Ep was the greatest in this study, followed by Eug, and least for Ec (Table 2). However, Eug possesses the largest sapwood area and DBH, and its trunk harbors the largest water conduction area. These factors, coupled to its maximum height, results in the longest water transport path and the greatest resistance to the upward movement of water due to friction and gravity [44], but vice versa for Ec, resulting in a starkly lower sap flow density for Eug than Ec. Hence, due to the combined effects of LAI, DBH, tree height, and SA, the sap flow density of the eucalyptus species was ranked thus: Ep > Ec > Eug.

In addition, our results uncovered considerable differences in the transpiration characteristics of the three eucalyptus species. The annual transpiration of Ep was 743.41 mm, which was 2.4 and 1.5 times higher than that of Ec (311.52 mm) and Eug (493.58 mm), respectively. These were all within the range of other records for Eucalyptus plantations [3,45–47]. The same species pattern was also observed for T accumulation by month and during the dry and rainy seasons. The average daily transpiration on all time scales examined had this pattern of variation: the mean daily transpiration of Eug exceeded that of Ec, but less than that of Ep (Table 2). Although the research consensus to date is that the magnitude of tree transpiration is influenced by a combination of species [44,48,49], stand structural characteristics [12,48,50,51], and environmental factors [12,25,42,52], the differences in transpiration between the three eucalyptus species in this study mainly arose from differences in species and tree stand structural characteristics because their close proximity precluded any effects of meteorological conditions.

Many previous studies have reported positive relationships of tree growth characteristics (e.g., height, DBH, SA and above-ground biomass) and transpiration [11,43,50]. However, the transpiration of the three eucalyptus species in this study was not consistent with their DBH, tree height, or SA, perhaps because the conclusions drawn in previous studies were for the same species; that is, the pattern may not be consistent between species due to differences in their physiological characteristics. Work by Hatton et al. [53] confirmed that leaf water efficiency differed negligibly between various eucalyptus species in four disparate environments: tropical woodlands with distinct wet and dry seasons, forests with a Mediterranean climate, a woodland system with an even distribution of precipitation throughout the year, and a woodland growing on a saline gradient. This finding, combined with the fact that the magnitude of LAI of the three eucalyptus species' plantations we studied remained consistent with the magnitude of measured transpiration, suggests that LAI may be the determining factor in the production of transpiration differences between the three eucalyptus species plantations.

From the previous analysis, it is clear that the differences in transpiration between eucalyptus species measured under the same climatic and stand conditions in this study are considerable, and such differences are mainly due to the species characteristics and the differences in stand structure under its influence. As such, it is necessary for regional eucalyptus transpiration estimates to take into account differences in transpiration characteristics between species. The annual transpiration of each eucalyptus species also differed significantly from that of other local species such as *Cunninghamia lanceolate* (Lamb.) Hook. (20 years old, 552.1 mm) [11], *Pinus elliottii* Engelm × *Pinus caribaea* Morelet (10 years old, 483.24 mm) [54] and *Acacia mangium* Willd. (19 years old, 244.5 mm) [55]; indeed, even the annual transpiration of Ep was much greater than theirs. Therefore, planting different tree species will have different effects on the water cycle processes in the planting area. Moreover, in this study, although the annual transpiration of Ep, Ec, and Eug amounted to only 50.8%, 21.3%, and 33.8% of the annual precipitation, respectively, the percentage increased to 117.7% (Ep), 49.1% (Ec), and 80.1% (Eug) in the dry season. This implies that Ep plantations will consume large amounts of soil water or groundwater to meet their evaporative demands during the dry season, which cannot be adequately supplemented by precipitation. This lower input of water than output during the dry season can place a large hydrological burden in the planting area. However, this hydrological burden will be attenuated in the Eug plantation, or even not occur in the Ec plantation. There are some limitations in this paper, including that we only measured the transpiration, not the total evapotranspiration, in plantations of three Eucalyptus species, which would reduce our ability to assess the impact of tree species selection on regional water resources. However, it is still possible to conclude from our results that it is necessary to take into account the differences in transpiration characteristics between species when selecting planting species. Of course, further studies on the total evapotranspiration of the three Eucalyptus species are needed to more completely assess the impact of species selection on regional water resources.

4.2. Factors Influencing Water Use by the Three Eucalyptus Species

Tree transpiration consists of a set of complex physiological responses to environmental change [56]. Therefore, the transpiration of a tree stand is mainly affected by meteorological factors [57,58] and available water [59]. In this study, the T of all three eucalyptus species clearly increased with an increasing Rs or VPD, but leveled off after reaching a certain threshold (Figure 5A,B). The threshold control of T by Rs and VPD has been observed in many studies [57,60,61] because the response of plant stomata to Rs and VPD is often the result of a trade-off between maximizing photosynthesis and minimizing transpiration [52,55]. As Rs and VPD exceed the thresholds, plants will prevent excessive water loss by closing their stomata. In addition, these threshold values can vary with time, environmental conditions, and tree species identity. For example, the thresholds for T corresponding to VPD and Rs in a coniferous forest in the Lesser Himalaya of Central Nepal were 0.4 kPa and 200 W m^{-2} [57], respectively, whereas in a *Robinia pseudocacia* L. plantation on the semi-arid Loess Plateau of China, they were 1.5 kPa and 250 W m^{-2}, respectively [62]. Ma et al. [41] found the threshold for sap flow corresponding to VPD in a *Robinia pseudocacia* L. forest on the Loess Plateau of China to be 1.9 kPa in 2015, but this value decreased to 1.6 kPa in 2016. Here, the calculated thresholds for Rs differed among the three eucalyptus species, being 403 W m^{-2} for Ep, 321 W m^{-2} for Ec, and 382 W m^{-2} for Eug, whose corresponding thresholds for VPD were 1.77 kPa for Ep, 1.73 kPa for Ec, and 2.40 kPa for Eug.

For the three eucalyptus species, their T increased markedly with a rising Ta, a result consistent with findings of O'Brien et al. [63] and Han et al. [25]. We also found that the relationship between T and RH for each species differed from many other studies that reported a significant negative linear relationship between T and RH [11]. We found that below a certain threshold of RH (i.e., 69.7% for Ep and 66.4% for Ec and Eug), the T of the three species increased with an increasing RH, while above the threshold, it decreased instead. This result may be explained by the fact that RH below the threshold value principally occurs essentially during the dry season, when higher RH may represent an increased environmental water supply, which would promote transpiration. Furthermore, the T of the three eucalyptus species also showed a significant linear decrease with an increasing wind speed (Figure 5E), consistent with the findings of Asbjornsen et al. [64], but contrary to those of O'Brien et al. [63] and Han et al. [25]. Still, some studies have argued that WS has little effect upon T. There is no consensus as to how wind speed affects transpiration, although there are reports of stomatal closure in direct response to high wind speeds [65–67], as well as reports of wind speed contributing to transpiration in accordance with stomatal and boundary layer theory [68–70].

In addition to meteorological factors, soil moisture availability also has a pivotal influence on tree transpiration. Soil water deficits tend to reduce leaf water potential, which reduces stomatal conductance and ultimately limits transpiration by trees [59,71]. Therefore, the more severe the soil water deficit is, the greater the limiting effect on transpiration. Previous studies have found that tree transpiration increased non-linearly with more soil moisture under conditions of soil moisture deficit [12,72,73], as also found in our study. Initially, the T of the three eucalyptus species increased rapidly and almost linearly at first, with increasing soil moisture, and then became progressively flatter until a threshold value was reached (Figure 5F). Once that threshold is exceeded (indicating an adequate soil moisture supply), soil moisture is no longer a limiting factor, at which point there is no clear relationship between SWC and T; this finding was also consistent with that of many studies [12,74]. Additionally, there is considerable variation in plant–soil moisture thresholds, depending on tree species, soil type, and climatic conditions. For example, Ungar et al. [75] obtained a value of 15% for the SWC threshold of an Aleppo pine (*Pinus halepensis* Mill) stand in Israel. In the semi-arid areas of the Liupan Mountains, the SWC thresholds were 19% and 28% for 28- and 33-year-old pure larch (*Larix principis-rupprechtii* Mayr) plantations, respectively [12,74]. The soil moisture thresholds for the three Eucalyptus species in our study also differed, at 19.3% for Ep, 18.6% for Eug, and 17.5% for Ec, likely due to their biological characteristics. These thresholds are consistent with the drought-tolerant nature of *Ec*.

The transpiration of forest stands is also closely related to their LAI, as confirmed by numerous studies [12,20,76,77]. Canopy LAI directly influences the water-conveying capacity of the forest and determines its transpiration potential [12], and therefore, forest transpiration tends to be positively correlated with LAI in the absence of other limiting factors [20]. For example, Forrester et al. [78] found that the transpiration of a *Eucalyptus nitens* (H. Deane & Maiden) Maiden plantation increased with a higher LAI in the range of 1.0 to 6.0. However, forest transpiration of the forest does not increase indefinitely with an increasing LAI, but rather, there is a threshold beyond which transpiration can no longer increase [12]. For example, Di et al. [79] found a sigmoidal function relationship between transpiration and LAI in *Populus tomentosa* Carr, and that transpiration no longer increased with increasing LAI after the threshold of 1.6. Similar findings were also found in studies by Wang et al. [12] and Bucci et al. [80]. This is probably due to the fact that when LAI reaches a certain value, the branches and leaves inside and outside the canopy will shade each other, and the radiation received by the total canopy foliage will not increase further. Similar to the results above, the T and LAI of the three eucalyptus species in this study were well described by the Gompertz model curve (Figure 6). This suggests the T of each species was strongly influenced by LAI that had a species-specific threshold value, beyond which LAI would no longer be a factor influencing their transpiration.

4.3. Daily Stand Transpiration (T) Predictive Models for the Three Eucalyptus Species

Although the stand-level transpiration can be measured in several ways, it is not practical to do so in every stand, nor is it plausible to predict future changes in transpiration for a particular stand [81]. To overcome this problem, the development of transpiration predictive models has become essential [82,83]. From the analysis reported here, it is clear that stand transpiration is usually influenced by a combination of the meteorological factors, SWC and LAI, in tandem; hence, a highly accurate model for predicting stand transpiration depends on at least three prevailing factors: soil moisture, canopy structure, and meteorology [12,20,84]. Many studies introduced such models for various tree species, but most only consider one or two of those key factors. For example, Han et al. [25]

developed a transpiration model for a *Larix principis-rupprechtii* Mayr plantation using only meteorological factors and SWC. Similarly, Jiao et al. [83] also developed a transpiration model for a *Robinia pseudocacia* L. plantation using only PET (potential evapotranspiration) and SWC. Neither of them considered the possible effect of LAI, which led to an inability to accurately predict transpiration when large changes in LAI occurred. Di et al. [79] built empirical models for predicting transpiration using LAI or LAI-coupled soil temperature for *Populus tomentosa* Carr plantations during periods of water abundance, but did so without taking into account the effect of soil moisture deficit. Thus, their models are no longer be applicable when soil moisture becomes a limiting factor. Many similar simplified models, which do not couple all the major influences (i.e., meteorological factors, SWC, and LAI), cannot be widely applied in the context of global climate change and under conditions where human management disturbances are changing the structure of forest stands.

In this study, the effects of the meteorological factors SWC and LAI on transpiration were fully considered, and models were accordingly developed to predict the transpiration of the three eucalyptus species. These models accurately predicted the daily transpiration of each species, with a predictive power of 91.1%, 87.2%, and 86.3% for Ep, Ec, and Eug, respectively (Table 3). These models could be used to reliably predict the transpiration following changes in canopy structure induced by forest growth and stand management practices, as well as the effects of future global climate change on stand transpiration, being more accurate with a wider range of potential applications than existing models that rely on only one or two of the three predominant factors [12,25]. Nonetheless, there are some limitations to our modeling, as it is based on data from only one test site, and the species-specific models still require optimization and enhancement with data from a wider range of areas and environments, which in turn will increase the reliability of their estimates and applicable areas [20,85]. Therefore, more research is needed to refine and improve the application of our models.

5. Conclusions

In plantations of eucalyptus in southern China, we found that the transpiration of all three species was influenced by a combination of the meteorological factors, SWC and LAI, and that how much T responded to each influencing factor differed among the three species. To understand and predict the variation in transpiration of the three eucalyptus species, we integrated meteorological factors, LAI and SWC to develop species-specific models for predicting their T, which explained 86.3%–91.1% of the highly dispersed variation in their daily transpiration. In addition, when controlled by the same environmental factors, all three eucalyptus species exhibited the same seasonal pattern of variation in sap flow density and transpiration. However, there were also significant differences in sap flow density and transpiration among the three eucalyptus species due to differences in their physiology and stand structural characteristics. We determined that the annual transpiration of Ep (743.41 mm) was 2.4 and 1.5 times higher than that of Ec (311.52 mm) and Eug (493.58 mm), respectively, while the mean annual sap flow density of Ec (0.08 cm min^{-1}) was significantly lower than that of Ep (0.11 cm min^{-1}), but significantly greater than that of Eug (0.042 cm min^{-1}). Furthermore, although the percentage of annual transpiration to precipitation was small for all three eucalyptus species, it increased to 117.7% (Ep), 49.1% (Ec), and 80.1% (Eug) during the dry season, with Ep imposing the highest hydrological burden on the planted area during the dry season, followed by Eug, and Ec the least. The results of this study can assist in the prediction of eucalyptus trees' transpiration, while the findings also imply that it is necessary to consider interspecific differences when estimating transpiration in Chinese eucalyptus plantations, and that evaluation of the impact of water use on regional water resources is necessary to select the most optimal regional eucalypt plantation species.

Author Contributions: Idea and study design, A.D. and Z.W.; data collection and analysis, Z.W., Y.X., W.Z., S.L. and A.D.; manuscript writing, Z.W. All authors have read and agreed to the published version of the manuscript.

Funding: This research was funded by the Natural Science Foundation of Guangdong Province (2021A1515010560), the Forestry Science and Technology Innovation Project of Guangdong Province (2018KJCX014), the Forestry Ecological Monitoring Network Platform Construction Project (2021CG535), and the Operation Project for Guangdong Zhanjiang Eucalyptus Forest Ecosystem National Positioning Observation and Research Station (2022132113).

Data Availability Statement: Some or all relevant data during the study are available from the corresponding author by request.

Acknowledgments: The authors also appreciate the South China Experiment Nursery for support during the selection of suitable plots for this study.

Conflicts of Interest: We declare that we have no financial and personal relationships with other people or organizations that can inappropriately influence our work, there is no professional or other personal interest of any nature or kind in any product, service and/or company that could be construed as influencing the position presented in, or the review of, the manuscript entitled "Differences in transpiration and water consumption characteristics among species of Eucalyptus on the Leizhou Peninsula, southern China".

References

1. National Forestry and Grassland Administration. *2014–2018 China Forest Resources Report*; China Forestry Publishing House: Beijing, China, 2019.
2. Morris, J.; Ningnan, Z.; Zengjiang, Y.; Collopy, J.; Daping, X. Water use by fast-growing Eucalyptus urophylla plantations in southern China. *Tree Physiol.* **2004**, *24*, 1035–1044. [CrossRef] [PubMed]
3. Ouyang, L.; Zhao, P.; Zhou, G.; Zhu, L.; Huang, Y.; Zhao, X.; Ni, G. Stand-scale transpiration of a Eucalyptus urophylla × Eucalyptus grandis plantation and its potential hydrological implication. *Ecohydrology* **2018**, *11*, e1938. [CrossRef]
4. Cabral, O.M.R.; Rocha, H.R.; Gash, J.H.C.; Ligo, M.A.V.; Freitas, H.C.; Tatsch, J.D. The energy and water balance of a Eucalyptus plantation in southeast Brazil. *J. Hydrol.* **2010**, *388*, 208–216. [CrossRef]
5. Jackson, R.B.; Jobbágy, E.G.; Avissar, R.; Roy, S.B.; Barrett, D.J.; Cook, C.W.; Farley, K.A.; Maitre, D.C.L.; McCarl, B.A.; Murray, B.C. Trading Water for Carbon with Biological Carbon Sequestrationy. *Science* **2005**, *310*, 1944–1947. [CrossRef] [PubMed]
6. Bosch, J.M.; Hewlett, J.D. A review of catchment experiments to determine the effects of vegetation changes on water yield and evapotranspiration. *J. Hydrol.* **1982**, *55*, 3–23. [CrossRef]
7. Scott, D.F.; Lesch, W. Streamflow responses to afforestation with *Eucalyptus grandis* and *Pinus patula* and to felling in the Mokobulaan experimental catchments, South Africa. *J. Hydrol.* **1997**, *199*, 360–377. [CrossRef]
8. Buckley, T.N.; Turnbull, T.L.; Pfautsch, S.; Gharun, M.; Adams, M.A. Differences in water use between mature and post-fire regrowth stands of subalpine Eucalyptus delegatensis R. Baker. *For. Ecol. Manag.* **2012**, *270*, 1–10. [CrossRef]
9. Lane, P.N.J.; Morris, J.; Zhang, N.; Zhou, G.; Zhou, G.; Xu, D. Water balance of tropical eucalypt plantations in south-eastern China. *Agric. For. Meteorol.* **2004**, *124*, 253–267. [CrossRef]
10. Soares, J.V.; Almeida, A.C. Modeling the water balance and soil water fluxes in a fast growing Eucalyptus plantation in Brazil. *J. Hydrol.* **2001**, *253*, 130–147. [CrossRef]
11. Ouyang, S.; Xiao, K.; Zhao, Z.; Xiang, W.; Xu, C.; Lei, P.; Deng, X.; Li, J. Stand transpiration estimates from recalibrated parameters for the granier equation in a Chinese Fir (*Cunninghamia lanceolata*) plantation in southern China. *Forests* **2018**, *9*, 162. [CrossRef]
12. Wang, Y.; Cao, G.; Wang, Y.; Webb, A.A.; Yu, P.; Wang, X. Response of the daily transpiration of a larch plantation to variation in potential evaporation, leaf area index and soil moisture. *Sci. Rep.* **2019**, *9*, 4697. [CrossRef]
13. Aranda, I.; Forner, A.; Cuestaa, B.; Valladares, F. Species-specific water use by forest tree species: From the tree to the stand. *Agric. Water Manag.* **2012**, *114*, 67–77. [CrossRef]
14. Hu, Y.; Zhao, P.; Huang, Y.; Zhu, L.; Ni, G.; Zhao, X.; Huang, Z. Hydrologic balance, net primary productivity and water use efficiency of the introduced exotic *Eucalyptus grandis* × *Eucalyptus urophylla* plantation in south-western China. *J. Plant Ecol.* **2019**, *12*, 982–992. [CrossRef]
15. Ouyang, L.; Wu, J.; Zhao, P.; Zhu, L.; Ni, G. Stand age rather than soil moisture gradient mainly regulates the compromise between plant growth and water use of *Eucalyptus urophylla* in hilly South China. *Land Degrad. Dev.* **2021**, *32*, 2423–2436. [CrossRef]
16. Wingfield, M.J.; Brockerhoff, E.G.; Wingfield, D.; Slippers, B. Planted forest health: The need for a global strategy. *Science* **2015**, *349*, 832–836. [CrossRef] [PubMed]
17. Ewers, B.E.; Mackay, D.S.; Gower, S.T.; Ahl, D.E.; Burrows, S.N.; Samanta, S.S. Tree species effects on stand transpiration in northern Wisconsin. *Water Resour. Res.* **2002**, *38*, 8-1–8-11. [CrossRef]
18. Siddiq, Z.; Zhang, Y.; Zhu, S.; Cao, K. Canopy water status and photosynthesis of tropical trees are associated with trunk sapwood hydraulic properties. *Plant Physiol. Biochem.* **2019**, *139*, 724–730. [CrossRef]

19. Siddiq, Z.; Cao, K.-F. Nocturnal transpiration in 18 broadleaf timber species under a tropical seasonal climate. *For. Ecol. Manag.* **2018**, *418*, 47–54. [CrossRef]
20. Wang, L.; Liu, Z.; Guo, J.; Wang, Y.; Ma, J.; Yu, S.; Yu, P.; Xu, L. Estimate canopy transpiration in larch plantations via the interactions among reference evapotranspiration, leaf area index, and soil moisture. *For. Ecol. Manag.* **2021**, *481*, 118749. [CrossRef]
21. Xu, Z.; Man, X.; Cai, T.; Shang, Y. How Potential Evapotranspiration Regulates the Response of Canopy Transpiration to Soil Moisture and Leaf Area Index of the Boreal Larch Forest in China. *Forests* **2022**, *13*, 571. [CrossRef]
22. Adil, B.; Hicham, F.; Ahmed, M.S.K.; Sanae, C.; Allal, S.; Mhamed, M.; Hassan, M. Study of Microclimate and Sapling Citrus Plant Transpiration in Tunnel Greenhouse Under Mediterranean Conditions. *Acta Tech. Agric.* **2022**, *2*, 61–66. [CrossRef]
23. Abreu, M.C.; Soares, A.A.V.; de Freitas, C.H.; Martins, F.B. Transpiration and growth responses by *Eucalyptus* species to progressive soil drying. *J. For. Res.* **2022**, *33*, 1529–1543. [CrossRef]
24. Lyu, J.; He, Q.; Chen, Q.; Cheng, R.; Li, G.; Otsuki, K.; Yamanaka, N.; Du, S. Distinct transpiration characteristics of black locust plantations acclimated to semiarid and subhumid sites in the Loess Plateau, China. *Agric. Water Manag.* **2022**, *262*, 107402. [CrossRef]
25. Han, C.; Chen, N.; Zhang, C.; Liu, Y.; Khan, S.; Lu, K.; Li, Y.; Dong, X.; Zhao, C. Sap flow and responses to meteorological about the Larix principis-rupprechtii plantation in Gansu Xinlong mountain, northwestern China. *For. Ecol. Manag.* **2019**, *451*, 117519. [CrossRef]
26. Tu, J.; Wei, X.; Huang, B.; Fan, H.; Jian, M.; Li, W. Improvement of sap flow estimation by including phenological index and time-lag effect in back-propagation neural network models. *Agric. For. Meteorol.* **2019**, *276*, 107608. [CrossRef]
27. Oogathoo, S.; Houle, D.; Duchesne, L.; Kneeshaw, D. Vapour pressure deficit and solar radiation are the major drivers of transpiration of balsam fir and black spruce tree species in humid boreal regions, even during a short-term drought. *Agric. For. Meteorol.* **2020**, *291*, 108063. [CrossRef]
28. Urban, J.; Rubtsov, A.V.; Urban, A.V.; Shashkin, A.V.; Vera, E.B. Canopy transpiration of a *Larix sibirica* and *Pinus sylvestris* forest in Central Siberia. *Agric. For. Meteorol.* **2019**, *271*, 64–72. [CrossRef]
29. Heidi, H.; Michael, H.A.; Susan, C.; Michael, C.; Sjirk, G.; Coert, J.G.; Graham, H.; Brett, P.H.; Wayne, J.; Jan-Hendrik, K.; et al. Eucalyptus camaldulensis in South Africa—Past, present, future. *Trans. R. Soc. South Afr.* **2020**, *75*, 1–22. [CrossRef]
30. Yang, G.; Lu, W.; Lin, Y.; Luo, J.; Wang, C.; Meder, R.; Warburton, P.M.; Arnold, R.J. Monitoring water potential and relative water content in eucalyptus camaldulensis using near infrared spectroscopy. *J. Trop. For. Sci.* **2017**, *29*, 121–128.
31. Saravanan, T.; Roger, J.A.; Luo, J.; Bala, R.T. Genomic studies reveal substantial dominant effects and improved genomic predictions in an open-pollinated breeding population of Eucalyptus pellita. *G3 Genes Genomes Genet.* **2020**, *10*, 3751–3763. [CrossRef]
32. Wirabuana, P.Y.A.P.; Sadono, R.; Juniarso, S. Fertilization Effects on Early Growth, Aboveground Biomass, Carbon Storage, and Leaf Characteristics of Eucalyptus pellita F. Muell. in South Sumatera. *J. Man. Hut. Trop.* **2019**, *25*, 154. [CrossRef]
33. Xu, Y.; Du, A.; Wang, Z.; Zhu, W.; Li, C.; Wu, L. Effects of different rotation periods of Eucalyptus plantations on soil physiochemical properties, enzyme activities, microbial biomass and microbial community structure and diversity. *For. Ecol. Manag.* **2020**, *456*, 117683. [CrossRef]
34. IUSS Working Group WRB. World reference base for soil resource 2006. In *World Soil Resources Reports No. 103*, 2nd ed.; FAO: Rome, Italy, 2006.
35. Campbell, G.S.; Norman, J. *An Introduction to Environmental Biophysics*, 2nd ed.; Springer: New York, NY, USA, 1998.
36. Granier, A. Evaluation of transpiration in a Douglas-fir stand by means of sap flow measurements. *Tree Physiol.* **1987**, *3*, 309–320. [CrossRef]
37. Lu, P.; Urban, L.; Zhao, P. Granier's Thermal Dissipation Probe (TDP) Method for Measuring Sap Flow in Trees: Theory and Practice. *Acta Bot. Sin.* **2004**, *46*, 631–646. [CrossRef]
38. Vertessy, R.A.; Benyon, R.G.; O'Sullivan, S.K.; Gribben, P.R. Relationships between stem diameter, sapwood area, leaf area and transpiration in a young mountain ash forest. *Tree Physiol.* **1995**, *15*, 559–567. [CrossRef] [PubMed]
39. Nadezhdina, N. Sap flow index as an indicator of plant water status. *Tree Physiol.* **1999**, *19*, 885–891. [CrossRef]
40. Peng, X.; Jun, F.; Wang, Q.; Warrington, D. Discrepancy of sap flow in *Salix matsudana* grown under different soil textures in the water-wind erosion crisscross region on the Loess Plateau. *Plant Soil* **2015**, *390*, 383–399. [CrossRef]
41. Ma, C.; Luo, Y.; Shao, M.; Li, X.; Sun, L.; Jia, X. Environmental controls on sap flow in black locust forest in Loess Plateau, China. *Sci. Rep.* **2017**, *7*, 13160. [CrossRef]
42. Tie, Q.; Hu, H.; Tian, F.; Guan, H.; Lin, H. Environmental and physiological controls on sap flow in a subhumid mountainous catchment in North China. *Agric. For. Meteorol.* **2017**, *240*, 46–57. [CrossRef]
43. Forrester, D.I.; Collopy, J.J.; Morris, J.D. Transpiration along an age series of Eucalyptus globulus plantations in southeastern Australia. *For. Ecol. Manag.* **2010**, *259*, 1754–1760. [CrossRef]
44. Cheng, J.; Ou, Y.; Huang, D.; Liu, S.; Li, Y. Sap Flow Characteristics of Four Dominant Tree Species in a Mixed Conifer-broadleaf Forest in Dinghushan. *Acta Ecol. Sin.* **2015**, *35*, 4097–4104. [CrossRef]
45. Mitchell, P.J.; Benyon, R.G.; Lane, P.N.J. Responses of evapotranspiration at different topographic positions and catchment water balance following a pronounced drought in a mixed species eucalypt forest, Australia. *J. Hydrol.* **2012**, *440*, 62–74. [CrossRef]

46. Macfarlane, C.; Bond, C.; White, D.A.; Grigg, A.H.; Ogden, G.N.; Silberstein, R. Transpiration and hydraulic traits of old and regrowth eucalypt forest in southwestern Australia. *For. Ecol. Manag.* **2010**, *260*, 96–105. [CrossRef]
47. Almeida, A.C.; Soares, J.V.; Landsberg, J.J.; Rezende, G.D. Growth and water balance of *Eucalyptus grandis* hybrid plantations in Brazil during a rotation for pulp production. *For. Ecol. Manag.* **2007**, *251*, 10–21. [CrossRef]
48. Hernandez-Santana, V.; Hernandez-Hernandez, A.; Vadeboncoeur, M.A.; Asbjornsen, H. Scaling from single-point sap velocity measurements to stand transpiration in a multispecies deciduous forest: Uncertainty sources, stand structure effect, and future scenarios. *Can. J. For. Res.* **2015**, *45*, 1489–1497. [CrossRef]
49. Link, P.; Simonin, K.A.; Maness, H.; Oshun, J.; Dawson, T.; Fung, I. Species differences in the seasonality of evergreen tree transpiration in a Mediterranean climate: Analysis of multiyear, half- hourly sap flow observations. *Water Resour. Res.* **2014**, *50*, 1869–1894. [CrossRef]
50. Yan, M.; Zhang, J.; He, Q.; Shi, W.; Otsuki, K.; Yamanaka, N.; Du, S. Sapflow-Based Stand Transpiration in a Semiarid Natural Oak Forest on China's Loess Plateau. *Forests* **2016**, *7*, 227. [CrossRef]
51. Horna, V.; Schuldt, B.; Brix, S.; Leuschner, C. Environment and tree size controlling stem sap flux in a perhumid tropical forest of Central Sulawesi, Indonesia. *Ann. For. Sci.* **2011**, *68*, 1027–1038. [CrossRef]
52. Di, N.; Xi, B.; Clothier, B.; Wang, Y.; Li, G.; Jia, L. Diurnal and nocturnal transpiration behaviors and their responses to groundwater-table fluctuations and meteorological factors of Populus tomentosa in the North China Plain. *For. Ecol. Manag.* **2019**, *448*, 445–456. [CrossRef]
53. Hatton, T.; Reece, P.; Taylor, P.; McEwan, K. Does leaf water efficiency vary among eucalypts in water-limited environments? *Tree Physiol.* **1998**, *18*, 529–536. [CrossRef] [PubMed]
54. Wang, Z.; Xu, Y.; Zhu, W.; Du, A. The transpiration water consumption of two common fast-growing forests in Leizhou Peninsula. *Acta Ecol. Sin.* **2019**, *39*, 2147–2155. [CrossRef]
55. Ma, L.; Lu, P.; Zhao, P.; Rao, X.; Cai, X.; Zeng, X. Diurnal, daily, seasonal and annual patterns of sap-flux-scaled transpiration from an *Acacia mangium* plantation in South China. *Ann. For. Sci.* **2008**, *65*, 402. [CrossRef]
56. Tian, Q.; He, Z.; Xiao, S.; Du, J. Growing Season Stem Water Status Assessment of Qinghai Spruce through the Sap Flow and Stem Radial Variations in the Qilian Mountains of China. *Forests* **2018**, *9*, 2. [CrossRef]
57. Ghimire, C.P.; Lubczynski, M.W.; Bruijnzeel, L.A.; Chavarro-Rincón, D. Transpiration and canopy conductance of two contrasting forest types in the Lesser Himalaya of Central Nepal. *Agric. For. Meteorol.* **2014**, *197*, 76–90. [CrossRef]
58. Wullschleger, S.D.; Wilson, K.B.; Hanson, P.J. Environmental control of whole-plant transpiration, canopy conductance and estimates of the decoupling coefficient for large red maple trees. *Agric. For. Meteorol.* **2000**, *104*, 157–168. [CrossRef]
59. White, D.A.; Beadle, C.L.; Worledge, D. Control of transpiration in an irrigated *Eucalyptus globulus* Labill. plantation. *Plant Cell Environ.* **2000**, *23*, 123–134. [CrossRef]
60. Chang, X.; Zhao, W.; Liu, H.; Wei, X.; Liu, B.; He, Z. Qinghai spruce (*Picea crassifolia*) forest transpiration and canopy conductance in the upper Heihe River Basin of arid northwestern China. *Agric. For. Meteorol.* **2014**, *198*, 209–220. [CrossRef]
61. Chen, D.; Wang, Y.; Liu, S.; Wei, X.; Wang, X. Response of relative sap flow to meteorological factors under different soil moisture conditions in rainfed jujube (*Ziziphus jujuba* Mill.) plantations in semiarid Northwest China. *Agri. Water Manag.* **2014**, *136*, 23–33. [CrossRef]
62. Jiao, L.; Lu, N.; Sun, G.; Ward, E.J.; Fu, B. Biophysical controls on canopy transpiration in a black locust (*Robinia pseudoacacia*) plantation on the semi-arid Loess Plateau, China. *Ecohydrology* **2016**, *9*, 1068–1081. [CrossRef]
63. O'Brien, J.J.; Oberbauer, S.F.; Clark, D.B. Whole tree xylem sap flow responses to multiple environmental variables in a wet tropical forest. *Plant Cell Environ.* **2004**, *27*, 551–567. [CrossRef]
64. Asbjornsen, H.; Tomer, M.D.; Gomez-Cardenas, M.; Brudvig, L.A.; Greenan, C.M.; Schilling, K. Tree and stand transpiration in a Midwestern bur oak savanna after elm encroachment and restoration thinning. *For. Ecol. Manag.* **2007**, *247*, 209–219. [CrossRef]
65. Meinzer, F.C.; Goldstein, G.; Jackson, P.; Holbrook, N.M.; Guti6rrez, M.V.; Cavelier, J. Environmental and physiological regulation of transpiration in tropical forest gap species: The influence of boundary layer and hydraulic properties. *Oecologia* **1995**, *101*, 514–522. [CrossRef] [PubMed]
66. Fisher, J.B.; Baldocchi, D.D.; Misson, L.; Dawson, T.E.; Goldstein, A.H. What the towers don't see at night: Nocturnal sap flow in trees and shrubs at two AmeriFlux sites in California. *Tree Physiol.* **2007**, *27*, 597–610. [CrossRef] [PubMed]
67. Zhao, C.; Si, J.; Qi, F.; Yu, T.; Li, P. Effect of wind speed on transpiration rate of Populus euphratica in extreme arid deserts. *J. Glaciol. Geocryol.* **2015**, *37*, 1104–1111. [CrossRef]
68. Campbell-Clause, J.M. Stomatal response of grapevines to wind. *Aust. J. Exp. Agric.* **1998**, *38*, 77–82. [CrossRef]
69. Gutirrez, M.V.; Meinzer, F.C.; Grantz, D.A. Regulation of transpiration in coffee hedgerows: Covariation of environmental variables and apparent responses of stomata to wind and humidity. *Plant Cell Environ.* **1994**, *17*, 1305–1313. [CrossRef]
70. Buckley, T.N.; Turnbull, T.L.; Pfautsch, S.; Adams, M.A. Nocturnal water loss in mature subalpine *Eucalyptus delegatensis* tall open forests and adjacent *E. pauciflora* woodlands. *Ecol. Evol.* **2011**, *1*, 435–450. [CrossRef]
71. Galmés, J.; Flexas, J.; Savé, R.; Medrano, H. Water relations and stomatal characteristics of Mediterranean plants with different growth forms and leaf habits: Responses to water stress and recovery. *Plant Soil* **2007**, *290*, 139–155. [CrossRef]
72. Oren, R.; Pataki, D.E. Transpiration in response to variation in microclimate and soil moisture in southeastern deciduous forests. *Oecologia* **2001**, *127*, 549–559. [CrossRef] [PubMed]

73. Bindi, M.; Bellesi, S.; Orlandini, S.; Fibbi, L.; Moriondo, M.; Sinclair, T. Influence of Water Deficit Stress on Leaf Area Development and Transpiration of Sangiovese Grapevines Grown in Pots. *Am. J. Enol. Vitic.* **2005**, *56*, 68–72. [CrossRef]
74. Li, Z.; Yu, P.; Wang, Y.; Webb, A.A.; He, C.; Wang, Y.; Yang, L. A model coupling the effects of soil moisture and potential evaporation on the tree transpiration of a semi-arid larch plantation. *Ecohydrology* **2016**, *10*, e1764. [CrossRef]
75. Ungar, E.D.; Rotenberg, E.; Raz-Yaseef, N.; Cohen, S.; Yakir, D.; Schiller, G. Transpiration and annual water balance of Aleppo pine in a semiarid region: Implications for forest management. *For. Ecol. Manag.* **2013**, *298*, 39–51. [CrossRef]
76. Ayyoub, A.; Er-Raki, S.; Khabba, S.; Merlin, O.; Ezzahar, J.; Rodriguez, J.C.; Bahlaoui, A.; Chehbouni, A. A simple and alternative approach based on reference evapotranspiration and leaf area index for estimating tree transpiration in semi-arid regions. *Agric. Water Manag.* **2017**, *188*, 61–68. [CrossRef]
77. Naithani, K.J.; Baldwin, D.C.; Gaines, K.P.; Lin, H.; Eissenstat, D.M. Spatial Distribution of Tree Species Governs the Spatio-Temporal Interaction of Leaf Area Index and Soil Moisture across a Forested Landscape. *PLoS ONE* **2013**, *8*, e58704. [CrossRef]
78. Forrester, D.I.; Collopy, J.J.; Beadle, C.L.; Warren, C.R.; Baker, T.G. Effect of thinning, pruning and nitrogen fertiliser application on transpiration, photosynthesis and water-use efficiency in a young *Eucalyptus nitens* plantation. *For. Ecol. Manag.* **2012**, *266*, 286–300. [CrossRef]
79. Di, N.; Wang, Y.; Clothier, B.; Liu, Y.; Jia, L.; Xi, B.; Shi, H. Modeling soil evaporation and the response of the crop coefficient to leaf area index in mature *Populus tomentosa* plantations growing under different soil water availabilities. *Agric. For. Meteorol.* **2019**, *264*, 125–137. [CrossRef]
80. Bucci, S.J.; Scholz, F.G.; Goldstein, G.; Hoffmann, W.A.; Meinzer, F.C.; Franco, A.C.; Giambelluca, T.; Miralles-Wilhelm, F. Controls on stand transpiration and soil water utilization along a tree density gradient in a *Neotropical savanna*. *Agric. For. Meteorol.* **2008**, *148*, 839–849. [CrossRef]
81. Li, L.; Chen, S.; Yang, C.; Meng, F.; Sigrimis, N. Prediction of plant transpiration from environmental parameters and relative leaf area index using the random forest regression algorithm. *J. Clean. Prod.* **2020**, *261*, 121136. [CrossRef]
82. Liu, N.; Buckley, T.; He, X.; Zhang, X.; Zhang, C.; Luo, Z.; Wang, H.; Sterling, N.; Guan, H. Improvement of a simplified process-based model for estimating transpiration under water-limited conditions. *Hydrol. Process.* **2019**, *33*, 1670–1685. [CrossRef]
83. Jiao, L.; Lu, N.; Fang, W.; Li, Z.; Wang, J.; Jin, Z. Determining the independent impact of soil water on forest transpiration: A case study of a black locust plantation in the Loess Plateau, China. *J. Hydrol.* **2019**, *572*, 671–681. [CrossRef]
84. Granier, A.; Loustaub, D.; Bréda, N. A generic model of forest canopy conductance dependent on climate, soil water availability and leaf area index. *Ann. For. Sci.* **2000**, *57*, 755–765. [CrossRef]
85. Sawano, S.; Hotta, N.; Tanaka, N.; Tsuboyama, Y.; Suzuki, M. Development of a simple forest evapotranspiration model using a process-oriented model as a reference to parameterize data from a wide range of environmental conditions. *Ecol. Modell.* **2015**, *309*, 93–109. [CrossRef]

forests

MDPI

Article

Transpiration Sensitivity to Drought in *Quercus wutaishansea* Mary Forests on Shady and Sunny Slopes in the Liupan Mountains, Northwestern China

Bingbing Liu [1], Pengtao Yu [1,*], Xue Zhang [2], Jiamei Li [1], Yipeng Yu [1], Yanfang Wan [1], Yanhui Wang [1], Xiao Wang [1], Zebin Liu [1], Lei Pan [3] and Lihong Xu [1]

[1] Key Laboratory of Forest Ecology and Environment of National Forestry and Grassland Administration, Liupan Mountains Forest Ecosystems National Positioning Observation and Research Station, Ecology and Nature Conservation Institute, Chinese Academy of Forestry, Beijing 100091, China
[2] School of Forestry, Northeast Forestry University, Harbin 150000, China
[3] Dalad Banner Forestry and Grassland Bureau, Erdos 301700, China
* Correspondence: yupt@caf.ac.cn; Tel.: +86-10-62889562

Abstract: Forests in water source areas are important factors for water supply security, soil, and water conservation, and their water consumption from transpiration is strongly affected by site conditions, including the slope aspect. However, the lack of research on how the slope aspect interferes with the response of stand transpiration to drought has hindered researchers from developing climate-resilient forest–water coordinated, sustainable development plans for different stand conditions. This study was conducted on *Quercus wutaishansea* forests in the southern part of Liupan Mountain in northwest China, and two sample plots were built on sunny and shady slopes. The responses of stand transpiration to various soil moisture and meteorological conditions on different slope orientations were analyzed. The results showed better-growing stands on shady slopes transpired more and consumed more soil moisture than those on sunny slopes. The soil moisture on shady slopes decreased rapidly below the threshold level during the drought, leading to a limitation of stand transpiration; however, its transpiration recovered rapidly after the drought. In contrast, stand transpiration on sunny slopes was not affected by this drought and maintained its pre-drought rate. Our results suggested that stands with higher water demand on shady slopes were more susceptible to drought when it occurred. This indicated that in the case of frequent droughts, the vegetation should be managed according to the vegetation-carrying capacities resulting from different site conditions.

Keywords: *Quercus wutaishansea*; transpiration; slope aspect; drought; relative extractable water

Citation: Liu, B.; Yu, P.; Zhang, X.; Li, J.; Yu, Y.; Wan, Y.; Wang, Y.; Wang, X.; Liu, Z.; Pan, L.; et al. Transpiration Sensitivity to Drought in *Quercus wutaishansea* Mary Forests on Shady and Sunny Slopes in the Liupan Mountains, Northwestern China. *Forests* **2022**, *13*, 1999. https://doi.org/10.3390/f13121999

Academic Editor: Romà Ogaya

Received: 28 September 2022
Accepted: 21 November 2022
Published: 25 November 2022

1. Introduction

Plant transpiration accounts for over 60% of evapotranspiration (ET), and it is an essential component of hydrological cycles [1–4]. It has been found to be strongly affected by increasing frequency of drought in the whole world [5–10]. The security level of the water supply in water source areas was also seriously threatened, especially for water-limited areas such as northwest China [11,12]. Therefore, it is essential to study the effect of drought on water consumption by stand transpiration to develop a reasonable integrated forest water management plan and to cope with climate change, especially in forested water source areas.

The studies on slope aspect changing drought effects on stand transpiration are scarce, although research focusing on the role of drought in stand transpiration is increasing [11,13,14]. It has been found that shaded and downslope slopes with sufficient soil moisture could enhance the resistance of trees to drought [15,16], and this effect would be further changed by tree size, i.e., the large trees would transpire more when faced with to drought [17,18]. Slope aspect also changes the site conditions and the soil moisture content [19,20]. Thus, slope aspect would indirectly affect the transpiration water consumption of forest stands [11,15,20,21]. However, most previous research has focused on shady slopes where trees are widely distributed and on stands with different slope positions. There is a lack of studies on the different responses of stand transpiration to drought between shady and sunny slopes.

Forest stands with different growth characteristics are severely threatened by drought [15,20,22–24], such as *Quercus wutaishansea* forests in the Liupan Mountains of northwestern China. In recent decades, the transpiration of *Quercus wutaishansea* trees in northwest China has been severely limited by the increased frequency of droughts [7,25]. Dominant trees under the same stand conditions are more susceptible to soil moisture deficits [17,18,26]. As an essential factor for stand growth, available soil moisture is essential for stand transpiration amounts [15,27], and, as a result, the trees grow better on shady slopes due to more available soil moisture [20]. However, increasing drought would lead to a decrease in the available soil moisture. This would lead to severe impacts on stands with high water requirements. Therefore, we hypothesized that the better-growing stands on shady slopes are more sensitive to drought than those on sunny slopes.

In this study, we analyzed how the slope aspect affected the transpiration of *Quercus wutaishansea* forest under drought conditions. The stand transpiration during the same drought period were compared among slopes and among different drought periods on the slope. The effects of slope, available soil moisture, and meteorological conditions on stand transpiration are furtherly partitioned. These results would help us understand the response mechanisms of the transpiration of *Quercus wutaishansea* forest stands with different slope aspects to drought under climate change and would provide a theoretical basis for integrated forest water management in water-limited areas.

2. Materials and Methods

2.1. Study Area

This study was conducted in the Qiuqianjia Forest farm (106°18′~106°29′ E, 35°31′~35°37′ N), located in the Natural Reserve of the Liupan Mountains in the midwestern part of the Loess Plateau in Northwest China (Figure 1). The forest farm area is 118.42 km^2, and the farm has an elevation range from 1612 to 2317 m a.s.l., of which 25% is covered by *Quercus wutaishansea* natural forest. This region has a semihumid continental climate with a mean annual temperature of 5.8 °C and mean annual precipitation of 618 mm, 87.6% of which is concentrated in the growing season (May–October) at the elevation of 1960 m a.s.l. The mean annual evaporation (1981–2010) is 1372 mm measured by an evaporation dish with a diameter of 20 cm, which is 2.22 times higher than the mean annual precipitation.

The soil in this region is dominated by the grey cinnamon soil type with a sandy loam texture, and the soil thickness is approximately 80 cm. The dominant tree species are *Quercus wutaishansea* pure forest and *Quercus wutaishansea* and *Populus davidiana* Dode mixed forest, among which *Quercus wutaishansea* pure forest is the most widely distributed, accounting for approximately 30.9% of the natural forest area. The shrub community consists mainly of *Crataegus pinnatifida* Bge., *Ostryopsis davidiana* Decne., and *Cotoneaster multiflora* Bunge. The herbaceous vegetation consists mainly of *Carex pediformis* C. A. Mey. and *Epimedium brevicornu* Maxim.

In this study, a northeast slope and a southwest slope were chosen as shady sample slope facing north and sunny sample slope facing south, respectively (Figure 1). Ten consecutive plots with an area of 30 m × 30 m were established on the sample slopes, i.e., the shady and sunny slopes as shown in Figure 1, and a detailed vegetation survey was made in May 2021. Parameters such as the height, diameter at breast height for each tree in the plots, and forest crown density were measured by the classical method in the vegetation survey. Then, two of the ten samples (Table 1) were selected as intensive plots, on which the sap flow and soil moisture were continuously observed in the growing season of 2021. These two intensive plots were similar in tree age and elevation (Table 1), and the soil thickness in them was greater than 80 cm, which did not limit the tree growth.

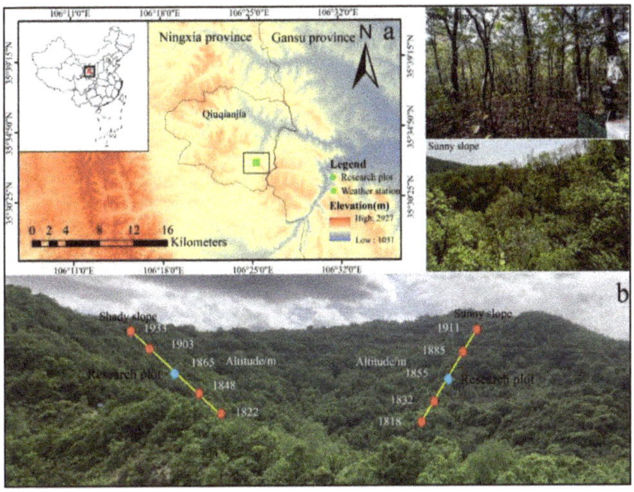

Figure 1. Geographical location of study plots in the Qiuqianjia Farm Forest: (**a**) relative positions of the study plots on the two slopes; (**b**) red dots indicate consecutive plots with the area of 30 m × 30 m on slopes; blue dots represent the intensive plots, in which the sap flow and soil moisture were continuously and simultaneously measured. (**c**,**d**) The pictures of *Quercus wutaishansea* forest on sunny and shady slopes are shown in the top right corner.

Table 1. The characteristics of sample plots on sunny and shady sample slopes. Plots P3 and P8 were selected as intensive plots. *DBH* indicates diameter at breast height; *LAI* indicates leaf area index.

Sample Plot No.	Aspect	Elevation (m a.s.l.)	Slope Gradient (°)	Stand Density (trees·ha^{-1})	Average Tree Age (a)	*LAI*	Average *DBH* (cm)	Average Tree Height (m)	Total Sapwood Area of the Stand (cm^2·ha^{-1})
P1		1822	29	800	54	2.95	13.85	8.02	74,071.4
P2		1848	34	667	52	3.11	19.95	11.63	63,422.0
P3	shady slope (5 degrees north by east)	1865	24	889	50	3.06	16.44	11.69	93,215.0
P4		1903	31	500	53	2.87	23.13	11.92	86,250.0
P5		1933	25	633	55	2.85	20.14	10.14	90,318.9
Average		-	28.6	697.8	52.8	2.97	19.67	10.68	81,455.5 ± 12,445.5
P6		1818	19	900	51	2.11	11.07	5.8	52,964.2
P7		1832	19	856	42	2.5	9.96	5.0	35,200.0
P8	sunny slope (60 degrees south by west)	1855	28	925	47	1.78	11.04	6.23	55,278.0
P9		1885	35	767	47	1.71	12.11	5.22	43,811.1
P10		1911	27	1244	50	2.3	8.08	3.9	42,448.0
Average		-	25.6	938.4	47.4	2.08	10.452	5.23	45,940.2 ± 8194.6

2.2. Sap Flow Measurement and Transpiration Calculation

All the tress in the sample plots were divided into different classes according to their *DBH*. For example, the trees in the shady slope plot were divided into six classes, i.e., with the *DBH* < 10 cm, 10–15 cm, 15–20 cm, 20–24 cm, 24–27 cm, and >27 cm; and those in the sunny slope plot into five classes, with the *DBH* < 9 cm, 9–14 cm, 14–18 cm, 18–22 cm, and >22 cm classes, respectively. Then, one tree, which had the *DBH* and *H* values close to the mean values of the corresponding *DBH* class, was selected from each *DBH* class as a sample tree, and eleven sample trees in total (Table 2) were observed and their stem sap flux was measured.

Table 2. Characteristics of the sample trees selected for sap flow measurement in intensive plots.

Intensive Plot No.	Sample Tree No.	*DBH* (cm)	Tree Height (m)	Measured Sapwood Thickness (cm)	Measured Sapwood Area (cm^2)	Sapwood Area Calculated by Equation (1) (cm^2)	Absolute Error (cm^2)/Relative Error (%) of Sapwood Area Calculation *
P3	1	9.1	7.9	1.3	32.65	31.70	−0.95/−2.91
	2	13	12.5	1.5	54.17	58.53	+4.36/+8.05
	3	17.9	18.8	2.1	104.19	101.42	−2.77/−2.66
	4	21.7	16.5	2.2	134.71	141.20	+6.49/+4.82
	5	24.6	17.3	2.6	179.61	175.18	−4.43/−2.47
	6	28.5	22.6	2.9	233.11	225.60	−7.51/−3.22
Average	-	-	-	-	-	-	−0.8/+0.27
P8	7	8.7	4.8	1.2	28.26	29.35	+1.08/+3.84
	8	12.7	7.5	1.6	55.77	56.23	+0.46/+0.82
	9	16.2	7.8	1.9	84.53	85.44	+0.91/+1.08
	10	20.1	11.1	2.0	113.67	123.78	+10.11/+8.89
	11	24.8	11.4	2.3	163.76	177.63	+13.87/+8.47
Average	-	-	-	-	-	-	+5.28/+4.62

* "+" indicates that the calculated value is higher than the measured value; "−" indicates that the calculated value is lower than the measured value. Total sapwood area represents the sum of the sapwood areas corresponding to the *DBH* classes of the sample plots.

The sap flow density of the sample trees was measured at breast height (1.3 m above ground) using a thermal diffusion probe (SF-L, Ecomatik, Munich, Germany). These probes consisted of two sensors 20 mm long and 2 mm in diameter (S0, a heated sensor powered by a constant current of 12 volts; S1, an unheated sensor). These sensors were inserted 20 mm outside the xylem at breast height on the north-facing side of the trunk (rather than the side exposed to sunlight). Before insertion, the outer bark was peeled off. Each probe was coated with silicone gel to ensure good thermal contact between the probe element and the sapwood. After insertion, the exposed bark was covered with silicone gel to reduce evaporation from the wood surface and then covered with aluminum foil to avoid physical damage and the thermal effects of solar radiation. Sap flow data were recorded every 5 min by a logger (CR1000x, Campbell Scientific Inc., Salt Lake City, UT, USA).

The sap flow density (J_S, mL·cm^{-2}·min^{-1}) (flow per unit of sapwood area) of an individual sample tree was calculated using Equation (1) [13]:

$$J_S = 0.714 \times \left(\frac{\Delta T_{max} - \Delta T}{\Delta T}\right)^{1.231} \quad (1)$$

where J_S is the sap flux density (ml·cm^{-2}·min^{-1}); ΔT is the temperature difference between the two needles; and ΔT_{max} is the maximum value of ΔT every night.

The transpiration of each *DBH* class was summed to obtain the transpiration of the stand per unit area. The daily stand transpiration (T, mm) was extrapolated using Equation (2):

$$T = \frac{\sum_{i=1}^{n} A_i \times J_S}{S} \times 60 \times 24 \div 1000 \quad (2)$$

where S is the projected area of the plot (m^2); n is the number of DBH class in the sample plots; A_i is the sum of the sapwood area of all trees in each DBH class (cm^2), which was calculated from the DBH of all trees in the sample plots by the relationship between sapwood area and DBH for *Quercus wutaishansea* forest using Equation (3):

$$A_S = 0.7122 \times DBH^{1.7189} \quad \left(R^2 = 0.95,\ n = 12\right) \tag{3}$$

This equation was established by us based on twelve core samples from our sample plots, where A_S (cm^2) is the sapwood area of the tree.

2.3. Weather and Soil Moisture Measurements

The weather conditions, including the precipitation (TE525MM, P (mm)), air temperature (HMP115, T (°C)), solar radiation (PQS1, R_S (W·m^{-2})), relative humidity (HMP155, Rh (%)), and wind speed (034E, U (m·s^{-1}) were monitored for the period from June to October, 2021, and recorded every 5 min with an automatic weather station (CR1000X, Campbell Scientific Inc., Salt Lake City, UT, USA) in an open area away 100 m from the sample plots.

The daily potential evapotranspiration (PET; mm·day^{-1}) was estimated using Equation (4) [28] based on the measured weather data:

$$PET = \frac{0.408\Delta(R_n - G) + \gamma\frac{900}{T+273}U(e_s - e_a)}{\Delta + \gamma(1 + 0.34U)} \tag{4}$$

where Δ (kPa·°C^{-1}) is the slope of the relationship between vapor pressure and air temperature; R_n (MJ·m^{-2}·day^{-1}) is the net radiation; G (MJ·m^{-2}·day^{-1}) is the soil heat density; γ (0.053 kPa·°C^{-1}) is the psychrometric constant; T (°C) is the mean air temperature at a 2 m height; U (m·s^{-1}) is the wind speed at a 2 m height; e_s (kPa) is the saturation vapor pressure; and e_a (kPa) is the actual vapor pressure.

Soil volumetric moisture in the root zone (soil layers 0–10, 10–20, 20–40, and 40–60 cm) was simultaneously monitored using soil moisture and temperature sensors (5-TE, Decagon, Pullman, WA, USA) in the two intensive plots. Data were collected every 5 min by a data logger (EM50, Decagon, Pullman, WA, USA). The relative extractable water (REW) was calculated as the ratio of the actual extractable water to the maximum extractable water [29]:

$$REW = \frac{VSM - VSM_w}{VSM_c - VSM_w} \tag{5}$$

where VSM is the volumetric soil moisture of the 0–60 cm soil layer (%); VSM_w is the wilting VSM of the 0–60 cm soil layer (%) (i.e., soil water potential equal to −1.56 MPa) [30]; and VSM_c is the field capacity VSM of the 0–60 cm soil layer (%).

2.4. Data Analysis

In this study, the upper boundary line method was applied to determine the response functions between T and REW data [13,31]. The corresponding thresholds of stand transpiration on REW are determined based on the inflection points of the outer boundary line [32].

3. Results

3.1. Meteorological Characteristics and the Occurrence of Drought

The summer of 2021 (June to August) was a dryer but hotter summer with low precipitation and high temperatures compared with that for the last 36 years. Although the precipitation from June to October (463.7 mm) was similar to the average precipitation levels of the past 36 years (482.9 ± 109.5 mm), the precipitation from June to August was only 221.7 mm, which was less than the average of the same period for the past 36 years (344.9 mm), with 36% less precipitation (Figure 2a). Unfortunately, this lower precipitation was accompanied by higher temperatures, with an average of 14.08 °C from June to October, which was higher than the 36-year average (8.83 ± 0.72 °C) (Figure 2b).

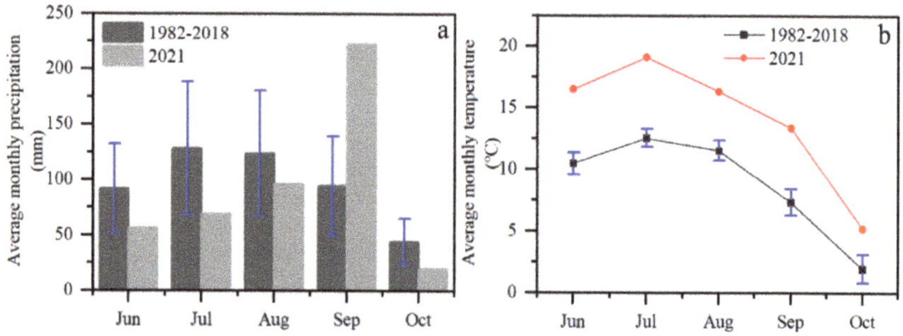

Figure 2. Multiyear average of the monthly cumulative precipitation from 1982 to 2018 and the monthly cumulative precipitation from 2021 (**a**) and multiyear average of the monthly mean temperature from 1982 to 2018 and the monthly mean temperature from 2021 (**b**). The error bars indicate one standard error.

From June to 17 August, the soil water content continued to decrease. In the period of from 4 July to 17 August, the *REW* reached its lowest level, and soil drought occurred (e.g., the shady slope *REW* varied from 0.37 to 0.47 with a mean value of 0.40). After 17 August, the *REW* on the shady slope gradually recovered above the threshold (0.43), which was determined by the shady slope *T* response *REW*. We defined the following four periods based on the threshold value (0.43) of the shady slope *T* response *REW*: for the pre-drought period (1 June–4 July), *REW* > 0.43; for the dry period (4 July–17 August), *REW* < 0.43; for the late dry period (17 August–19 September), *REW* > 0.43; and for the late growing season (19 September–31 October), *REW* > 0.43. The *REW* trend on the sunny slope was similar to that on the shady slope, but its value was lower than that on the shady slope for all periods. For example, in the dry period, the *REW* on the sunny slope varied in the range 0.31–0.45 with a mean value of 0.37, which was 7.5% lower than that on the shady slope.

PET was significantly higher on the sunny slope than on the shady slope (personal correspondence of Li Jiamei) (Figure 3). The *PET* on sunny slopes was 0.2–4.8 mm·day^{-1}, with a mean value of 2.3 mm·day^{-1}, which was 1.2 times higher than on shady slopes (the range of 0.03–4.1 mm·day^{-1}, mean value of 1.9 mm·day^{-1}). The *PET* difference between shady and sunny slopes was significant before and after the drought. For example, low precipitation and high air temperature before and during the drought resulted in higher *PET* with mean values of 3.26 mm·day^{-1} and 3.27 mm·day^{-1} on sunny slopes, respectively, and 2.73 mm·day^{-1} and 2.74 mm·day^{-1} on shady slopes. During the latter part of the drought, *PET* gradually declined, with mean values of 1.95 mm·day^{-1} and 1.56 mm·day^{-1} on sunny and shady slopes, respectively. Late in the growing season, *PET* declined to its lowest value, with mean values of 0.90 mm·day^{-1} and 0.61 mm·day^{-1}, respectively.

3.2. *Sap Flow Density*

The average daily sap flow density was higher on sunny slopes than on shady slopes for all drought periods, i.e., the pre-drought, drought, and late drought periods (Figure 4a). Moreover, the change of daily sap flow density according to drought from the pre-drought period to late drought were very different between the sunny slope and shady slope. On the sunny slope, the sap flow density maintained a stable trend and did not decrease with the development of drought, i.e., there was no big difference of daily sap flow density among three drought periods. Conversely, on the shady slope, it sharply declined significantly ($p < 0.05$) during the drought period.

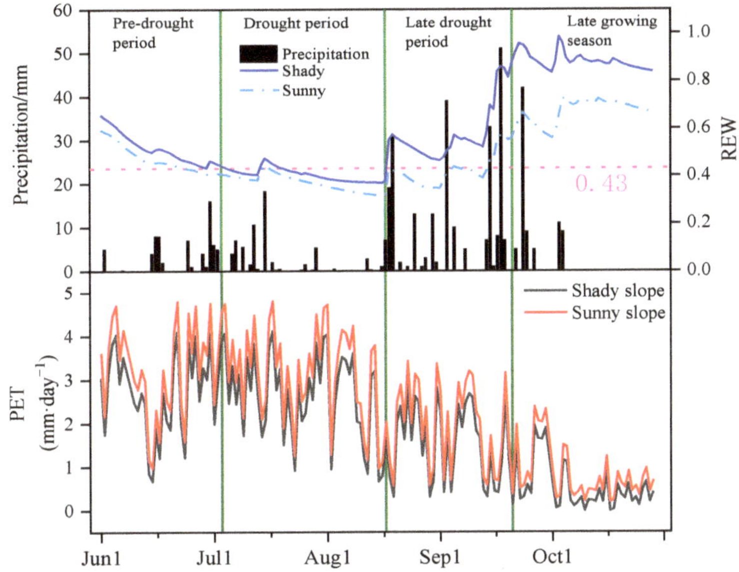

Figure 3. Precipitation variation, *REW* variation, and *PET* variation for shady and sunny slope sample plots from June to October 2021 (the *PET* of both slopes was calculated from unpublished data). A value of 0.43 was determined based on the stand transpiration response *REW* threshold value in 3.3. A *REW* above 0.43 indicated adequate soil moisture, and a *REW* below 0.43 indicated inadequate soil moisture.

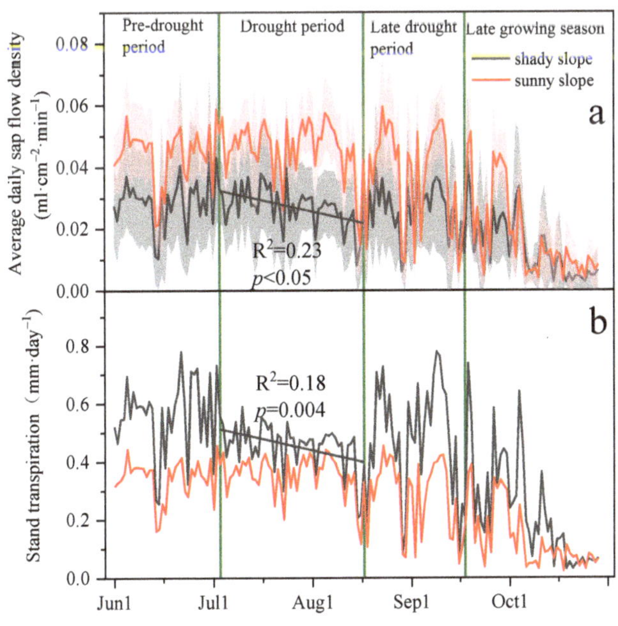

Figure 4. Daily variation of sap flow density (**a**), and stand transpiration (**b**) on shady and sunny slopes. The shaded area indicates the standard deviation of the sap flow density.

3.3. Stand Transpiration Differences between Shady and Sunny Slope Forests

In contrast to the sap flow density situation, stand transpiration on the sunny slope was lower than that on the shady slope for the whole growing season (Figure 4b). The stand transpiration on the sunny slope varied from 0.30 mm·day^{-1} to 0.36 mm·day^{-1} on average, and no dramatic drop of stand transpiration occurred in response to the onset of drought. Surprisingly, on the shady slope, stand transpiration not only decreased significantly during the drought period ($p = 0.004$, mean 0.42 mm·day^{-1}), which was 19.1% and 17.9% lower compared to that in the pre-drought and late drought periods, respectively, but also quickly recovered to the pre-drought level after drought (Figure 4b). However, in any case, the stand transpiration on the sunny slope was 23.4%–40% lower than that on the shady slope.

3.4. Responses of Stand Transpiration to REW and PET in Forest Stands with Different Slope Aspects

For both shady and sunny slopes, *T* increases as a saturated exponential function of *REW* (Figure 5). The effect of the *REW* on *T* remains stable when *REW* reaches a certain level, i.e., a threshold value. However, *T* responded to *REW* at different thresholds among different slope aspects, i.e., the response threshold of stand transpiration to *REW* was 0.43 for the shady slope and 0.33 for the sunny slope.

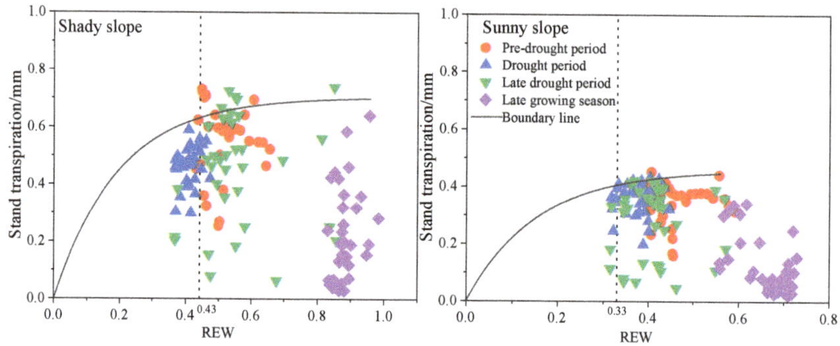

Figure 5. Transpiration responses to *REW* in shady and sunny slope stands.

The response of T to *PET* always followed an increasing saturation exponential function, i.e., T increased rapidly with increasing *PET* and gradually stabilized after it reached a threshold value. However, the response of T to *PET* under sufficient soil moisture was significantly stronger than that under insufficient soil moisture. For the sunny slope, this difference of T among different soil moisture levels was not significant (Figure 6).

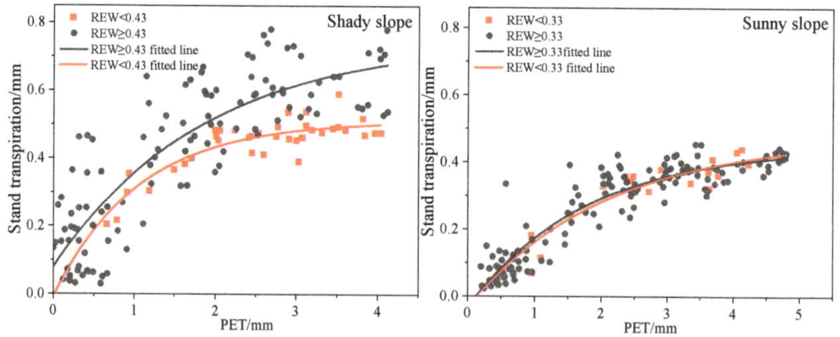

Figure 6. Effects of *PET* on stand transpiration under sufficient and insufficient *REW* conditions.

4. Discussion

4.1. Stand Transpiration Is More Sensitive to Drought on Shady Slope

It has been widely recognized that slope aspect affects stand structure [33] and the forests on shady slopes grow better, such as with larger *DBH*, height, and *LAI*. Additionally, there is less solar radiation and lower *PET* on shady slopes, which accounts for 73.7% and 81.3% of that on sunny slopes in the Liupan Mountains, Northwest China, respectively (personal correspondence of Li Jiamei), and lower soil evaporation [20]. Thus, the stands on shady slopes have more available soil water for tree growth compared with those on sunny slopes (Figure 3).

Our results showed that stand transpiration on shady slopes decreased significantly when soil drought occurred, but recovered quickly after the drought. At the same time, there was no significant change of stand transpiration on sunny slopes regardless of whether the drought occurred. This verified our hypothesis that stand transpiration on shady slopes was more susceptible to drought than that on sunny slopes. Larger trees on shady slopes were more susceptible to drought that occurred frequently [17,18,34,35]. Past studies can indirectly support our results, i.e., that tall trees must lift water to greater heights against gravity and pathlength-associated resistance; therefore, tall trees face greater hydraulic challenges [36–38]. Therefore, the tree growth situation of the stands and the amounts of available soil moisture on different slope aspects are critical in influencing stand transpiration.

4.2. Stand Growth Characteristics Affect Transpiration Response to Drought

The slope aspect shapes the stand growth characteristics [39]. In general, shady slope stands consume more soil water because of their larger sapwood areas, tree heights, and stand transpiration amounts [15,20,26,40]. Thus, large stand transpiration would accelerate the *REW* decline to the threshold, and the *REW* is insufficient to resist gravitational and distance-based transportation to higher tree tops, resulting in a rapid reduction in stand transpiration (Figure 7) [41]. The results of the present study validate that the *T* response to *PET* is significantly higher in the case of adequate *REW* than in the case of insufficient *REW* (Figure 6). As in past studies, better-growing stands have higher *REW* requirements, and the transpiration begins to be limited when soil moisture is insufficient to meet stand transpiration needs [8,42,43].

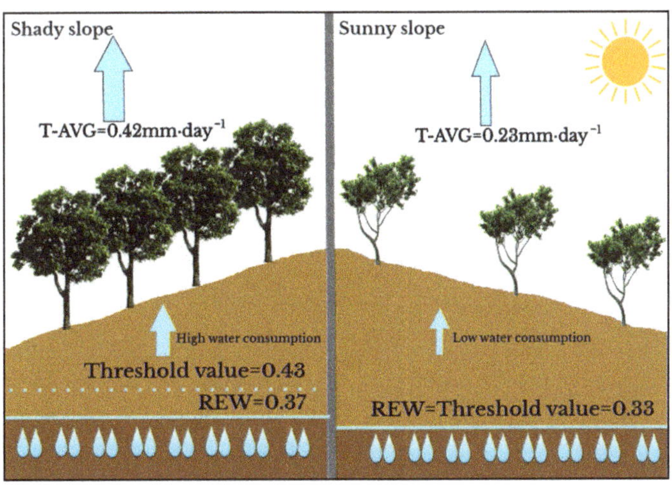

Figure 7. Vegetation profile and response to soil moisture on shady and sunny slopes.

Stands with lower tree height and *LAI* would consume less soil moisture [17,34]. In our study, the sapwood areas and tree heights of the sunny slope stands are smaller [20], and the stand transpiration amounts are lower (only 42% of that of the shady slope) (Figure 4), thus consuming less soil moisture [40]. Under extreme drought, soil moisture may not drop below or near the threshold due to low water consumption by stand transpiration (Figure 5). In the present study, the *LAI* and tree height on the sunny slope are very low to accommodate higher *PET* and lower *REW* [20,33], thus consuming less soil water. Our results demonstrate that there is a small threshold of the *T* response to the *REW* (0.33) and no difference in the *PET* response under different *REW* levels in this drought of 2021 (Figures 6 and 7).

Quercus wutaishansea is the main natural vegetation in forest water areas and grows mainly on shady slopes. Our research shows that the stand transpiration on shady slopes has already been affected by drought. With increasing drought, forests in water source areas will face more serious threats. This reminds us to pay more attention to the forest dynamics related to drought.

5. Conclusions

We analyzed the transpiration characteristics of forest stands on shady and sunny slopes and their responses to drought. We concluded that on shady slopes with better stand conditions, the trees grew better, consumed more water, and were more susceptible to drought fluctuations than on sunny slopes. On sunny slopes with poorer stand conditions, the trees exhibited long-term drought adaptation strategies, such as low LAI and tree heights. These results highlight that in future forest management, more attention should be paid to the forests, especially on shady slopes where forests are predominantly located.

Author Contributions: Conceptualization, B.L., P.Y., and Y.W. (Yanhui Wang); instrument installation, B.L., X.Z., L.P., and J.L.; investigation, B.L., Z.L., and X.Z.; formal analysis, B.L.; writing—original draft preparation, B.L. and P.Y.; writing—review and editing, B.L., P.Y., Y.W. (Yanfang Wan), Y.Y., X.W., B.L., and L.X. All authors have read and agreed to the published version of the manuscript.

Funding: This work was financially supported by the Central Public-Interest Scientific Institution Basal Research Fund of Chinese Academy of Forestry (CAFYBB2022XD003, CAFYBB2021ZW002), the National Natural Science Foundation of China (U21A2005, U20A2085, 42161144008), and the National Key Research & Development Program of China (2022YFF0801804, 2022YFF0801803).

Acknowledgments: We thank Yongqiang Hu and Jun Zhang of Liupanshan Forestry Bureau for their support in field work.

Conflicts of Interest: The authors declare no conflict of interest.

References

1. Good, S.P.; Noone, D.; Bowen, G. Hydrologic connectivity constrains partitioning of global terrestrial water fluxes. *Science* **2015**, *349*, 175–177. [CrossRef] [PubMed]
2. Jasechko, S.; Sharp, Z.D.; Gibson, J.J.; Birks, S.J.; Yi, Y.; Fawcett, P.J. Terrestrial water fluxes dominated by transpiration. *Nature* **2013**, *496*, 347–350. [CrossRef] [PubMed]
3. Lian, X.; Piao, S.; Huntingford, C.; Li, Y.; Zeng, Z.; Wang, X.; Ciais, P.; McVicar, T.R.; Peng, S.; Ottlé, C.; et al. Partitioning global land evapotranspiration using CMIP5 models constrained by observations. *Nat. Clim. Change* **2018**, *8*, 640–646. [CrossRef]
4. Schlesinger, W.H.; Jasechko, S. Transpiration in the global water cycle. *Agric. For. Meteorol.* **2014**, *189–190*, 115–117. [CrossRef]
5. Adams, H.D.; Luce, C.H.; Breshears, D.D.; Allen, C.D.; Weiler, M.; Hale, V.C.; Smith, A.M.S.; Huxman, T.E. Ecohydrological consequences of drought- and infestation-triggered tree die-off: Insights and hypotheses. *Ecohydrology* **2012**, *5*, 145–159. [CrossRef]
6. Sun, S.; Sun, G.; Caldwell, P.; McNulty, S.; Cohen, E.; Xiao, J.; Zhang, Y. Drought impacts on ecosystem functions of the US National Forests and Grasslands: Part II assessment results and management implications. *For. Ecol. Manag.* **2015**, *353*, 269–279. [CrossRef]
7. Lyu, J.; He, Q.-Y.; Chen, Q.-W.; Cheng, R.-R.; Li, G.; Otsuki, K.; Yamanaka, N.; Du, S. Distinct transpiration characteristics of black locust plantations acclimated to semiarid and subhumid sites in the Loess Plateau, China. *Agric. Water Manag.* **2022**, *262*, 107402. [CrossRef]
8. Song, X.; Lyu, S.; Wen, X. Limitation of soil moisture on the response of transpiration to vapor pressure deficit in a subtropical coniferous plantation subjected to seasonal drought. *J. Hydrol.* **2020**, *591*, 125301. [CrossRef]

9. Gutierrez Lopez, J.; Tor-Ngern, P.; Oren, R.; Kozii, N.; Laudon, H.; Hasselquist, N.J. How tree species, tree size, and topographical location influenced tree transpiration in northern boreal forests during the historic 2018 drought. *Glob. Change Biol.* **2021**, *27*, 3066–3078. [CrossRef]
10. Wang, W.; McDowell, N.G.; Pennington, S.; Grossiord, C.; Leff, R.T.; Sengupta, A.; Ward, N.D.; Sezen, U.U.; Rich, R.; Megonigal, J.P.; et al. Tree growth, transpiration, and water-use efficiency between shoreline and upland red maple (*Acer rubrum*) trees in a coastal forest. *Agric. For. Meteorol.* **2020**, *295*, 108163. [CrossRef]
11. Tian, A.; Wang, Y.; Webb, A.A.; Liu, Z.; Pengtao, Y.; Xiong, W.; Wang, X. Partitioning the causes of spatiotemporal variation in the sunny day sap flux density of a larch plantation on a hillslope in northwest China. *J. Hydrol.* **2019**, *571*, 503–515. [CrossRef]
12. Chang, X.; Zhao, W.; Liu, H.; Wei, X.; Liu, B.; He, Z. Qinghai spruce (*Picea crassifolia*) forest transpiration and canopy conductance in the upper Heihe River Basin of arid northwestern China. *Agric. For. Meteorol.* **2014**, *198–199*, 209–220. [CrossRef]
13. Liu, Z.; Wang, Y.; Tian, A.; Webb, A.A.; Yu, P.; Xiong, W.; Xu, L.; Wang, Y. Modeling the Response of Daily Evapotranspiration and its Components of a Larch Plantation to the Variation of Weather, Soil Moisture, and Canopy Leaf Area Index. *J. Geophys. Res. Atmos.* **2018**, *123*, 7354–7374. [CrossRef]
14. Xiong, W.; Oren, R.; Wang, Y.; Yu, P.; Liu, H.; Cao, G.; Xu, L.; Wang, Y.; Zuo, H. Heterogeneity of competition at decameter scale: Patches of high canopy leaf area in a shade-intolerant larch stand transpire less yet are more sensitive to drought. *Tree Physiol.* **2015**, *35*, 470–484. [CrossRef]
15. Hawthorne, S.; Miniat, C.F. Topography may mitigate drought effects on vegetation along a hillslope gradient. *Ecohydrology* **2017**, *11*, e1825. [CrossRef]
16. Shi, L.; Liu, H.; Xu, C.; Liang, B.; Cao, J.; Cressey, E.L.; Quine, T.A.; Zhou, M.; Zhao, P. Decoupled heatwave-tree growth in large forest patches of Larix sibirica in northern Mongolian Plateau. *Agric. For. Meteorol.* **2021**, *311*, 108667. [CrossRef]
17. Lüttschwager, D.; Jochheim, H. Drought Primarily Reduces Canopy Transpiration of Exposed Beech Trees and Decreases the Share of Water Uptake from Deeper Soil Layers. *Forests* **2020**, *11*, 537. [CrossRef]
18. Colangelo, M.; Camarero, J.J.; Borghetti, M.; Gazol, A.; Gentilesca, T.; Ripullone, F. Size Matters a Lot: Drought-Affected Italian Oaks Are Smaller and Show Lower Growth Prior to Tree Death. *Front. Plant Sci.* **2017**, *8*, 135. [CrossRef]
19. Yang, J.; El-Kassaby, Y.A.; Guan, W. The effect of slope aspect on vegetation attributes in a mountainous dry valley, Southwest China. *Sci. Rep.* **2020**, *10*, 16465. [CrossRef]
20. Metzen, D.; Sheridan, G.J.; Benyon, R.G.; Bolstad, P.V.; Griebel, A.; Lane, P.N.J. Spatio-temporal transpiration patterns reflect vegetation structure in complex upland terrain. *Sci. Total Environ.* **2019**, *694*, 133551. [CrossRef]
21. Wang, L.; Liu, Z.; Guo, J.; Wang, Y.; Ma, J.; Yu, S.; Yu, P.; Xu, L. Estimate canopy transpiration in larch plantations via the interactions among reference evapotranspiration, leaf area index, and soil moisture. *For. Ecol. Manag.* **2021**, *481*, 118749. [CrossRef]
22. Kume, T.; Tsuruta, K.; Komatsu, H.; Shinohara, Y.; Katayama, A.; Ide, J.; Otsuki, K. Differences in sap flux-based stand transpiration between upper and lower slope positions in a Japanese cypress plantation watershed. *Ecohydrology* **2016**, *9*, 1105–1116. [CrossRef]
23. Renner, M.; Hassler, S.K.; Blume, T.; Weiler, M.; Hildebrandt, A.; Guderle, M.; Schymanski, S.J.; Kleidon, A. Dominant controls of transpiration along a hillslope transect inferred from ecohydrological measurements and thermodynamic limits. *Hydrol. Earth Syst. Sci.* **2016**, *20*, 2063–2083. [CrossRef]
24. Dudley, B.D.; Marttila, H.; Graham, S.L.; Evison, R.; Srinivasan, M.S. Water sources for woody shrubs on hillslopes: An investigation using isotopic and sapflow methods. *Ecohydrology* **2018**, *11*, e1926. [CrossRef]
25. Lyu, J.; He, Q.-Y.; Yang, J.; Chen, Q.-W.; Cheng, R.-R.; Yan, M.-J.; Yamanaka, N.; Du, S. Sap flow characteristics in growing and non-growing seasons in three tree species in the semiarid Loess Plateau region of China. *Trees* **2020**, *34*, 943–955. [CrossRef]
26. Wan, Y.; Yu, P.; Wang, Y.; Wang, B.; Yu, Y.; Wang, X.; Liu, Z.; Liu, X.; Wang, S.; Xiong, W. The Variation in Water Consumption by Transpiration of Qinghai Spruce among Canopy Layers in the Qilian Mountains, Northwestern China. *Forests* **2020**, *11*, 845. [CrossRef]
27. Szatniewska, J.; Zavadilova, I.; Nezval, O.; Krejza, J.; Petrik, P.; Čater, M.; Stojanović, M. Species-specific growth and transpiration response to changing environmental conditions in floodplain forest. *For. Ecol. Manag.* **2022**, *516*, 120248. [CrossRef]
28. Allen, R.; Pereira, L.; Raes, D.; Smith, M.; Allen, R.G.; Pereira, L.S.; Martin, S. *Crop Evapotranspiration: Guidelines for Computing Crop Water Requirements*; FAO Irrigation and Drainage Paper 56; FAO: Rome, Italy, 1998.
29. Granier, A.; Loustau, D.; Brda, N. A generic model of forest canopy conductance dependent on climate, soil water availability and leaf area index. *Ann. For. Sci.* **2000**, *57*, 755–765. [CrossRef]
30. Zhang, Y.P.; Wang, J.; Qu, Y.T.; Liu, S.P.; Wang, Y.; Shanggune, Z.P. Soil Water Characteristic Curves and Soil Water Constants at Different Depths in the Abandoned Sloping Field of Loess Hilly Region. *J. Irrig. Drain.* **2020**, *39*, 6. [CrossRef]
31. Li, Z.; Yu, P.; Wang, Y.; Webb, A.A.; He, C.; Wang, Y.; Yang, L. A model coupling the effects of soil moisture and potential evaporation on the tree transpiration of a semi-arid larch plantation. *Ecohydrology* **2016**, *10*, e1764. [CrossRef]
32. Wang, Y.; Wang, Y.; Pengtao, Y.U.; Xiong, W.; Apeng, D.U.; Zhenhua, L.I.; Liu, Z.; Ren, L.; Lihong, X.U.; Zuo, H. Simulated responses of evapotranspiration and runoff to changes in the leaf area index of a *Larix principis-rupprechtii* plantation. *Acta Ecol. Sin.* **2016**, *36*, 6928–6938. [CrossRef]
33. Renninger, H.J.; Carlo, N.J.; Clark, K.L.; Schafer, K.V. Resource use and efficiency, and stomatal responses to environmental drivers of oak and pine species in an Atlantic Coastal Plain forest. *Front. Plant Sci.* **2015**, *6*, 297. [CrossRef] [PubMed]

34. Bennett, A.C.; McDowell, N.G.; Allen, C.D.; Anderson-Teixeira, K.J. Larger trees suffer most during drought in forests worldwide. *Nat. Plants* **2015**, *1*, 15139. [CrossRef]
35. Zang, C.; Pretzsch, H.; Rothe, A. Size-dependent responses to summer drought in Scots pine, Norway spruce and common oak. *Trees* **2011**, *26*, 557–569. [CrossRef]
36. Ryan, M.G.; Phillips, N.; Bond, B.J. The hydraulic limitation hypothesis revisited. *Plant Cell Environ.* **2006**, *29*, 367–381. [CrossRef]
37. Zhang, Y.J.; Meinzer, F.C.; Hao, G.Y.; Scholz, F.G.; Bucci, S.J.; Takahashi, F.; Villalobos-Vega, R.; Giraldo, J.P.; Cao, K.F.; Hoffmann, W.A. Size-dependent mortality in a Neotropical savanna tree: The role of height-related adjustments in hydraulic architecture and carbon allocation. *Plant Cell Environ.* **2009**, *32*, 1456–1466. [CrossRef] [PubMed]
38. Mcdowell, N.G.; Allen, C.D. Darcy's law predicts widespread forest mortality under climate warming. *Nat. Clim. Change* **2015**, *5*, 669–672. [CrossRef]
39. Zalloni, E.; Battipaglia, G.; Cherubini, P.; Saurer, M.; De Micco, V. Wood Growth in Pure and Mixed *Quercus ilex* L. Forests: Drought Influence Depends on Site Conditions. *Front. Plant Sci.* **2019**, *10*, 397. [CrossRef]
40. Yu, S.; Guo, J.; Liu, Z.; Wang, Y.; Ma, J.; Li, J.; Liu, F. Assessing the Impact of Soil Moisture on Canopy Transpiration Using a Modified Jarvis-Stewart Model. *Water* **2021**, *13*, 2720. [CrossRef]
41. Ouyang, L.; Gao, J.; Zhao, P.; Rao, X. Species-specific transpiration and water use patterns of two pioneer dominant tree species under manipulated rainfall in a low-subtropical secondary evergreen forest. *Ecohydrology* **2020**, *13*, e2234. [CrossRef]
42. Song, L.; Zhu, J.; Zhang, T.; Wang, K.; Wang, G.; Liu, J. Higher canopy transpiration rates induced dieback in poplar (Populus × xiaozhuanica) plantations in a semiarid sandy region of Northeast China. *Agric. Water Manag.* **2021**, *243*, 106414. [CrossRef]
43. Fang, W.; Lu, N.; Liu, J.; Jiao, L.; Zhang, Y.; Wang, M.; Fu, B. Canopy transpiration and stand water balance between two contrasting hydrological years in three typical shrub communities on the semiarid Loess Plateau of China. *Ecohydrology* **2019**, *12*, 106414. [CrossRef]

forests

MDPI

Article

Water Uptake Pattern by Coniferous Forests in Two Habitats Linked to Precipitation Changes in Subtropical Monsoon Climate Region, China

Jianbo Jia [1,2,3], Yu Chen [1,2], Jia Lu [1,2] and Wende Yan [1,3,*]

1 Central South University of Forestry and Technology, Changsha 410004, China; jotham880303@163.com (J.J.); cchenyu0702@163.com (Y.C.); t20162294@csuft.edu.cn (J.L.)
2 Key Laboratory of Soil and Water Conservation and Desertification Combating in Hunan Province, Changsha 410004, China
3 Lutou National Station for Scientific Observation and Research of Forest Ecosystem in Hunan Province, Yueyang 414000, China
* Correspondence: csfuywd@hotmail.com

Abstract: Variations in precipitation patterns under climate changes influence water availability, which has important implications for plants' water use and the sustainability of vegetation. However, the water uptake patterns of the main forest species under different temporal spatial conditions of water availability remain poorly understood, especially in areas of high temporal spatial heterogeneity, such as the subtropical monsoon climate region of China. We investigated the water uptake patterns and physiological factors of the most widespread and coniferous forest species, *Cunninghamia lanceolata* L. and *Pinus massoniana* L., in the early wet season with short drought (NP), high antecedent precipitation (HP), and low antecedent precipitation (LP), as well as in the early dry season (DP), in edaphic and rocky habitats. The results showed that the two species mainly absorbed soil water from shallow layers, even in the short drought period in the wet season and switched to deeper layers in the early dry season in both habitats. It was noted that the trees utilized deep layers water in edaphic habitats when the antecedent rainfall was high. The two species showed no significant differences in water uptake depth, but exhibited notably distinct leaf water potential behavior. *C. lanceolata* maintained less negative predawn and midday water potential, whereas *P. massoniana* showed higher diurnal water potential ranges. Moreover, the water potential of *P. massoniana* was negatively associated with the antecedent precipitation amount. These results indicate that for co-existing species in these communities, there is significant eco-physiological niche segregation but no eco-hydrological segregation. For tree species in two habitats, the water uptake depth was influenced by the available soil water but the physiological factors were unchanged, and were determined by the species' genes. Furthermore, during the long drought in the growing season, we observed probable divergent responses of *C. lanceolata* and *P. massoniana*, such as growth restriction for the former and hydraulic failure for the latter. However, when the precipitation was heavy and long, these natural species were able to increase the ecohydrological linkages between the ecosystem and the deep-layer system in this edaphic habitat.

Keywords: plant water source; habitat; stable isotope technology; leaf water potential; water use efficiency

Citation: Jia, J.; Chen, Y.; Lu, J.; Yan, W. Water Uptake Pattern by Coniferous Forests in Two Habitats Linked to Precipitation Changes in Subtropical Monsoon Climate Region, China. *Forests* **2022**, *13*, 708. https://doi.org/10.3390/f13050708

Academic Editor: Steven McNulty

Received: 1 April 2022
Accepted: 29 April 2022
Published: 30 April 2022

1. Introduction

Increases in vegetation greenness have been reported around the world over the last three decades, manifested as the expansion of afforestation and reforestation [1–3]. However, forests may be vulnerable to degradation due to global climate changes with new precipitation patterns [4–6]. Changes in the characteristics of precipitation may result in changes in water availability, which have implications for plants' water uses in ecosystems [7,8]. The variations in plants' water use responses to precipitation and water availability plays important role in the sustainability of the restored vegetation and the promotion of the water cycle in critical zones [9–11].

The temporal–spatial heterogeneity of precipitation and water availability affect plant water use strategies [12,13]. At the point scale, the water source variability along the soil profile is one of the most important factors for water uptake by plants [14]. At the surface scale, the aquifer storage is distinct in different habitats, such as deep soil habitats [15], outcrop habitats [16], and soil with rock fragments habitats [17], which is related to the soil properties and plant water consumption. At different stages of the same season, the plant water uptake depth may also differ with changes in rhizosphere water availability [15,18]. Meanwhile, the amount of precipitation may be a critical factor affecting the water sources of trees. The plant water uptake can be identified by contrasting the δD and $\delta^{18}O$ of xylem water and all the potential water sources. Previous studies have shown that tree species may switch their water sources from shallow layers in the wet season with sufficient precipitation to stable layers in the dry season using stable isotope techniques [19–21] Liu [22] found that following rainfall events, *Platycladus orientalis* L. trees with a dense and shallow fine root system absorbed more water from the soil surface layers and precipitation. Other plants, however, mostly take up water from deep and stable layers regardless of seasonal changes or precipitation events in the semi-arid regions [16,23]. In contrast, in subtropical regions, evergreen species use shallow soil water with a drought-avoidance strategy even under seasonal drought conditions [24].

The divergent response of plant water uptake to changes in precipitation and water availability has been related to physiological characteristics. It has been suggested that the predawn and midday leaf water potential can be used to describe the daily patterns of plant-water relations, coupling water among the root zone, the plant itself and the atmosphere [25]. Previous studies have shown that plants relying on shallower water sources exhibited a larger diurnal range of leaf water potential, and on the contrary, narrower diurnal ranges are usually linked with deep and stable water sources [26–28]. Moreover, the plant water efficiency (WUE) has attracted attention as a means of reflecting plant water use characteristics, together with plant water uptake [29,30]. Nie [31] explored leaf WUE based on $\delta^{13}C$ values and found that the high WUE corresponded with the use of deep water sources, indicating more conservative water-use strategies in a subtropical monsoon climate region. The plant water uptake pattern was found to be influenced by water availability and physiological traits in different ecosystems [32]. However, the relationship between these two factors affecting plant water uptake is unclear, especially in complex and fragile forest ecosystems, which limits the understanding of restored vegetation adaptability and rock-soil-water-plant-atmosphere interactions in critical zones.

Subtropical China, which is characterized by a monsoon climate, is an ecologically sensitive area that is affected by global changes [33]. The precipitation in this region is abundant and the alternation of dry and wet is obvious. The change in precipitation patterns has led to a reduction in the available water in the ecosystem, and the risk of drought stress and drought death has significantly increased [4]. The distribution of the plantations is a clustered distribution with heterogeneous habitats (such as thin soil habitats with rock fragments, and outcrop habitats with soil fragments) [34]. Different rock and soil structures could influence hydrological processes and the amount of soil water available. Plant water use strategies in different habitats are critically important for the evaluation of vegetation adaptation. A number of previous studies have primarily focused on the water sources of different types of plantations or natural vegetation in one specific habitat [16,35]. Few have paid attention to differences in plant water uptake patterns in different habitats, which has limited our understanding of plant water adaptation and the evaluation of sustainable vegetation restoration. With changes in the global precipitation pattern, short-term drought and rainstorms have become more frequent, especially in the wet/growing season. However, it is unclear how the water uptake of plants in the different habitats responds to these precipitation changes.

Based on the above analysis, we applied stable isotope techniques (δD and $\delta^{18}O$) and measurements of leaf water potential to determine coniferous-leaved forest water uptake patterns in two habitats (edaphic and rocky habitats) with different antecedent precipitation during the growing season in the subtropical monsoon climate region of China. The main objectives of this study were: (i) to investigate the responses of belowground water use patterns of *Cunninghamia lanceolata* L. and *Pinus massoniana* L. to the temporal–spatial heterogeneity of water availability, as shown in conditions of different antecedent precipitation levels and edaphic and rocky habitats; and (ii) to understand the aboveground physiological responses to varied water availability of two species, analyzed by examining the variations in leaf water potential behavior and water use efficiency. Our first hypothesis was that soil water availability could have an effect on the plant water uptake depth, and that the two species may show similar water sources, and the second was that the plants' physiological factors may vary with the changes in water availability and species types.

2. Materials and Methods

2.1. Study Sites

This study was conducted at the Hunan Lutou forest ecosystem observation and research Station (28°31′7″–28°38′ N, 113°51′52″–113°58′24″ E) (Figure 1). The region has a subtropical mountainous monsoon climate, with a mean annual precipitation of 1450.8 mm and an annual temperature of 18.5 °C. The wet season, which receives more than 60% of the annual rainfall, lasts from late April to late September, and the dry season extends from December to February [28]. The growing season spans from April to October. The study area is dominated by *Cunninghamia lanceolata* and *Pinus massoniana* secondary forests. The understory contains species such as *Fortunearia sinensis* Rehd, *Ilex cornuta* and *Asparagus cochinchinensis*, and the forest coverage rate is more than 90%. Soil in the study area is predominantly red soil, having a general soil layer that is 80–100 cm thick. The other part of the slope has a high exposed rock ratio whereas the soil occurs discontinuously, only in rock gaps. Thus, the habitats were variable, with the different outcrop ratios, such as an edaphic habitat with a low outcrop ratio, a continuous broken rock habitat with patches of soil, an isolated outcrop habitat, and so on. Springs sometimes appear at the bottom of hillslopes during the rainy season or after rains in the drought season.

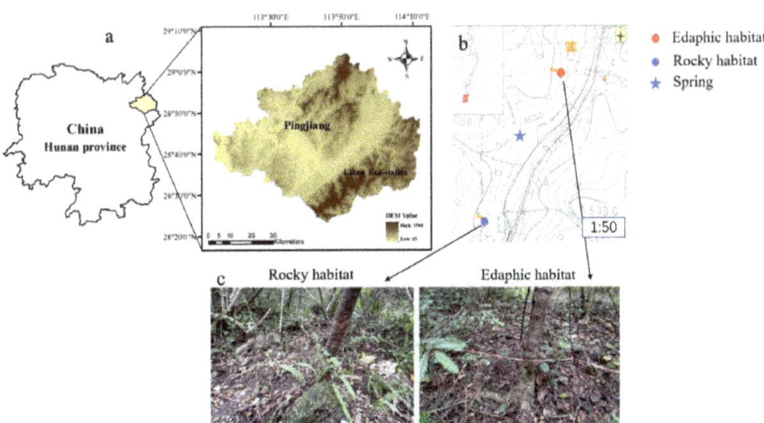

Figure 1. Map of the study and the field area. (**a**): The location of Lutou forest ecosystem observation and research Station in Hunan province; (**b**): The sampling site location in the study area; (**c**) photographs of the two habitats.

According to the distribution of these typical habitats, one habitat consisting of thick soil with rock fragments habitat (the "edaphic habitat" for short) and the another habitat consisting of continuous stone outcrops with soil fragments (the "rocky habitat" for short) were chosen at the foot of the Northwest-facing hillslope in two 20 m × 20 m sample plots. The two habitats were 250 m apart whereas the elevation difference was about five meters. In the edaphic habitat, the soil was relatively thick (about 90 cm deep), horizontally interrupted by small outcrops, and vertically interrupted by small rocks. Along the soil profile, the upper layer soil (0–30 cm) was well-drained whereas the lower layers (30–70 cm) were sticky with a low saturated hydraulic conductivity (K_s). Underneath the soil was a high-weathered dolomite bedrock zone (70–90 cm). The outcrop ratio was about 20% in this habitat. In the rocky habitat, the outcrop ratio was more than 80%, and the range of height from the top of the outcrop to the soil in the rock gaps was from 0.3 m to 3 m. The soil was inlaid in the rock in a spotty pattern and was discontinuous (average 30 cm deep on average). Similarly, a high-weathered bedrock zone was present under the soil. The vegetation was sparse in this habitat. There was an intermittent spring outflow near the two habitats at the bottom of the hillslope.

For the subsequent analysis and comparison, the plant water sources were divided into shallow (0–30 cm), middle (30–70 cm in the edaphic habitat and 30–50 cm in the rocky habitat), and deep (70–90 cm in the edaphic habitat and 50–70 cm in the rocky habitat) layers and spring according to the soil texture and fluctuations and patterns of isotopic ratios in the soil water, VWC, and the impact of the rainfall pulse. (1) Shallow soil layer: The variability of soil water isotopic compositions and VWC in this layer was larger, and it was vulnerable to rainfall pulses and evaporation with seasons. (2) Middle soil layer: The variability of soil water isotopic compositions and SWC in this layer was lower than that of the 0–30 cm soil layer. The impacts of rainfall pulses and evaporation were moderate. Both the clay content and soil bulk density were higher than the shallow layers. (3) Deep soil layer: This layer was high-weathered bedrock with high leakage and low water holding capacity in the rocky habitat and high water moisture in the edaphic habitat, respectively.

2.2. Plant and Soil Sampling

In order to explore the relationship between plant physiological traits and water uptake patterns for adapting to precipitation change, plant and soil sampling were conducted simultaneously at the two habitats bimonthly on 12 June (wet season with high antecedent precipitation, HP), 5 August (wet season with low antecedent precipitation, LP), and 18 October (early dry season, DP) 2020. We also sampled on 18 May in the early wet season with a 20 day drought (no rain, NP). Two coniferous species, *Cunninghamia lanceolata* (DBH of from 5 to 11cm, average DBH was 7.9 cm) and *Pinus massoniana* (DBH of from 6 to 12cm, average DBH was 8.5cm) in each of the habitats were selected for the study. We selected four individuals per species for analysis, and the DBH of sampled trees were used to represent the average DBH in the stands. The leaf and plant xylem samples from every selected plant were collected in each habitat. Every selected plant was collected in each stand-age tree per month. The fully sun-exposed, mature and healthy leaves in the upper canopy from each selected plant were collected in different directions on each sampling date. The leaves were mixed and packed into craft paper bags and brought them back to the laboratory for the measurement of the plant leaves' $\delta^{13}C$ levels. Shoots ranging from 0.3 to 0.5 cm in diameter and 3 to 5cm in length were collected at mid-day from stems that were more than 2 years old [28]; the outer bark and phloem of the shoots were removed to obtain the xylem sample.

Soil samples were obtained in two habitats from six depth intervals (0–10, 10–20, 20–30, 30–50, 50–70, 70–90 cm) with an auger (sampling only at 70 cm deep in the rocky habitat) and five replicates were collected at each layer. Among them, the high-weathered bedrock samples were collected between 70–90 cm in the edaphic habitat and 50–70 cm in the rocky habitat. A subsample of the soil samples was stored at −20 °C for isotopic analysis, whereas the remainder of the samples were sealed for the measurement of gravimetric soil water content, obtained by oven drying for one day. The volumetric water content (VWC) was converted according to gravimetric water content and bulk density (Table 1) of each layer.

Table 1. The soil bulk density of two the habitats.

Soil Depth (cm)	Edaphic Habitat (g·cm^{-3})	Rocky Habitat (g·cm^{-3})
0–10	0.90	0.88
10–20	0.97	0.85
20–30	1.09	0.91
30–50	1.13	1.03
50–70	1.01	0.88
70–90	0.91	-

2.3. Precipitation and Spring Sampling

Water samples were routinely collected for each rain event above 5 mm from May 2020 to December 2020. The isotopic values of precipitation were not collected from January to April due to the COVID-19 pandemic. The collection equipment was designed based on the new device for monthly rainfall sampling developed for the Global Network of Isotopes in Precipitation [36]. The rainwater samples were stored in cap vials, wrapped in parafilm, and stored in a freezer until the analysis of stable isotopes. Data on temporal distribution of rainfall data and other meteorological data were collected at a meteorological station located in the middle of the same small catchment. Spring water discharged from1 June to 29 November, but were cut off between 25 July to 29 August. The spring was sampled regularly during the outflow period. Both rainwater and spring water were stored in cap vials, wrapped in parafilm, and frozen until stable isotope analysis.

2.4. Isotopic Analyses

The water was extracted from the xylem and soil using an automatic cryogenic vacuum distillation water extraction system (LI-2100, LICA, Beijing, China) [37,38]. The δD and δ^{18}O in the xylem and soil water samples were measured with liquid water isotope ratio infrared spectroscopy (IRIS, DLT-100, Los Gatos Research, Mountain View, CA, USA) at the Key Laboratory for Agro-Ecological Processes in Subtropical Region, Chinese Academy of Sciences. The δ^{13}C level in the plant leaves were analyzed using an isotope ratio mass spectrometer (IRMS, MAT253, Thermo Fisher Scientific, Bremen, Germany).

The isotope composition is reported in δ notation relative to V-SMOW as

$$\delta X = (R_{sample}/R_{standard} - 1) \times 1000 \quad (1)$$

where X represents D, ^{18}O, or ^{13}C. R_{sample} and $R_{standard}$ are the ratios of D/H, $^{18}O/^{16}O$, or $^{13}C/^{12}C$ ratio of a measured sample and a standard sample, respectively. The standard deviation for repeat measurements was ±1‰ for δD, ±0.2‰ for δ^{18}O, and ±0.15‰ for δ^{13}C.

Extracting water from the plant xylem using cryogenic vacuum distillation can result in the mixing of organic materials (e.g., methanol and ethanol), which may affect the spectroscopy and lead to erroneous stable isotope values when analyzing them with IRIS [39,40]. We have corrected the isotopic values of the xylem according to Liu [28].

2.5. Leaf Water Potential

Predawn and midday water potentials (Ψ_{pd} and Ψ_{md}, respectively) of leaves were measured in the wet seasons (simultaneously with isotope sampling) with a pressure chamber (PMS Instruments Co., Corvallis, OR, USA). Samples (n = 5 per species) were collected from branches that were fully exposed to the sun, at places where branches were 2/3 of the way up of the canopy, at least 2 m above ground. The measurements were performed between 4:00 to 6:00 h for predawn water potential and between 12:00 and 14:00 h for midday water potential on the same day.

2.6. Data Analysis

Plant water source partitioning was determined by means of the Bayesian mixing model MixSIAR (version 3.1.7) [41]. The raw isotopic ratios of the xylem water were input into MixSIAR as the mixture data. The averages and standard deviations of the soil water isotopes in the different soil layers were the source data. The discrimination was set to zero for both δD and $\delta^{18}O$ because there is generally no isotopic discrimination of water during plant water uptake by roots [42].

Independent-samples *t*-tests and One-way ANOVA were used to detect the differences in plant water sources and water potential among the species, habitats and their seasonal differences. Post hoc comparisons were based on Tukey's HSD. Moreover, Pearson correlation was used to conduct the correlation analysis, and the figures were plotted with Origin 9.0 software (Origin, Origin Lab, Farmington, ME, USA).

3. Results

3.1. Isotopic Compositions of Precipitation, Soil Water, and Springs

The total precipitation was approximately 2121 mm in 2020 and the distribution of rainfall was temporally uneven, with 79.32% of the rainfall occurring during the wet season (Figure 2). It was noted that there are two extreme precipitation events occurred—on 7 in September (282.2 mm) and on 7 June (115.2 mm). Except for the NP sampling with a 20 day drought, the accumulated precipitation amounts ten days before the last three samplings were 283.6 mm, 49.4 mm, and 55.4 mm, respectively.

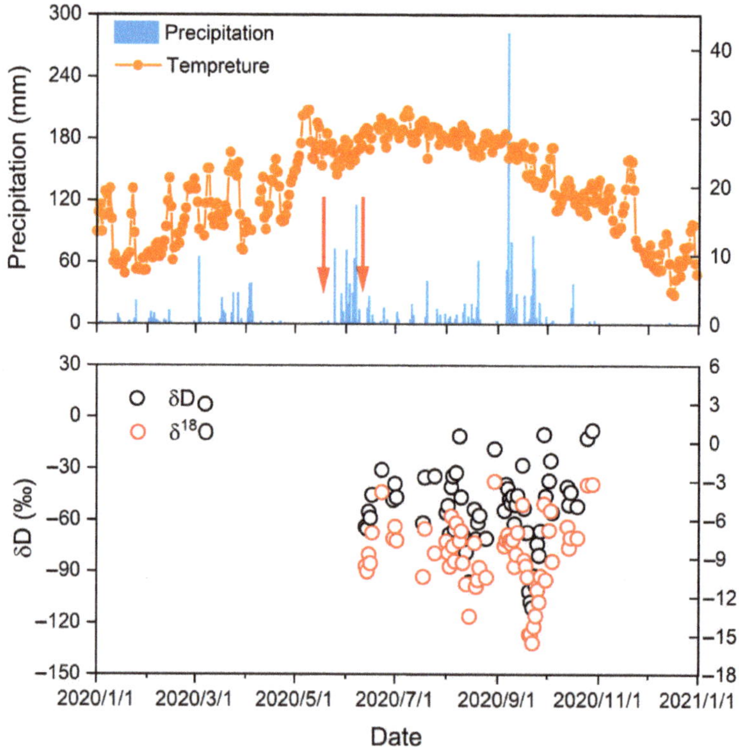

Figure 2. Variations in precipitation, mean air temperature, and isotopic values (δD, $\delta^{18}O$) in precipitation at a daily timescale in 2020. Arrows indicate sampling dates. (The isotopic values of precipitation were not collected from January to April due to the COVID-19 pandemic impacting).

The isotopic compositions of the precipitation showed a large fluctuation during the study period (Figure 2). The mean δD of the precipitation was −48.69 ‰, the mean $\delta^{18}O$ of the precipitation was −7.88 ‰. The relatively depleted isotopic values of precipitation occurred when it rained continuously for a long time, with high precipitation. The δD level obtained for ten days of precipitation before the three samplings were ranged from −23.55‰ to −57.52‰, −34.54‰ to −68.36‰, −40.76‰ to −51.02‰, respectively. The $\delta^{18}O$ of precipitation before three samplings were ranging from −5.27‰ to −8.15 ‰, −7.6 ‰ to −9.65 ‰, −6.54‰ to −7.4‰, respectively.

The δD and $\delta^{18}O$ values of soil water in the different habitats varied with soil depth and season (Figures 3 and 4). In the edaphic habitat, the average δD value of the soil water was −45.56‰ ± 16.05 ‰ (mean ± S.D.), and the average $\delta^{18}O$ value was −6.55‰ ± 1.73‰. The average δD and $\delta^{18}O$ values of soil water in the rocky habitat were −44.6‰ ± 16.58‰ and −6.7‰ ± 1.96 ‰, respectively. There were no significant differences ($p = 0.84$ for δD, $p = 0.79$ for $\delta^{18}O$) in the soils' isotopic compositions in the different habitats. In NP, the soil water isotopes were observed to be depleted with soil depth (Figures 3a and 4a). In HP, the δD and $\delta^{18}O$ values of water along the soil profile were consistent with recent rainfall values (Figures 3b and 4b). In the two late two samplings, the soil water isotope composition converged at the top and bottom layers, which were similar to recent rainfall values (Figure 3c,d, and Figure 4c,d). The middle-layer soil water showed more enriched values in LP and depleted isotopic values in DP and exhibited less variation with soil depth. There were no significant differences obtained for soil water ($p = 1.28$ for δD, $p = 0.93$ for $\delta^{18}O$) in different sample layers (i.e., 0–10 cm, 10–30 cm, 30–50 cm, 50–70 cm, 70-90 cm). However, when merging sample layers into shallow, middle and deep layers (see Methods), the soil water isotope was significant different in the two habitats ($p < 0.05$).

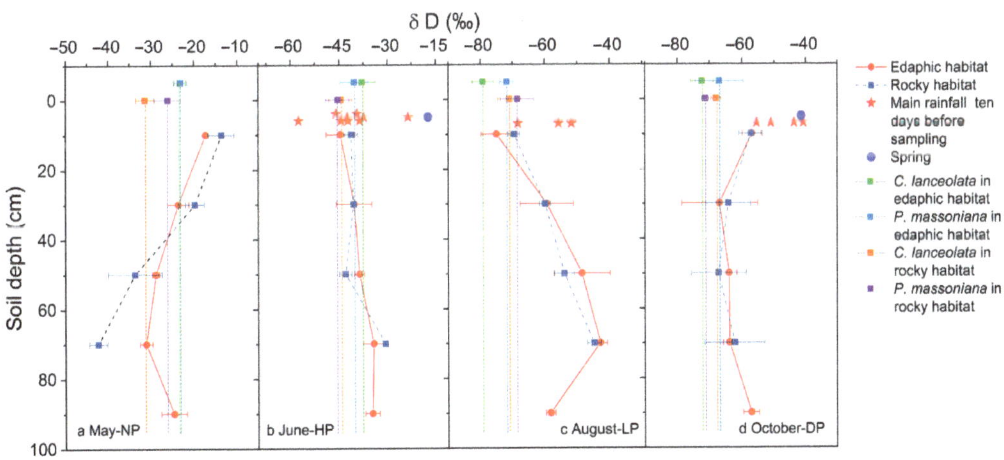

Figure 3. Variation in mean (±S.D.) δD of soil water with the soil profile, precipitation, spring, and xylem water during the wet season (**a**) May sampling; (**b**) June sampling; (**c**) August sampling; (**d**) October sampling.

The isotopic composition of springs changed across the sampling time. The isotopic values were less negative in HP than in DP. The δD and $\delta^{18}O$ values of xylem water were less negative in NP and became more negative with the seasonal changes. There were no significant differences in isotopic composition between species types and habitats ($p > 0.05$), except in June-HP, when the xylem water isotope was more negative in the rocky habitat than that in the edephic habitat ($p < 0.05$).

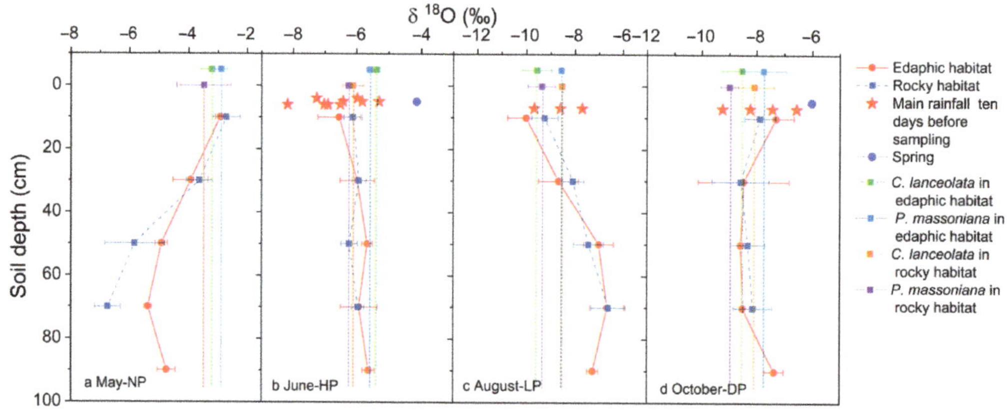

Figure 4. Variation in mean (±S.D.) $\delta^{18}O$ of soil water with the soil profile, precipitation, spring, and xylem water during the wet season (**a**) May sampling; (**b**) June sampling; (**c**) August sampling; (**d**) October sampling.

3.2. Variations in Soil Water Content, Water Uptake Patterns, and Their Linkage with Precipitation

The VWCs of the two habitats displayed clear vertical and seasonal variations (Figure 5). The average VWCs were 43.42% ± 7.68% in the edaphic habitat and 38.24% ± 8.42% in the rocky habitat, with no significant differences ($p = 0.07$) during the study periods. However, the VWCs of shallow soil layers in the two habitats differed significantly ($p < 0.001$). In NP, the VWC of the shallow layer was the lowest in the two habitats and the soil moisture increased with depth (Figure 5a). Furthermore, the VWC exhibited a slightly increasing tendency along the soil profile in the edaphic habitat but a decreasing tendency in the rocky habitat with the seasonal changes. It was noted that the soil moisture in the edaphic habitat was significantly higher than that in the rocky habitat in LP, especially in the middle layers, which may be related to the different soil texture and plant transpiration characteristic (Figure 5c).

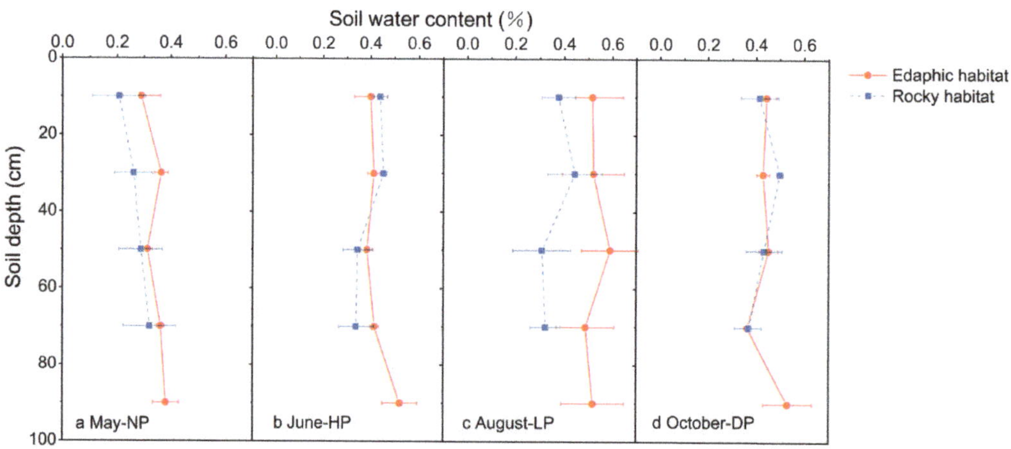

Figure 5. Variation in mean (±S.D.) soil water content along the soil profile during the wet season (**a**) May sampling; (**b**) June sampling; (**c**) August sampling; (**d**) October sampling.

The two tree species mainly took up soil moisture throughout the wet season in two habitats, and the proportions of water sources used by the two species exhibited no significant differences ($p > 0.05$) (Figure 6). However, the plant water uptake depth differed between habitats and across seasons. In HP, trees in the rocky habitat absorbed more than 67.14% of their water from shallow soil layers, whereas the mean water uptake ratio of the two tree species in the edaphic habitat were 64.45% for the middle and deep soil layers. In DP, the *C. lanceolata* and *P. massoniana* in edaphic habitat obtained more than 74.82% of their water from the shallow and deep soil layers. On the other hand, in the rocky habitat, the two species mainly extracted soil water from shallow and middle layers (82.13%). In NP and LP, both *C. lanceolata* and *P. massoniana* in the two habitats utilized the largest proportion of shallow soil water (64.97%, 0–30cm).

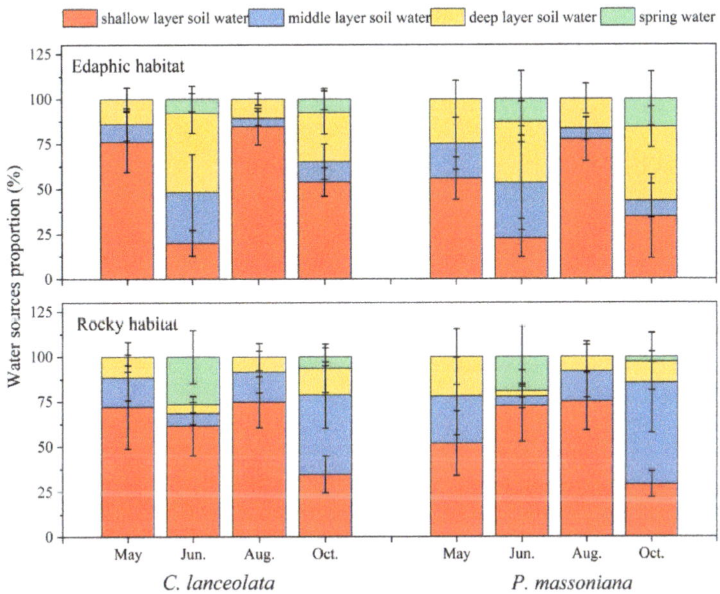

Figure 6. Variation in mean (±S.D.) water source proportions for *C. lanceolata and P. massoniana* during the wet season.

The responses of the proportion of plant water sources used in each soil layers to the amount of precipitation ten days before sampling were distinct in the two habitats (Figure 7). In the edaphic habitat, tree species absorbed less water from shallow layers and absorbed more deep soil water with the increases in precipitation (Figure 7a,c). On the other hand, the trees maintained a high water uptake from shallow layers in the rocky habitat regardless of precipitation variations (Figure 7d). Meanwhile, there were significant negative linear relationships between the water source proportions of the middle and deep soil layers and precipitation (Figure 7e,f).

3.3. Variation in Leaf Water Potential and Its Linkage with Precipitation

The ψ_{pd} of the two species was found to be less negative (>−1MPa) in the two habitats, indicating no severe water stress during the study period, whereas the ψ_{md} was lower than ψ_{pd}, away from the 1:1 line, and exhibited profoundly seasonal variations ($p < 0.01$) (Figure 8). Both of the two species showed lower ψ_{md} values in NP and DP than that in HP and LP ($p < 0.05$). Furthermore, *P. massoniana* showed significantly more negative ψ_{md} values than *C. lanceolata*, especially in NP and DP ($p < 0.05$). However, there were no significant differences in ψ_{pd} and ψ_{md} between the edaphic and rocky habitats for the two species ($p > 0.05$).

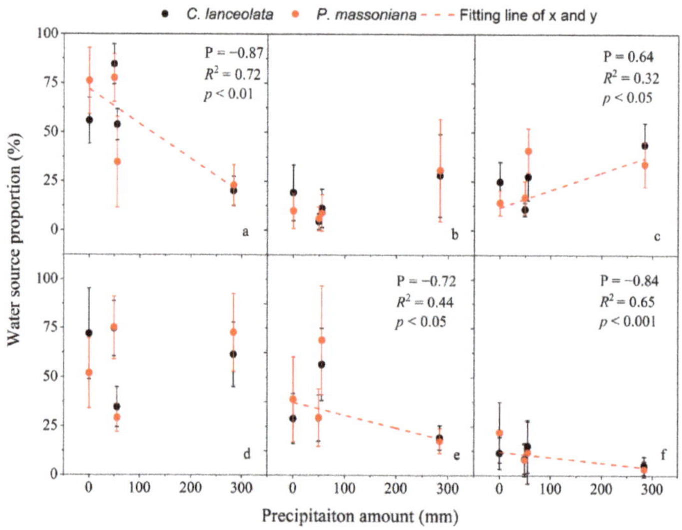

Figure 7. Relationships between water source proportions for each soil layers (mean ± S.D.) and the precipitation amount ten days before sampling. P is Pearson correlation, R^2 represents the fitting degree of the relationship between the water source proportion and the precipitation amount; p is the p-value of the fitting ((**a–c**) plant water sources from the shallow, middle, and deep layers in the edaphic habitat, respectively; (**d–f**) plant water sources from the shallow, middle, and deep layers in the rocky habitat, respectively). The black dots represented *C. lanceolata*, and the red dots represent *P. massoniana*.

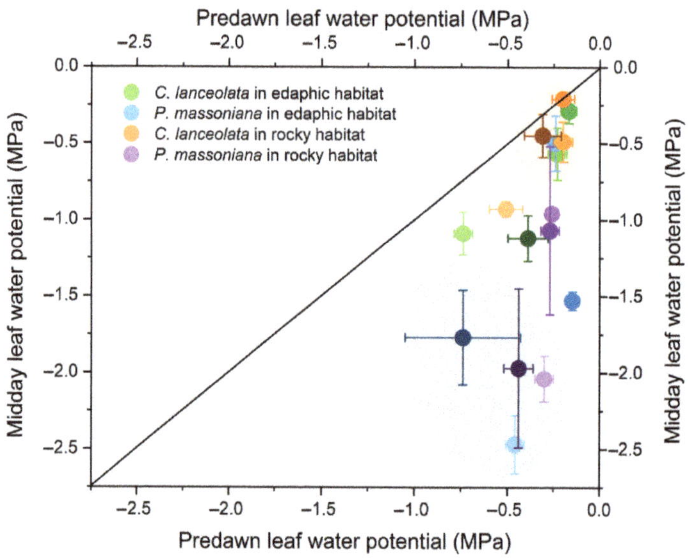

Figure 8. Plot of mean (±S.D.) predawn water potential and midday water potential of two tree species in the edaphic and rocky habitats. The black line is the 1:1 line of predawn water potential vs. midday water potential. The light to dark colors for each species represent the sampling dates as May-NP, June-HP, August-LP, and October-DP. The light orange shadow represents the cluster of trees close to the 1:1 line in HP and LP. The light blue shadow represents the cluster of trees away from the 1:1 line in NP and DP.

The diurnal ranges of water potential ($\Delta\psi$) exhibited significant variation in different species with seasonal changes ($p < 0.01$) (Figure 9). *C. lanceolata* showed significantly lower $\Delta\psi$ values than *P. massoniana* ($p < 0.001$). The $\Delta\psi_{max}$ value was the highest in NP for *P. massoniana* (-1.84 ± 0.19 MPa) and in DP for *C. lanceolata* (-0.45 ± 0.34 MPa). Both of the two tree species displayed the minimum $\Delta\psi$ (-0.48 ± 0.11 MPa and 0.09 ± 0.06 MPa, respectively) in HP and LP. Both of the two species showed significantly higher diurnal ranges of water potential in the edaphic habitat than those in the rocky habitat ($p < 0.001$) during the sampling period, except for *P. massoniana* in LP and DP. Furthermore, there was no significant correlation between the $\Delta\psi$ values and water uptake depth for *C. lanceolata* or *P. massoniana* in the two habitats.

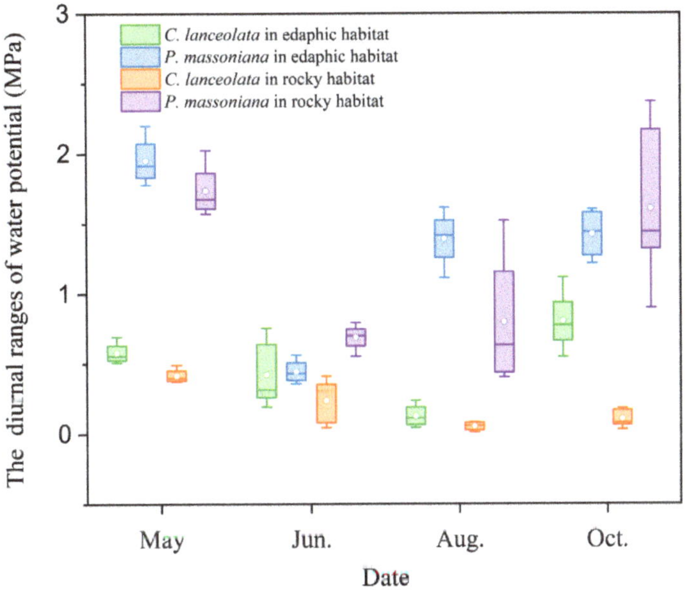

Figure 9. Variation in mean (±S.D.) diurnal ranges of water potential for *C. lanceolata and P. massoniana* during the wet season.

The high vapor pressure deficit and solar radiation showed a strong atmospheric evaporation force (AEF) (Table 2) that influenced transpiration and the diurnal changes in water potential. The meteorological factors during the sample period showed no significant variation ($p > 0.05$), except for DP with lower AEF. On the other hand the $\Delta\psi$ values exhibited seasonal changes and showed the highest values in NP and DP for the two species. These changes may be affected by the available of soil water. The responses of the $\Delta\psi$ to the precipitation amount ten days before sampling were different in the two tree species (Figure 10). The $\Delta\psi$ values of *C. lanceolata* did not increase with the change in conditions from no rain to high rainfall in the two habitats. However, the $\Delta\psi$ values of *P. massoniana* showed lower values with the precipitation increases in the edaphic and rocky habitats. Moreover, the plant water uptake depth was not correlated with the diurnal range of water potential (Table 3).

Table 2. The variations in meteorological factors during the sampling dates.

	T (°C)	RH (%)	VPD (KPa)	PAR_{max} ($umol{\cdot}m^{-2}{\cdot}s^{-1}$)
May-NP	26.6	63	1.29	1336.4
June-HP	30.1	74	1.11	1074.8
August-LP	31.1	66	1.54	1503.7
October-DP	20.9	68	0.79	1194.1

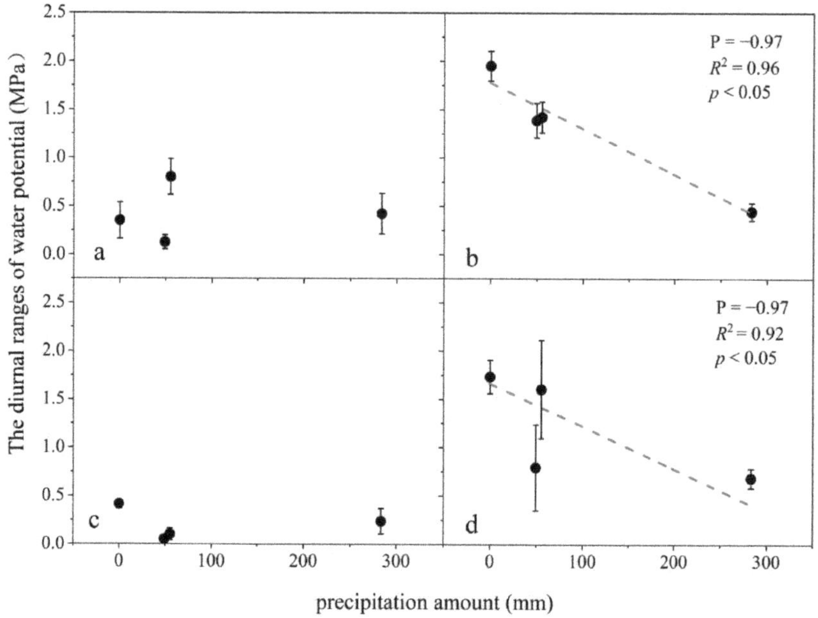

Figure 10. Relationships between the diurnal ranges of water potential (mean ± S.D.) and the precipitation amount ten days before sampling. P is Pearson correlation, R^2 represents the fitting degree of the relationship between the diurnal ranges of water potential and the precipitation amount; p is the p-value of the fitting ((**a**) *C. lanceolata* in the edaphic habitat; (**b**) *P. massoniana* in the edaphic habitat; (**c**) *C. lanceolata* in the rocky habitat; (**d**) *P. massoniana* in the rocky habitat).

Table 3. Relationship between water uptake depth and the diurnal ranges of water potential for *C. lanceolata* and *P. massoniana* in edaphic and rocky habitats.

Pearson Correlation	Edaphic Habitat (−MPa)		Rocky Habitat (−MPa)	
	C. lanceolata	*P. massoniana*	*C. lanceolata*	*P. massoniana*
Shallow layer	−0.41, $p > 0.05$	0.77, $p > 0.05$	0.328, $p > 0.05$	−0.825, $p > 0.05$
Middle layer	0.152, $p > 0.05$	−0.837, $p > 0.05$	−0.411, $p > 0.05$	−0.78, $p > 0.05$
Deep layer	0.441, $p > 0.05$	−0.562, $p > 0.05$	−0.064, $p > 0.05$	−0.886, $p > 0.05$

3.4. Variation in Leaf Water Use Efficiency

The $\delta^{13}C$ values of the two species were significantly more negative in the middle wet season and early dry season than that in May-NP ($p < 0.05$) (Table 4). The leaf water use efficiency was higher in the drought stage in the wet season. There were no significant differences ($p > 0.05$) in $\delta^{13}C$ values between the two habitats. Furthermore, with the exception of in DP in the edaphic habitat and in LP in the rocky habitat, *C. lanceolata* and *P. massoniana* showed no significant differences in $\delta^{13}C$.

Table 4. Comparisons of $\delta^{13}C$ values of *C. lanceolata* and *P. massoniana* in edaphic and rocky habitats.

	Edaphic Habitat (‰)		Rocky Habitat (‰)	
	C. lanceolata	*P. massoniana*	*C. lanceolata*	*P. massoniana*
May-NP	−25.78 ± 0.85Aa	−26.67 ± 0.73Aa	−25.66 ± 0.57Aa	−26.93 ± 0.05Aa
June-HP	−27.07 ± 0.41Ba	−27.75 ± 0.85Ba	−27.95 ± 1.26Ba	−28.31 ± 0.18Aa
August-LP	−27.65 ± 0.69Ba	−28.8 ± 0.42Ba	−27.43 ± 0.18Ba	−29.51 ± 0.27Bb
October-DP	−27.22 ± 0.47Ba	−28.23 ± 0.18Bb	−27.86 ± 0.64Ba	−28.86 ± 0.25Ba

Note: Capital letters represent significant differences of the same tree species among different sampling dates at the 0.05 level; lowercase letters represent significant differences between *C. lanceolata and P. massoniana* in the same habitat at the 0.05 level.

4. Discussion

4.1. Water Uptake of Tree Species in Two Habitats

The variation of plant water uptake depth in the two habitats was consistent, except in June-HP. These two species, growing at the foot of the slope, mainly absorbed soil water from shallow layers in the early and middle wet season, and switched to deeper layers in the late wet season. This water uptake pattern has also been observed in other natural species and plantations in the similar study areas [16,28]. However, it was noted that the plants utilized shallow soil water rather than deep water (no springs flowing) in the early wet season with a 20 day drought, which was inconsistent with other studies in this climate region [16,22]. Although the mean soil moisture was lower compared to other samplings, the VWC was still higher than that observed in semiarid climate regions in the wet season [15,43]. Meanwhile, with a relatively lower wilting coefficient and high spatial heterogeneity [44], the shallow layers could also provide enough available water for plants. Previous studies showed that the plant species adjusted their physiological factors, such as water potential behavior, water use efficiency, in response to the environment changes [25,45,46]. In our study, *C. lanceolata* and *P. massoniana* exhibited the highest leaf $\Delta\psi$ and $\delta^{13}C$ values in NP, indicating that they tried their best to absorbed enough shallow soil water with lower midday leaf water potential to balance carbon-water relations in tandem with high leaf-level intrinsic water use efficiency (iWUE). Moreover, this water uptake pattern is an adaptation, enabling plants to save more energy for growth in the early wet season. Both *C. lanceolata* and *P. massoniana* grow quickly, showing high energy consumption in May, as well as in the early growing season. Although the deep soil layer has a higher VWC, the energy required to take up water from the deep layer is greater than that of the upper layers [47,48]. Thus, the trees extracted shallow soil water to avoid excessive energy consumption through physiological adjustments [49–51]. In the middle and late wet season, plants water uptake depth shifted from shallow to deeper layers. Soil water availability may be the main reason for this water uptake pattern [52,53].

When the antecedent precipitation was much higher in the middle wet season, the plants still absorbed water from shallow layers in the rocky habitat, but in the edaphic habitat they switched to deep layers of soil water. Water availability is the most important factor influencing the plants water uptake depth [26,54]. Soil structure, such as soil texture, bulk density, affected water holding capacity, and migration, along with soil profiles, thus regulated plant water use [28,55]. The bulk density in the rocky habitat was lower than that in the edaphic habitat, promoting the high water holding capacity, whereas in the thin deep layers with large cracks and crevices in the rocky habitat, moisture leaked into the layer, flowing through the springs. In the thick deep layer with fine cracks in the edaphic habitat, the amount of stored water was higher than that in the shallow layer after large and continuous precipitation. Therefore, discrepancies in soil properties are the main reasons underlying the different soil water availability along the profiles in the two habitats. Furthermore, the low diurnal ranges of water potential of *C. lanceolata* and *P. massoniana* also demonstrated that they were both had sufficient water supplies in the two habitats.

4.2. Water Uptake Pattern and Physiological Factors Change in the Different Tree Species

The two coexisting plants—both in the edaphic and the rocky habitat—exhibited no significant differences in water uptake patterns, indicating that they had the same eco-hydrological niche and no water source segregation. This result was consistent with a previous study in a similar climate region, which showed that the six mixed plantations had similar water sources, using the 0–30 cm soil layers in the wet season [28]. Studies in other regions also showed that coexisting species usually exhibited water competition in mixed stand [56,57]. Nie [58] investigated three communities on adjacent rocky hill slopes, and found that different species within each community all exhibited the use of a similar water source. Du [59] studied three karst climate forest communities of a typical hill, and obtained the same results. The similar root distribution of *C. lanceolata* and *P. massoniana* may be the main reason that they exhibited the same water uptake pattern [60,61]. Hence, the interspecific difference in community was relatively low in the subtropical monsoon climate region. However, as per the above analysis, the water uptake pattern was different between the edaphic and rocky habitats for the same species. This suggested that the habitats may have more of an influence on plant water use than the interspecific differences in the community, especially when the antecedent precipitation is high.

Although the water uptake depth was similar for the two species, the two species had different physiological responses to the water uptake. In our study, *C. lanceolata* maintained small diurnal ranges of water potential, high leaf $\delta^{13}C$ values, and a large amount of branching from the base of the trunk, whereas *P. massoniana* showed the inverse characteristics. Meanwhile, the $\Delta\psi$ values of *P. massoniana* in the two habitats were negatively associated with antecedent precipitation, but a significant relationship was not observed in *C. lanceolata*. Wang [15] found similar results for two species in a mixed plantation in the Loess Plateau. Moreno-Gutiérrez [50] observed the existence of species-specific eco-physiological niche segregation in dryland plant communities. A possible explanation was that the inter-specific competition in the same habitat caused each tree species to establish different hydrological niches for water uptake [48,62]. Unlike the previous findings, in our study, there was significant eco-physiological niche segregation but no ecohydrological segregation for the two species in the same habitat. The plant water uptake depth was not correlated with the diurnal range of water potential. In other words, the aboveground physiological parameters showed significantly differences between two species, whereas the belowground water uptake was consistent among the two species. This discrepancy may be attributed to sufficient precipitation and soil water availability for ecohydrological non-segregation [63] and interspecific differences in terms of eco-physiological segregation [54].

4.3. Implications for Plant Water Adaptation under Precipitation Changes

With the increasing temperatures, precipitation patterns change seasonally and become more variable [8], which could lead to an increase in either the severity of drought or extreme precipitation, especially in the growing season [64–66]. When drought or extreme precipitation occurs, soil water availability may influence the plants' water use strategies.

In our study, plants absorbed soil water from shallow layers by increasing the diurnal ranges of water potential and water use efficiency in the early wet season with a 20 day drought. The tree species sought a balance between water uptake and growth through their relatively high water use efficiency [67]. However, if the drought was prolonged, soil moisture would decline and fail to supply water for plants. Ding [26] conducted a 135 day rainfall exclusion experiment in a catchment, and found two adverse responses, according to different physiological characteristics, to the severe water limitation: canopy defoliation and/or mortality and survival. In our study, *P. massoniana*, as the species exhibiting profligate water use exhibited larger $\Delta\psi$ and lower ψ_{md} values for absorbing water sources [26]. Once the ψ_{md} values beconmes lower than the hydraulic trait values, the species may suffer from the risk of hydraulic failure, such as xylem cavitation and leaf turgor loss [56,68]. On the contrary, *C. lanceolata* displayed stable $\Delta\psi$ values in the sampling period, indicating the rigorous stomatal control [69]. The tree growth rate of *C. lanceolata* may slow due to the reduction in shallow soil water sources and the advanced stomatal closure.

Except for the drought in the growing season, the frequency of rainstorms and extreme precipitation also has also been increasing recently [70,71]. Plants are the main conduit for returning terrestrial water to the atmosphere, thereby exerting a strong effect on hydrologic fluxes of the terrestrial-atmospheric system [63]. In our study, the plants that mainly utilized for deep layer soil water in the edaphic habitat and the $\Delta\psi$ values of *P. massoniana* were lower when the precipitation was extremely high. These results illustrated that the tree species could adjust their water use strategies and increase the eco-hydrological linkages between the ecosystem and the deep-layer system [59].

5. Conclusions

In this study, the stable isotope technique and a pressure chamber were applied to detect the seasonal water uptake patterns of two coniferous species in edaphic and rocky habitats in a subtropical monsoon climate region. The results showed that the two species mainly absorbed soil water from shallow layers, even in the short drought period in the wet season and switched to deeper layers in the early dry season. It was noted that the trees utilized deep-layers water in edaphic habitats when the antecedent rainfall was high. The two species showed no significant differences in water uptake depth, but notable differences in their leaf water potential behaviors. *C. lanceolata* displayed narrow and stable $\Delta\psi$ values whereas the $\Delta\psi$ values of *P. massoniana* were negatively associated with antecedent precipitation. Thus, for co-existing species in communities, there was significant eco-physiological niche segregation but no eco-hydrological segregation. For the tree species in the two habitats, the water uptake depth was influenced by the soil water availability, but the physiological factors were unchanged, determined by the species genetics. Furthermore, during a long drought in the growing season, *C. lanceolata* and *P. massoniana* probably show divergent responses, such as growth restriction and hydraulic failure. However, when the precipitation is heavy and long, these species could increase the ecohydrological linkages between the ecosystem and the deep-layer system in the edaphic habitat.

Author Contributions: J.J. and Y.C. conducted field experiment, performed data analysis, and wrote the draft manuscript. J.L. and W.Y. conceived the study, designed the experiment. All authors contributed to discussion and interpretation of resulting data. All authors have read and agreed to the published version of the manuscript.

Funding: This research was funded by the key research and development program in Hunan province (2020NK2022), the National Natural Science Foundation of China (41807162) and the Hunan Province Natural Science Foundation (2019JJ50994).

Data Availability Statement: All relevant data are within the manuscript.

Conflicts of Interest: We declare that we have no financial and personal relationships with other people or organizations that can inappropriately influence our work, there is no professional or other personal interest of any nature or kind in any product, service and/or company that could be construed as influencing the position presented in, or the review of, the manuscript entitled "Water uptake pattern by coniferous forests in two habitats linked to precipitation changes in subtropical monsoon climate region, China".

References

1. Forzieri, G.; Miralles, D.G.; Ciais, P.; Alkama, R.; Ryu, Y.; Duveiller, G.; Zhang, K.; Robertson, E.; Kautz, M.; Martens, B. Increased control of vegetation on global terrestrial energy fluxes. *Nat. Clim. Chang.* **2020**, *10*, 356–362. [CrossRef]
2. Piao, S.; Yin, G.; Tan, J.; Cheng, L.; Huang, M.; Li, Y.; Liu, R.; Mao, J.; Myneni, R.B.; Peng, S. Detection and attribution of vegetation greening trend in China over the last 30 years. *Glob. Chang. Biol.* **2015**, *21*, 1601–1609. [CrossRef] [PubMed]
3. Zhu, Z.C.; Piao, S.L.; Myneni, R.B.; Huang, M.; Ning, Z. Greening of the Earth and its drivers. *Nat. Clim. Chang.* **2016**, *6*, 182. [CrossRef]

4. Aguirre, G.J. Drier tropical forests are susceptible to functional changes in response to a long-term drought. *Ecol. Lett.* **2019**, *22*, 855–865. [CrossRef] [PubMed]
5. Gu, L.H.; Pallardy, S.G.; Hosman, K.P.; Sun, Y. Impacts of precipitation variability on plant species and community water stress in a temperate deciduous forest in the central US. *Agr. For. Meteorol.* **2016**, *217*, 120–136. [CrossRef]
6. Malhi, Y.; Roberts, T.J.; Betts, R.A.; Killeen, T.J.; Li, W.; Nobre, C.A. Climate change, deforestation, and the fate of the Amazon. *Science* **2008**, *319*, 169–172. [CrossRef]
7. Dietrich, L.; Kahmen, A. Water relations of drought-stressed temperate trees benefit from short drought-intermitting rainfall events. *Agric. For. Meteorol.* **2019**, *265*, 70–77. [CrossRef]
8. Konapala, G.; Mishra, A.K.; Wada, Y.; Mann, M.E. Climate change will affect global water availability through compounding changes in seasonal precipitation and evaporation. *Nat. Commun.* **2020**, *11*, 3044. [CrossRef]
9. Anderegg, W.; Konings, A.G.; Trugman, A.T.; Yu, K.; Bowling, D.R.; Gabbitas, R.; Karp, D.S.; Pacala, S.; Sperry, J.S.; Sulman, B.N. Hydraulic diversity of forests regulates ecosystem resilience during drought. *Nature* **2018**, *561*, 538–541. [CrossRef]
10. Chi, C.; Park, T.; Wang, X.; Piao, S.; Xu, B.; Chaturvedi, R.K.; Fuchs, R.; Brovkin, V.; Ciais, P.; Fensholt, R. China and India lead in greening of the world through land-use management. *Nat. Sustain.* **2019**, *2*, 122–129.
11. Macias, F.M. Satellite images show China going green. *Nature* **2018**, *553*, 411–413. [CrossRef] [PubMed]
12. Liang, X.; He, P.; Liu, H.; Zhu, S.; Uyehara, I.K.; Hou, H.; Wu, G.L.; Zhang, H.; You, Z.T.; Xiao, Y.Y. Precipitation has dominant influences on the variation of plant hydraulics of the native Castanopsis fargesii (Fagaceae) in subtropical China. *Agric. For. Meteorol.* **2019**, *271*, 83–91. [CrossRef]
13. Zhao, Y.; Wang, L. Plant water use strategy in response to spatial and temporal variation in precipitation patterns in China: A stable isotope analysis. *Forests* **2018**, *9*, 123. [CrossRef]
14. Kühnhammer, K.; Kübert, A.; Brüggemann, N.; Diaz, P.D.; Dusschoten, D.V.; Javaux, M.; Merz, S.; Vereecken, H.; Dubbert, M.; Rothfuss, Y. Investigating the root plasticity response of Centaurea jacea to soil water availability changes from isotopic analysis. *New Phytol.* **2020**, *226*, 98–110. [CrossRef] [PubMed]
15. Wang, J.; Fu, B.J.; Wang, L.X.; Lu, N.; Li, J.Y. Water use characteristics of the common tree species in different plantation types in the Loess Plateau of China. *Agric. For. Meteorol.* **2020**, *288–289*, 108020. [CrossRef]
16. Deng, Y.; Ke, J.; Wu, S.; Jiang, G.; Zhu, A. Responses of plant water uptake to groundwater depth in limestone outcrops. *J. Hydrol.* **2020**, *590*, 125377. [CrossRef]
17. Ceacero, C.J.; Díaz-Hernández, J.L.; Campo, A.D.; Navarro-Cerrillo, R.M. Soil rock fragment is stronger driver of spatio-temporal soil water dynamics and efficiency of water use than cultural management in holm oak plantations. *Soil Till Res.* **2020**, *197*, 104495. [CrossRef]
18. Nie, Y.P.; Chen, H.S.; Wang, K.L.; Yang, J. Water source utilization by woody plants growing on dolomite outcrops and nearby soils during dry seasons in karst region of Southwest China. *J. Hydrol.* **2012**, *420*, 264–274. [CrossRef]
19. Drake, P.L.; Franks, P.J. Water resource partitioning, stem xylem hydraulic properties, and plant water use strategies in a seasonally dry riparian tropical rainforest. *Oecologia* **2013**, *137*, 321–329. [CrossRef]
20. Yang, B.; Wen, X.F.; Sun, X.M. Seasonal variations in depth of water uptake for a subtropical coniferous plantation subjected to drought in an East Asian monsoon region. *Agric. For. Meteorol.* **2015**, *201*, 218–228. [CrossRef]
21. Yang, F.T.; Feng, Z.M.; Wang, H.; Dai, X.Q.; Fu, X.L. Deep soil water extraction helps to drought avoidance but shallow soil water uptake during dry season controls the inter-annual variation in tree growth in four subtropical plantations. *Agric. For. Meteorol.* **2017**, *234–235*, 106–114. [CrossRef]
22. Liu, Z.Q.; Yu, X.X.; Jia, G.D. Water uptake by coniferous and broad-leaved forest in a rocky mountainous area of northern China. *Agric. For. Meteorol.* **2019**, *265*, 381–389. [CrossRef]
23. Zunzunegui, M.; Boutaleb, S.; Díaz Barradas, M.C.; Esquivias, M.P.; Valera, J.; Jáuregui, J.; Tagma, T.; Ain-Lhout, F. Reliance on deep soil water in the tree species Argania spinosa. *Tree Physiol.* **2018**, *38*, 678–689. [CrossRef] [PubMed]
24. Ouyang, L.; Gao, J.G.; Zhao, P.; Rao, X.Q. Species-specific transpiration and water use patterns of two pioneer dominant tree species under manipulated rainfall in a low-subtropical secondary evergreen forest. *Ecohydrology* **2020**, *13*, e2234. [CrossRef]
25. Hochberg, U.; Rockwell, F.E.; Holbrook, N.M.; Cochard, H. Iso/Anisohydry: A Plant-Environment Interaction Rather Than a Simple Hydraulic Trait. *Trends Plant. Sci.* **2018**, *23*, 112–120. [CrossRef]
26. Ding, Y.L.; Nie, Y.P.; Chen, H.S.; Wang, K.L.; Querejeta, J.I. Water uptake depth is coordinated with leaf water potential, water-use efficiency and drought vulnerability in karst vegetation. *New Phytol.* **2020**, *229*, 1339–1353. [CrossRef]
27. Eggemeyer, K.D.; Tala, A.; Edwin, H.F.; David, A.W.; Zhou, X.H.; William, Z.C. Seasonal changes in depth of water uptake for encroaching trees Juniperus virginiana and Pinus ponderosa and two dominant C4 grasses in a semiarid grassland. *Tree Physiol.* **2009**, *29*, 157–169. [CrossRef]
28. Liu, W.N.; Chen, H.S.; Zou, Q.Y.; Nie, Y.P. Divergent root water uptake depth and coordinated hydraulic traits among typical karst plantations of subtropical China: Implication for plant water adaptation under precipitation changes. *Agric. Water Manag.* **2021**, *249*, 106798. [CrossRef]
29. Craven, D.; Hall, J.S.; Ashton, M.S.; Berlyn, G.P. Water-use efficiency and whole-plant performance of nine tropical tree species at two sites with contrasting water availability in Panama. *Trees* **2013**, *27*, 639–653. [CrossRef]
30. Hasselquist, N.J.; Allen, M.F.; Santiago, L.S. Water relations of evergreen and drought-deciduous trees along a seasonally dry tropical forest chronosequence. *Oecologia* **2010**, *164*, 881–890. [CrossRef]

31. Nie, Y.P.; Chen, H.S.; Wang, K.L.; Ding, Y.L. Seasonal variations in leaf δ13C values: Implications for different water-use strategies among species growing on continuous dolomite outcrops in subtropical China. *Acta Physiolo Plant.* **2014**, *36*, 2571–2579. [CrossRef]
32. Volkmann, T.; Haberer, K.; Gessler, A.; Weiler, M. High-resolution isotope measurements resolve rapid ecohydrological dynamics at the soil–plant interface. *New Phytol.* **2016**, *210*, 839–849. [CrossRef] [PubMed]
33. Bruelheide, H.; Nadrowski, K.; Assmann, T.; Bauhus, J.; Both, S.; Buscot, F.; Chen, X.Y.; Ding, B.Y.; Durka, W.; Erfmeier, A. Designing forest biodiversity experiments: General considerations illustrated by a new large experiment in subtropical china. *Meth. Ecol Evol.* **2014**, *5*, 74–89. [CrossRef]
34. Liu, H.Y.; Jiang, Z.H.; Dai, J.Y.; Wu, X.C.; Peng, J. Rock crevices determine woody and herbaceous plant cover in the karst critical zone. *Sci. China Earth Sci.* **2019**, *62*, 1756–1763. [CrossRef]
35. Ding, Y.L.; Nie, Y.P.; Susanne, S.; Chen, H.S.; Yang, J.; Zhang, W.; Wang, K.L. A novel approach for estimating groundwater use by plants in rock-dominated habitats. *J. Hydrol.* **2018**, *565*, 760–769. [CrossRef]
36. Agency, I.A.E. A new device for monthly rainfall sampling for GNIP. *Water Environ. News* **2002**, *16*, 5.
37. Ehleringer, J.R.; Roden, J.; Dawson, T.E. Assessing Ecosystem-Level Water Relations Through Stable Isotope Ratio Analyses. In *Methods in Ecosystem Science*; Sala, O.E., Jackson, R.B., Mooney, H.A., Howarth, R.W., Eds.; Springer: New York, NY, USA, 2000. [CrossRef]
38. Li, S.G.; Romero-Saltos, H.; Tsujimura, M.; Sugimoto, A.; Sasaki, L.; Davaa, G.; Oyunbaatar, D. Plant water sources in the cold semiarid ecosystem of the upper Kherlen River catchment in Mongolia: A stable isotope approach. *J. Hydrol.* **2007**, *333*, 109–117. [CrossRef]
39. Liu, W.R.; Peng, X.H.; Shen, Y.J.; Chen, X.M. Measurements of hydrogen and oxygen isotopes in liquid water by isotope ratio infrared spectroscopy (IRIS) and their spectral contamination corrections. *Chin. J. Ecol.* **2013**, *32*, 1181–1186. (In Chinese with English abstract)
40. Schultz, N.M.; Griffis, T.J.; Lee, X.; Baker, J.M. Identification and correction of spectral contamination in $^{2}H/^{1}H$ and $^{18}O/^{16}O$ measured in leaf, stem, and soil water. *Rapid Commun. Mass Sp.* **2011**, *25*, 3360–3368. [CrossRef]
41. Stock, B.C.; Semmens, B.X. MixSIAR GUI User Manual, Version 3.1. Available online: https://cran.r-project.org/web/packages/MixSIAR/readme/README.html (accessed on 14 May 2020).
42. Ehleringer, J.R.; Dawson, T.E. Water uptake by plants: Perspectives from stable isotope composition. *Plant. Cell Environ.* **2010**, *15*, 1073–1082. [CrossRef]
43. Tokumoto, I.; Heilman, J.L.; Schwinning, S.; Mcinnes, K.J.; Litvak, M.E.; Morgan, C.; Kamps, R.H. Small-scale variability in water storage and plant available water in shallow, rocky soils. *Plant. Soil.* **2014**, *385*, 193–204. [CrossRef]
44. Fu, T.G.; Chen, H.S.; Fu, Z.Y.; Wang, K.L. Surface soil water content and its controlling factors in a small karst catchment. *Environ. Earth Sci.* **2016**, *75*, 1406. [CrossRef]
45. Liu, C.C.; Liu, Y.G.; Fan, D.Y.; Guo, K. Plant drought tolerance assessment for re-vegetation in heterogeneous karst landscapes of southwestern China. *Flora* **2012**, *207*, 30–38. [CrossRef]
46. Nie, Y.P.; Chen, H.S.; Ding, Y.L.; Zou, Q.Y.; Ma, X.Y.; Wang, K.L. Qualitative identification of hydrologically different water sources used by plants in rock-dominated environments. *J. Hydrol.* **2019**, *573*, 386–394. [CrossRef]
47. Liu, Z.Q.; Jia, G.D.; Yu, X.X. Variation of water uptake in degradation agroforestry shelterbelts on the North China Plain. *Agr. Ecosyst. Environ.* **2020**, *287*, 106697. [CrossRef]
48. Williams, D.G.; Ehleringer, J.R. Intra- and Interspecific Variation for Summer Precipitation Use in Pinyon-Juniper Woodlands. *Ecol Monogr.* **2000**, *70*, 517–537.
49. Li, D.D.; Liu, J.Q.; Anne, V.; Xi, B.Y.; Virginia, H.S. Understanding the relationship between biomass production and water use of Populus tomentosa trees throughout an entire short-rotation. *Agric. Water Manag.* **2021**, *246*, 106710. [CrossRef]
50. Moreno-Gutiérrez, C.; Dawson, T.E.; Nicolas, E.; Querejeta, J.I. Isotopes reveal contrasting water use strategies among coexisting plant species in a Mediterranean ecosystem. *New Phytol.* **2012**, *196*, 489–496. [CrossRef]
51. Renninger, H.J.; Nicholas, C.; Clark, K.L.; Schfer, K.V.R. Physiological strategies of co-occurring oaks in a water- and nutrient-limited ecosystem. *Tree Physiol.* **2014**, *2*, 159–173. [CrossRef]
52. Gaines, K.P.; Stanley, J.W.; Meinzer, F.C.; Mcculloh, K.A.; Woodruff, D.R.; Weile, C.; Adams, T.S.; Lin, H.; Eissenstat, D.M. Reliance on shallow soil water in a mixed-hardwood forest in central Pennsylvania. *Tree Physiol.* **2016**, *36*, 444–458. [CrossRef]
53. Zhan, L.C.; Chen, J.S.; Li, L.; Xin, P. Plant water use strategies indicated by isotopic signatures of leaf water: Observations in southern and northern China. *Agric. For. Meteorol.* **2019**, *276–277*, 107624. [CrossRef]
54. Sánchez-Costa, E.; Poyatos, R.; Sabaté, S. Contrasting growth and water use strategies in four co-occurring Mediterranean tree species revealed by concurrent measurements of sap flow and stem diameter variations. *Agric. For. Meteorol.* **2015**, *207*, 24–37. [CrossRef]
55. Yang, J.; Chen, H.S.; Nie, Y.P.; Wang, K.L. Dynamic variations in profile soil water on karst hillslopes in Southwest China. *Catena* **2019**, *172*, 655–663. [CrossRef]
56. Magh, R.K.; Eiferle, C.; Burzlaff, T.; Dannenmann, M.; Dubbert, M. Competition for water rather than facilitation in mixed beech-fir forests after drying-wetting cycle. *J. Hydrol.* **2020**, *587*, 124944. [CrossRef]
57. Liu, Z.Q.; Jia, G.D.; Yu, X.X. Water uptake and WUE of Apple tree-Corn Agroforestry in the Loess hilly region of China. *Agric. Water Manag.* **2020**, *234*, 106138. [CrossRef]

58. Nie, Y.P.; Chen, H.S.; Ding, Y.L.; Wang, K.L.; Pugnaire, F. Water source segregation along successional stages in a degraded karst region of subtropical China. *J. Veg. Sci.* **2018**, *29*, 933–942. [CrossRef]
59. Du, H.; Zeng, F.; Song, T.; Liu, K.; Liu, M. Water depletion of climax forests over humid karst terrain: Patterns, controlling factors and implications. *Agric. Water Manag.* **2021**, *244*. [CrossRef]
60. Ellsworth, P.Z.; Sternberg, L.S.L. Seasonal water use by deciduous and evergreen woody species in a scrub community is based on water availability and root distribution. *Ecohydrology* **2015**, *8*, 538–551. [CrossRef]
61. Schwinning, S. The ecohydrology of roots in rocks. *Ecohydrology* **2010**, *3*, 238–245. [CrossRef]
62. Gonzalez, D.; Andres, E.; Camarero, J.J.; Blanco, J.A.; Bosco, I.; Yueh-Hsin, S.-B.; Gabriel, C.; Federico, J. Tree-to-tree competition in mixed European beech-Scots pine forests has different impacts on growth and water-use efficiency depending on site conditions. *J. Ecol.* **2018**, *106*, 59–75. [CrossRef]
63. Asbjornsen, H. Ecohydrological advances and applications in plant-water relations research: A review. *J. Plant. Ecol.* **2011**, *4*, 3–22. [CrossRef]
64. Hogg, E.H. A global overview of drought and heat-induced tree mortality reveals emerging climate change risks for forests. *For. Ecol. Manag.* **2010**, *259*, 660–684.
65. Fan, Z.X.; Thomas, A. Spatiotemporal variability of reference evapotranspiration and its contributing climatic factors in Yunnan Province, SW China, 1961–2004. *Clim. Chang.* **2013**, *116*, 309–325. [CrossRef]
66. Messmer, M.; Simmonds, I. Global analysis of cyclone-induced compound precipitation and wind extreme events. *Weather. Clim. Extrem.* **2021**, *32*, 100324. [CrossRef]
67. Keep, T. To grow or survive: Which are the strategies of a perennial grass to face severe seasonal stress? *Funct. Ecol.* **2021**, *35*, 1145–1158. [CrossRef]
68. Choat, B. Global convergence in the vulnerability of forests to drought. *Nature* **2012**, *491*, 752–756. [CrossRef]
69. Renninger, H.J.; Carlo, N.J.; Clark, K.L.; Schafer, K.V. Resource use and efficiency, and stomatal responses to environmental drivers of oak and pine species in an Atlantic Coastal Plain forest. *Front. Plant. Sci.* **2015**, *6*. [CrossRef]
70. Berg, P.; Moseley, C.; Haerter, J.O. Strong increase in convective precipitation in response to higher temperatures. *Nat. Geosci.* **2013**, *6*, 181–185. [CrossRef]
71. Min, S.K.; Zhang, X.; Zwiers, F.W.; Hegerl, G.C. Human contribution to more-intense precipitation extremes. *Nature* **2011**, *470*, 378–381. [CrossRef]

forests

MDPI

Article

Water Balance Components of Sub-Mediterranean Downy Oak Landscapes of Southeastern Crimea

Roman Gorbunov *, Vladimir Tabunshchik, Tatiana Gorbunova and Mariia Safonova

A.O. Kovalevsky Institute of Biology of the Southern Seas of RAS, 299011 Sevastopol, Russia
* Correspondence: karadag_station@mail.ru

Abstract: This article discusses the processes of moisture intake, redistribution, and consumption within the downy oak forest community, along with their interannual and interseasonal water balance dynamics. The study of the water balance components was conducted using a combination of field research methods and geoinformation modeling on the territory of the Karadag landscape and ecological station of the Karadag Nature Reserve for the period from 2010 to 2020. The study of the water balance of downy oak forests located at the furthest extent of their range represents an important problem, whose solution will further scientific understanding by uncovering individual patterns of the internal organization of such systems. The indicators having the most tangible impact on the water balance are the amount of precipitation and evapotranspiration. The average annual precipitation on the territory of Karadag for the analyzed period was 448 mm; in recent years, a decrease in the amount of precipitation has been recorded. The evapotranspiration values within the downy oak forests approximately coincide with the values of this indicator in the Mediterranean region to average 450 mm per year. The influence of stemflow and relief features on the redistribution of moisture within the landscape is described. The analyzed water balance components' dynamics form conditions conducive to the displacement of steppe communities by forest species.

Keywords: water balance components; downy oak forests; landscape; Southeastern Crimea; precipitation; evapotranspiration; runoff

Citation: Gorbunov, R.; Tabunshchik, V.; Gorbunova, T.; Safonova, M. Water Balance Components of Sub-Mediterranean Downy Oak Landscapes of Southeastern Crimea. *Forests* **2022**, *13*, 1370. https://doi.org/10.3390/f13091370

Academic Editor: Tim Martin

Received: 26 July 2022
Accepted: 25 August 2022
Published: 27 August 2022

1. Introduction

The study of the Mediterranean climate and the features of its radiation, heat, and water balance are highly topical issues [1–5], since this region is located on the border of an arid climate and a temperate and rainy climate [6]. The theoretical foundations for studying the water balance of the territory of the Crimean Peninsula are laid out in the works of [7–13]. An example of evaluating the regional and local water balance characteristics in Crimea can be found in the works of [9,14–17]. At the same time, very little attention has been paid to the water balance of certain types of plant communities growing at the local level on the territory of the Crimean Peninsula.

Despite the Crimean Peninsula being located far beyond the Mediterranean Basin, the southern coast of Crimea, along with other regions of the Caucasus, is defined as an exclave of Mediterranean vegetation [18–20].

On the territory of the Crimean Peninsula, subtropical forest landscapes are mainly represented by downy oak and juniper formations with an admixture of Christ's thorn, smoke tree, and Oriental hornbeam. These are represented from the coast to an altitude of 300–350 m [9], where they are either replaced by typical downy oak forests and their derivatives with an admixture of hornbeam, higher rock-oak forests with an admixture of hornbeam and ash forests, or rise to the edge of the steep southern limestone slope of the yayla (upland pasture). These communities are typically represented by shiblyak thickets, comprising a dense low-growing closed forest area, or, conversely, sparse woodlands. The Crimean shiblyak and phrygana biomes are often identified with the Mediterranean maquis and garrigues, respectively, although they differ in some important respects [14]. In contrast to the predominantly sclerophyllous maquis, deciduous species predominate in shiblyak biomes [20]. These forests formed at the furthest extent of their range due to the effect of the subtropical climate, which occurs here due to the circulatory position of the Southern coast of Crimea. Cold air masses from the north tend not to penetrate here, while those from the southern and southwestern directions linger on the south-facing slopes of the Main Ridge. At the same time, Southeastern Crimea is located in the circulatory and insolation shadow formed by the Main Ridge of the Crimean Mountains, which determines the region's high aridity and decreased radiation balance values. Thus, the sub-Mediterranean (or semi-subtropical) climate formed on the territory of the Southeastern Crimea is characterized by warm winters, maximum precipitation during the cold period of the year (late autumn-winter), and hot, mostly dry summers.

In this regard, the study of the water balance of downy oak forests located at the furthest extent of their range represents an important problem, whose solution will further scientific understanding by uncovering individual patterns of the internal organization of such systems. Despite the relatively small areas that they cover, the importance of studying the water balance of downy oak forests in Southeastern Crimea is also due to the ecosystem functions that these forests perform (preservation of moisture reserves in the soil, prevention of negative physical and geographical processes on the slopes, etc.).

The aim of this work is to study the temporal variability and identify trends in the components of the water balance of the sub-Mediterranean downy oak landscapes of the Southeastern Crimea on the example of the Karadag landscape and ecological station. The strong suit of this study is that almost all data are obtained from long-term stationary studies. To achieve this goal, the following tasks were solved: we described in detail the conditions of the forest formation with regard to the influence of landscape components on the water balance of the territory; we revealed the dynamics of water balance components of sub-Mediterranean downy oak landscapes.

Identification of dynamics features and formation of water balance components on the basis of long-term stationary observations, and not only by means of modeling, makes it possible to more accurately identify dependencies and patterns in climate change for Mediterranean territories vulnerable to external changes in environmental conditions in the future.

2. Study Area

The area of study encompasses the southeastern part of the Crimean Peninsula. The concept of "South-Eastern Crimea" is often encountered in scientific publications. However, no clear boundaries of this region have yet been defined. The analysis of publications shows that the authors most often refer to the South-Eastern Crimea as the territory comprising the municipalities of Sudak and Feodosia. However, this designation is associated not with administrative units, but rather with landscape features. The natural boundaries of the studied area take the form of watersheds. The western border of the district runs along the Goller Ridge watershed between the Raven and Shelen Rivers, including the peaks of Piyakya, Biyuk-Krizh, and Livaz-Kaya. The northern border runs along the main watershed that delimits the basins of rivers flowing into the Black Sea (i.e., in a southerly direction) and the Sea of Azov (in a northerly direction). The northern border runs along the cliffs of Yuvan-Kaya, the Khambal ridge, the peaks of Kukushlu-Oba, Kazas-Olan and Apaly, Tuar-Alan Ridge, and Jady-Kaya Mountain. The northeastern and eastern borders run along the line from Kara-Oba along the Uzun-Syrt and Tepe-Oba ridges to end in the area of Cape Ilya.

The relief of Southeastern Crimea, which is represented by mountain ranges in the central and western parts, is gradually transformed into a plain towards the east. The elevation of the area varies from 0 to 917 m (Figure 1). Forest communities are mainly represented in the most elevated western and central parts of Southeastern Crimea.

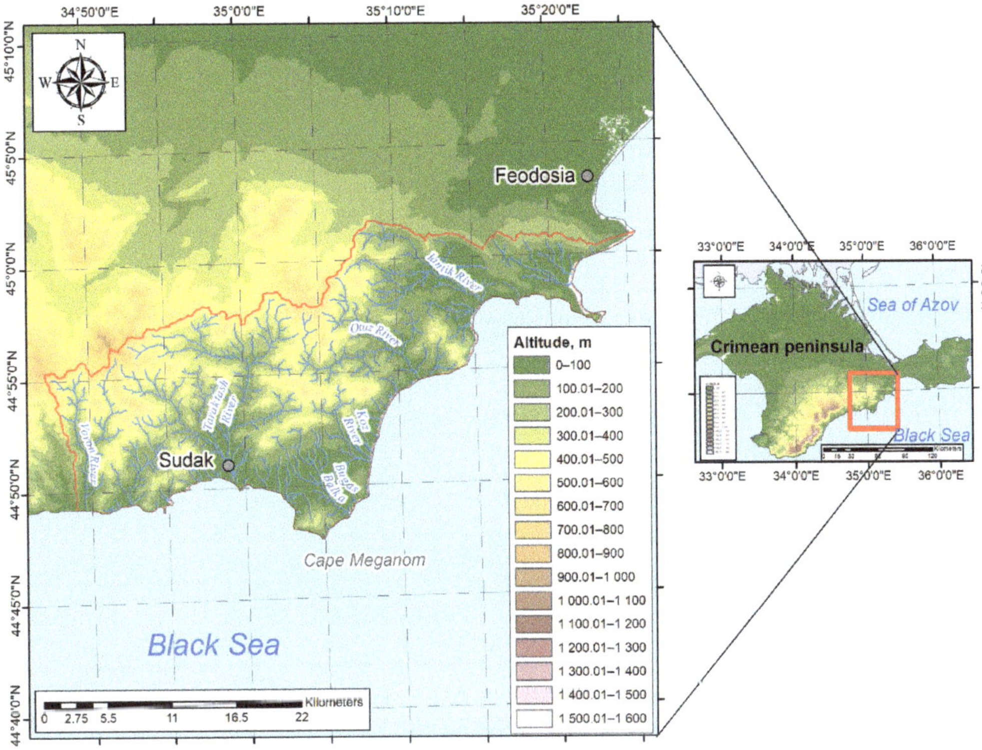

Figure 1. Territory of Southeastern Crimea.

The distribution of plant communities throughout the region is influenced by its geomorphological structure. Steppe communities within Southeastern Crimea can be described as occupying the lowest areas (average elevation—about 100 m); these are replaced by oak shiblyaks with an average elevation of 230 m, then downy oak forests (average elevation—285 m), rock-oak forests (average elevation—340 m) and hornbeam-beech forests (average elevation—470 m). Within the area under consideration, territories with minimal terrain dissection are covered by steppe, psammophytic, and halophytic communities, while territories with the greatest relief dissection are covered by forest communities.

The slopes of Southeastern Crimea are subject to various erosion processes, including ravine formation, which determines the redistribution of forests along the slopes. Snow catchments play an important role in the spatial differentiation of forests. The linear erosion Stream Power Index (SPI) calculated for the Southeastern Crimea shows that most of the area is susceptible to strong erosion processes.

The territory of the Southeastern Crimea is characterized by Mediterranean climatic features. The average annual air temperature within the Southeastern Crimea area, which varies from the northwest to the southeast, ranges from +9° to +13°. Southeastern Crimea experiences between 100 and 300 mm of precipitation during the winter period. The precipitation field decreases from west to east (Figure 2a), with the greatest amount falling in the northwestern part of the study area. In summer, 80 to 160 mm of precipitation falls over the territory of the Southeastern Crimea, with the precipitation field decreasing from the northwest to the southeast (Figure 2b). Both in winter and in summer, the coastal areas of Southeastern Crimea are the most arid. The annual distribution of precipitation varies from 700 to 350 mm in the west-to-east and northwest-to-northeast directions.

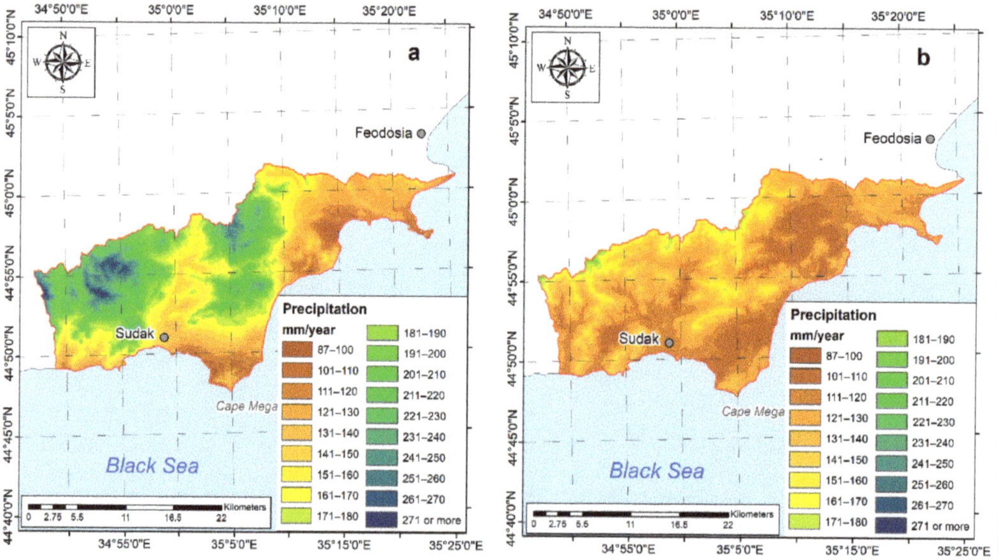

Figure 2. Average precipitation on the territory of Southeastern Crimea for the winter (**a**) and summer (**b**) periods.

The formation of vegetation cover is affected by the distribution fields of climatic factors. Both the greatest amount of precipitation and lowest temperatures prevail in the most elevated areas of Southeastern Crimea; it is here that hornbeam and beech forests are mainly developed. Along with a decrease in the amount of precipitation, woody vegetation is gradually replaced by shrub and steppe communities. Among other formations, oak shiblyaks and tipchak steppes are characteristic of the areas where the least amount of precipitation will fall.

The density of the river network within Southeastern Crimea decreases from the southwest to the northeast from 0.3 to 0.1 km/km^2 [21]. This indicator is indirectly associated with increased climate aridity and decreased precipitation. The watercourses of Southeastern Crimea fully belong to the rivers of the southern macro-slope of the Crimean Mountains. Nevertheless, the river network of Southeastern Crimea is quite complex; in general, the area is characterized by a combination of river valleys, interfluves, and watersheds (Figure 1).

Conditions for the existence of separate small catchments of small gullies and other erosive forms on the coast of Southeastern Crimea are due to the complex rugged terrain, which creates unique topological conditions for the existence of plant communities. For example, local factors disrupt the manifestation of quasi-zonal processes within the valleys of the watercourses of Southeastern Crimea. In addition to oak communities, alder, poplar, and willow are also found throughout the area.

Southeastern Crimea is characterized by a predominance of brown soils (up to 80% of the total area) [22]. Mainly weak and underdeveloped soils are formed on the destruction products of limestones, sandstones, and conglomerates composing individual mountain massifs and ridges, as well as on volcanic rocks of the Karadag massif. In the lower parts of the slopes and at their feet, full-profile and drift soil types are formed. Alluvial (17%) and meadow (2.5%) soil types are commonly encountered in the valleys. Brown mountain–forest soils within the study area have an extremely scanty distribution (about 0.3%).

In certain areas of Southeastern Crimea, vegetation cover has undergone a fundamental transformation. Here, the vegetation is represented either by agricultural crops, or goes through stages of degradation and restoration, which in turn directly affects the water balance of the area due to the capacity of vegetation to retain and redistribute moisture. The vegetation and soil cover, along with the water balance of soils, are additionally affected by the widespread rooting activity of wild boar, which leads to desiccation of the soil and increased evaporation from its surface [23].

3. Materials and Methods

In the classical form, the water balance equation [24] can be represented in the following form (Figure 3):

$$x + y_1 + w_1 + z_1 = y_2 + w_2 + z_2 \pm \Delta u, \tag{1}$$

where x is precipitation on the surface of the object; y_1 is surface inflow of water from outside; w_1 is underground inflow of water from outside; z_1 is condensation of water vapor; y_2 is surface outflow of water outside the object; w_2 is underground outflow of water outside; z_2 is evaporation; Δu is change in the volume of water within the object (contour).

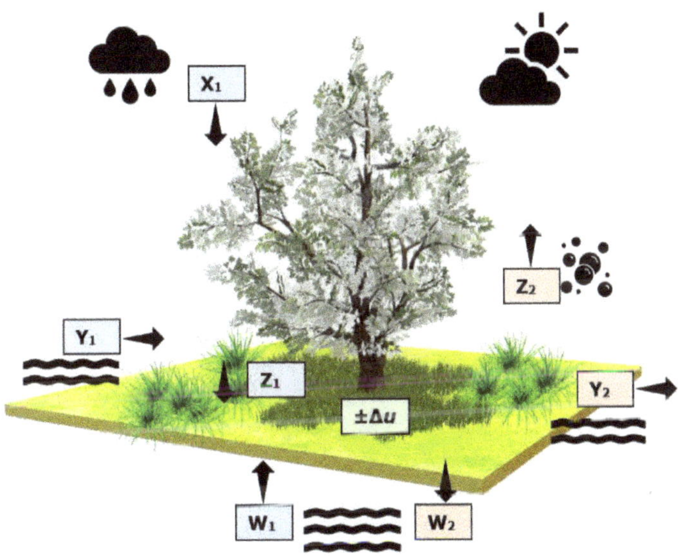

Figure 3. Scheme of the water balance equation. Black arrows show the flows direction.

At the local level of the study, the structure of the water balance becomes an even more complex phenomenon. The structure of the water balance of forests comprises several elements: the flow of water with precipitation and surface runoff from neighboring areas, canopy interception, stem flow, transpiration and evaporation from tree crowns, water flow to the forest litter surface, evaporation from the forest litter surface, infiltration, evaporation of water from the soil, its retention by precipitation (formation of soil moisture capacity and precipitation), surface runoff, subsurface runoff, and water transfer to groundwater runoff.

The study of the water balance components of downy oak forests was conducted under the auspices of the Karadag Landscape and Ecological Station (KLES) of the Karadag Nature Reserve. The KLES is located in the redivision of the major catchment gully (0.6 km long, 17 ha area) on the eastern slope of the Besh-Tash ridge, 1.5 km from the village of Kurortnoe. There, the precipitation and soil moisture parameters are measured in a meteorological site located in an open area in the lower part of the catchment. A precipitation collector and soil lysimeters are also installed in the middle part of the catchment area under the canopy of the forest. Observations of soil moisture are carried out at 30 points located in areas with forest, shrub, and steppe vegetation. Monitoring of the runoff is carried out on two runoff sites located within the KLES area. Figure 4 shows the network of monitoring stations.

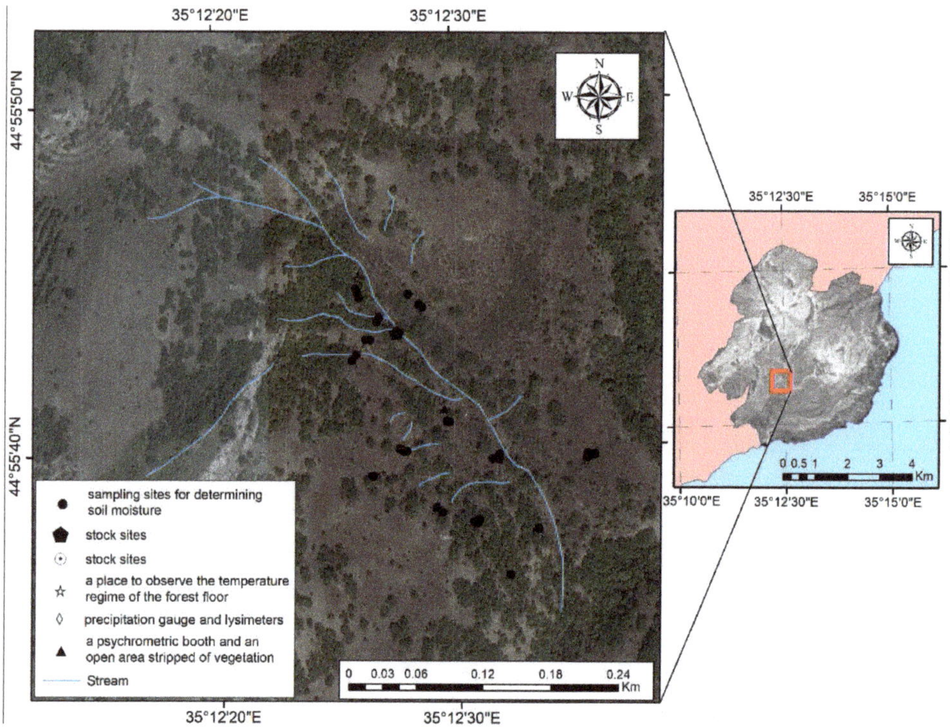

Figure 4. Area covered by the Karadag Landscape and Ecological Station (KLES) and monitoring points.

The KLES area is representative of the entire Karadag Nature Reserve, since almost all types of plant communities are represented here. The vegetation cover is characterized by a strong mosaicism, which is due, among other things, to the relief dissection. More than 20% of the KLES area is occupied by forest communities. They are mainly represented by a downy oak formation (*Querceta pubescentis*) with the addition of *Acereta campestrii*, *Pistacieta muticae*, and *Pineta pallasianae*. Shrub communities, which occupy about a tenth of the study center area, are represented by formations of thorny shrubs—*Paliureta spinae-christi*, *Rosaeteta corymbiferae*, *Pruneta spinosae*, and *Rubeta taurici*. Herbaceous communities are quite diverse and occupy more than 60% of the hospital territory [9].

In order to obtain values for building evapotranspiration maps, data from the open database of the ERA5 reanalysis of the Copernicus Climate Change Service project [25], implemented by the European Center for Medium-Range Weather Forecasts (ECMWF), were used. As a result, the average monthly data for 1978–2021 were obtained. The data are presented in the netCDF4 format and cover the territory in question with a regular grid with a cell size of $0.25° \times 0.25°$.

4. Results and Discussion

The water balance of any territory depends on climatic factors and the nature of the underlying surface. The literature provides various values of the average annual precipitation falling on the territory of the Karadag reserve. On average, 357 mm of precipitation falls on Karadag (for the period from 1930 to 1980) [26]; according to other data, 389 mm of precipitation (for the period from 1920 to 2006) [27], 501 mm of precipitation (for the period from 2000 to 2011), and 467 mm of precipitation (for the period from 2012 to 2018) according to [28].

According to our observations, an average of 448 mm of precipitation fell annually on the territory of Karadag from 2010 to 2020. This is significantly less than in a typical Mediterranean region. Due to the presence of downy oak forests in Northeastern Crimea, environmental conditions can be considered as transitional from optimal to pessimal. For example, within the limits of downy oak forests growing about 15 km north of Montpellier in southern France at an altitude of 250 m, the average annual precipitation is 1311 mm [29]. In Italy, downy oak forests receive about 850 mm of precipitation; in the area of the city of Evora in Portugal, the corresponding figure is 665 mm [30].

In terms of the interannual quantitative dynamics of incoming precipitation, a wide range of values is observed: in 2020, an average of 275 mm of precipitation fell on Karadag, while in 2010, the corresponding amount was 715 mm. Thus, the amplitude reaches 440 mm of precipitation. In 2010–2020, there is a tendency towards a reduction in the amount of precipitation: 2019 and 2020 are both characterized by extremely low average annual precipitation values.

In the intra-annual dynamics, a constant winter maximum of precipitation is observed within the downy oak forests. However, in some years (2013, 2014), absolute precipitation maxima were observed in the study area in June (45% of the annual precipitation), which were catastrophic for the environment. July and August are characterized by stable arid conditions, with an average of 31 mm precipitation. The winter maximum is characterized by a predominance or uniform distribution of precipitation during the late autumn and winter months. The average amount of precipitation in December is 45 mm; in January, 48 mm; in February, 33 mm. Spring and autumn precipitation minima are characterized by a decrease in the amount of precipitation: on average, 27 mm of precipitation falls in spring within the downy oak forests, while in autumn, average precipitation is 34 mm.

In downy oak communities, retention of precipitation in crowns is around 24% of precipitation, comprising an average of 104 mm. Considering their relatively small amount of precipitation, the forest communities of Karadag can retain up to 100% of precipitation [28]. At the same time, there is an intra-annual dynamic to this process, associated both with the vegetation stages of plants and with the intensity of precipitation. Thus, the maximum values of precipitation retention by crowns are typical for the second half of spring to early summer due to plants achieving a maximum of green phytomass during this period. For example, in July, an average of 9 mm (about 41%) of precipitation of precipitation is retained. The retention of precipitation by tree crowns is also affected by the precipitation intensity. In [28] it is indicated that up to 20% of precipitation is retained when more than 20 mm precipitation falls on Karadag, while when precipitation is less than 2 mm, over 45% is retained. Data analysis shows that the increase in precipitation observed in Southeastern Crimea is formed due to summer heavy rainfalls [16]; this in turn leads to most of the water penetrating under the canopy of the forest, with little retention in the crown. In addition, large drops characteristic of heavy rainfall are poorly retained on the leaves, quickly draining to the surface of the forest floor.

A significant role in the water balance structure of downy oak forests is played by stem flow. This effect is interesting due to its concentration of water flow directly under the roots of trees. However, under conditions of sloping terrain, this process can intensify sheet-, drip-, and fine-jet erosion of soils, exposing the roots of trees and accelerating the processes of badland formation. In [9], a methodology for assessing stem runoff in the territory of Karadag is given; however, quantitative characteristics of the experimental results are lacking. A.A. Klyukin in [31] notes that in an oak forest with a canopy density of 0.8, growing on a slope with a steepness of 20–25° and a length of 100 m, the average accumulation of erosion near tree trunk is about 0.004 mm/year. At the same time, A.A. Klyukin notes that the average rate of erosion can reach 2.0–3.9 mm/year on steep and short slopes that are adjacent to river valley beds and talvegs of temporary watercourses, accompanied by exposure of the parent rock. Thus, such processes are characteristic of steep slopes. On slopes of medium and low steepness, such processes are not significantly manifested, with the retentiveness of the soil being provided by an herbaceous layer. A.A. Klyukin in [31] notes that the trunks of an adult oak can run off 5–6 L of precipitation per year; in our opinion, this figure is significantly overestimated for the forests under study. The more objective data in [7] indicate that runoff from oak trunks reaches 3%–5% of the annual precipitation layer, which is equivalent to approximately 20–40 mm/year. Stem runoff is also actively affected by the roughness of the bark of tree trunks, which demonstrates that the real area of the runoff is larger than the ideal cylindrical shape of the trunk. In oak forests, this indicator, according to [9], reaches 1.2, allowing them to be attributed to medium rough type. By comparison, stem runoff in oak forests in Spain accounts for slightly more than 10% of precipitation [32].

According to measured data on sample trees, throughfall accounts for about 75% of the total precipitation amount. Comparable results were obtained by V.A. Bokov's expeditions in this area in 1997 and 1998. [9]. According to his research, a dense downy oak forest intercepts about 25%–35% of precipitation during the growing season from May to October, which is about 103 mm in a wet summer (1997) and 55 mm in a dry one (1998). However, this moisture, which enters the soil surface more gradually than heavy rainfall, is generally well-retained by the forest floor, helping to form its moisture reserve. This corresponds to around 28 and 35%, respectively, of the precipitation depth that fell during the growing season, or 18 and 16% of the annual precipitation layer. During winter, when vegetation is absent in deciduous forests and there is no foliage on the trees, precipitation almost completely penetrates under the canopy of the forest, only being slightly delayed by tree branches. According to the results of measurements carried out in the winter season, precipitation retention of slightly more than 20% does not exceed the annual average. However, given that the winter months account for the maximum precipitation and taking into account the seepage through the crowns, this indicator is much higher than the summer maximum and accounts for no more than 15 mm of precipitation. These data are partially confirmed in the work [14], where it is indicated that during the period from 2000 to 2008, 162 mm was retained within the downy oak forests of Southeastern Crimea, which is approximately 30% of the amount of precipitation. At the same time, observations show that heavy rains are not so characteristic of the winter precipitation maximum, resulting in the possibility of a satisfactory soaking of the forest litter and soil. M.A. Kochkin notes [7] that the forest floor of oak forests, consisting of leaves and dead herbaceous vegetation, absorbs and retains water from precipitation, thereby reducing surface runoff.

For the period from 2010 to 2020, there has been a significant decrease in soil moisture: in average annual terms, by about 1.6 times. The inter-seasonal dynamics of decreasing soil moisture shows that almost all seasons of the year are characterized by a drop in average relative humidity values close to the average annual value, with the exception of autumn, when relative humidity decreases by 1.4 times. Considering the interannual dynamics, it is important to highlight a significant reduction in soil moisture in December (by more than 2.5 times), as well as minor reductions in September. The maximum relative values of soil moisture are experienced during the winter months, as well as in March, when the average value does not fall below 25%, while the minimum values occur at the end of summer and beginning of autumn (August and September), when the relative soil moisture values drop to 15%–18%.

During the period from 2010 to 2020, evapotranspiration within the downy oak forest community averages 450 mm. These values are generally equivalent to the evapotranspiration of oak forests in the Mediterranean region: 343 ± 37 mm in the Mediterranean region as a whole [30], 458 mm in the province of Tarragona in Spain [32], 823 mm in the eastern Pyrenees [33]. Maximum evapotranspiration values in the Mediterranean can reach 1300 mm per year [34].

Analyzing the interannual evapotranspiration dynamic within the oak forests of Karadag, its values can be seen to form a steadily increasing trend (Figure 5).

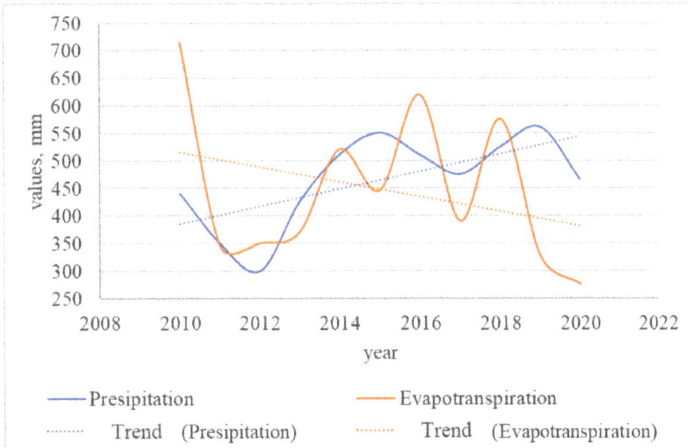

Figure 5. Average annual values showing the relationship between evapotranspiration and precipitation in the downy oak forest in the period 2010–2020.

When considering the seasonal evapotranspiration values, the fact of the predominance of precipitation over evapotranspiration in the winter and autumn periods becomes clearly evident. During winter, the average evapotranspiration value is 11 mm, which is approximately 26% of the amount of precipitation. In autumn, the average evapotranspiration is 22 mm, which is approximately 63% of the average amount of precipitation for this time of year. In the remaining seasons of the year, there is an excess of evapotranspiration values over precipitation. In spring, the average evapotranspiration of 56 mm is 2 times higher than the average amount of precipitation for this season; in summer, the equivalent amount is 67 mm, which is 1.5 times more than the average amount of precipitation at this time of year. At the same time, a steady tendency towards increased average annual values of evapotranspiration is evident at all times of year.

When considering the distribution of evapotranspiration by month, the highest values of evapotranspiration are characteristic of May, June, and July, when the average values of evapotranspiration exceed 70 mm. The lowest values are experienced for December, January, and February, when average evapotranspiration does not exceed 15 mm. In spring, there is a gradual increase in evapotranspiration values, while in autumn there is a gradual decrease. Nevertheless, data analysis also demonstrates a significant increase in evapotranspiration occurring in recent years during late summer and early autumn.

A significant role in the spatial differentiation of downy oak forests is played by accumulations of snow. Around 5% of precipitation falls in the form of snow in the Karadag area [26]. According to [9], in the Southeastern Crimea in the area of Karadag and Echkidag, between 50% to 90% of snow can be blown off the windward northeastern slopes. Although winter precipitation is generally not retained by tree crowns due to the absence of foliage, it can be carried as a result wind transport. For example, in 2019, a layer of snow with a 15 cm high was formed as a result of wind activity. For the period from 2011 to 2020, the average annual precipitation in solid form was 38 mm, accounting for approximately 9% of the total. However, in terms of the interannual dynamics of the indicators under consideration, the years with the minimum amount of solid precipitation (2013, 2020) can be distinguished, when the amount of precipitation in solid form did not exceed 10 mm, as well as the years of maximum solid precipitation—for example, 2015 and 2016—when the corresponding amount was more than 60 mm. Despite the duration of snow in Southeastern Crimea lasting no more than a week, the phenomenon of snow catchments is very characteristic; where the accumulated snow is blown onto the slopes of the southern exposures, more hydrophilic conditions are formed on these slopes than those with northern exposures. For example, within one of the gullies on the eastern slope of the Besh-Tash Range, the slope of the southern exposure is characterized by better moisturization than the opposite slope. This is due to the extensive snow collection adjacent to the slope of the south-southwestern exposure, from which the snow is blown by northeasterly winds to the slope of the south-southwestern exposure (Figure 6) [35]. In this regard, the slope of the south-southwestern exposure is better forested and characterized by greater erosion indentation as compared with the opposite slope of the north-northeastern exposure and slopes of the south-southwestern exposure, where areas with heavily washed soil cover are noted.

Figure 6. Snow-covered slopes of the south-southwest exposure of the gully on the eastern slope of the Besh-Tash ridge [35].

The formation of surface runoff depends on a whole complex of factors. A slope runoff characterized by insignificant indicators is identified on the basis of observations carried out at the runoff sites. In most cases, runoff is caused either by melting snow or heavy precipitation. In 1996–1997, the average runoff was approximately 0.16 L/km^2 [9]. The average value of the runoff from 2010 to 2020 recorded by the measurement results was 0.04 L/km^2. The highest average runoff values, which were recorded in 2010, amount to 0.09 L/km^2, while the lowest figures recorded in 2011 and 2020 were slightly less than 0.02 L/km^2. Thus, these values have no significant effect on the final equation of the water balance of downy oak forests. A decrease in the values of surface runoff in the interannual dynamics is noted for the period under review.

Horizontal precipitation (condensation from fog, frost, frost, ice, etc.) plays a prominent role in the formation of the water balance of downy oak forests. Although we unfortunately did not have the opportunity to conduct real measurements of their contribution to the formation of the water balance of forests, according to the data [36], their contribution can reach 20%–25% of the annual precipitation layer. From September to March, no more than 20 days with foggy conditions can be observed within the downy oak forests; on an annual basis, the number of foggy days varies from 20 to 40 [37]. According to the Feodosiya weather station [8], there is an average of 24 foggy days per year. In 2020, 36 days with fog were recorded within the downy oak forest, of which 75% occurred during the winter period, significantly affecting the incoming part of the water balance.

All the processes described above take place against a background of the displacement of steppe communities by forest communities on the upper parts of the slopes (Figure 7). The growth of heavy rainfall in summer leads to a rapid flushing of water from the upper parts of the slopes, which are characterized by petrophytous steppe communities, along with an increase in soil moisture in the lower parts of the slopes, which creates conditions for forest communities to rise up the slope, reducing the number of slope microzones. This phenomenon is clearly visible when calculating the normalized difference vegetation index

(NDVI) (Figures 7 and 8). When analyzing the average NDVI values for August, which were obtained from Sentinel-2 satellite images in the period from 2016 to 2018 and from 2019 to 2021, an increase in the values can be clearly observed, along with a corresponding increase in the quantity of green phytomass. Figure 8 shows that forest vegetation increases up to 158 m on the profile AB, with the NDVI values not significantly changing from one period to the next. Conversely, above the 158 m mark, where there is initially less woody vegetation, and in August, respectively, less green phytomass, there is an increase in NDVI values and, accordingly, the amount of green phytomass.

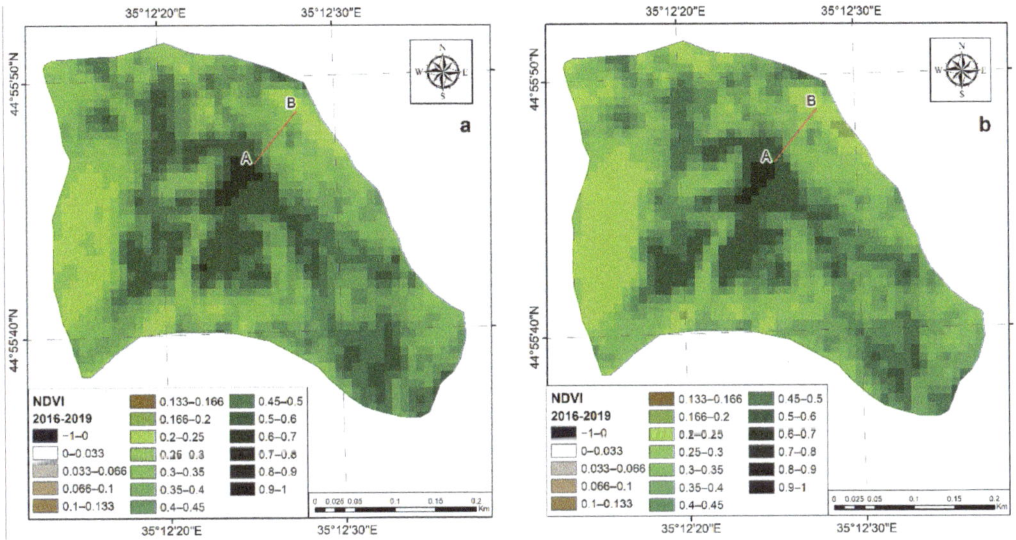

Figure 7. Comparison of NDVI values in 2016–2018 (**a**) and 2019–2021 (**b**). AB—slope profile.

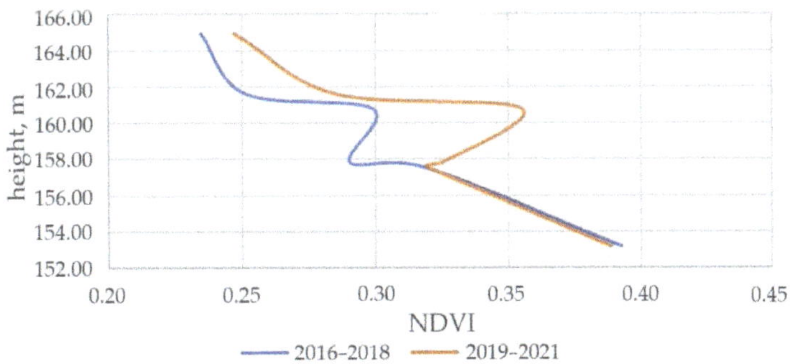

Figure 8. Change of NDVI values for the period from 2016 to 2018 and from 2019 to 2021 from the foot of the slope to the top along profile AB.

5. Conclusions

When studying the forests water balance, it is extremely important to describe in detail the conditions its formation: geological structure, relief, hydrological features of the territory. It is necessary to give a description, characterizing their influence on the spatial differentiation of forests. So, summarizing the above, the slopes of the Southeastern Crimea are subject to ravine formation, which determines the redistribution of forests along the slopes. A significant role in the spatial differentiation of downy oak forests is played by accumulations of snow. In general, the Southeastern Crimea is the territory of river valleys and interfluves; it is a mudflow area. The passage of mudflows in the strongest way transforms the appearance of forests. The mechanical composition and gritty consistence of soils determine the water regime of forest communities. In the study area, we are dealing with very lithosolic soils formed on the weathering products of flysch, limestone, marl, intrusive, and effusive rocks. It is extremely diverse in terms of parent rock material, and, accordingly, in terms of forest conditions. Forests of the Southeastern Crimea have a different appearance and combines with different types of steppe communities. The study area includes forests, woodlands, and shrub communities. A vivid example of the influence of the animal world on the formation of the water balance of forest ecosystems is the burrowing activity of the wild boar, which is ubiquitous and leads to the drying up of the soil, increasing evaporation from its surface.

Based on field research conducted in the Karadag Nature Reserve and analysis of literature data, the main components of the water balance of the downy oak forests of Southeastern Crimea are considered for the period from 2010 to 2020.

The average annual precipitation on the territory of Karadag for the analyzed period was 448 mm; in recent years, a decrease in the amount of precipitation has been recorded. On average, the crowns of downy oak communities retain up to 24% of precipitation. In terms of seasonal dynamics, the maximum retention of precipitation (up to 41%) occurs in the period from late spring to early summer; this is due to the achievement of maximum volumes of phytomass during this period.

According to the literature data, stem runoff intercepts 3%–5% of the annual precipitation; on steeper slopes only, this can lead to negative consequences in the form of planar soil flushing. Runoff formed from the crowns of trees reaches the surface of the soil more gradually to forms the moisture reserve of the forest floor. During the study period, soil moisture decreased significantly by 1.6 times. The maximum relative soil moisture values are experienced during the winter months and in March, when the average value does not fall below 25%. The features of the relief of the Southeastern Crimea created conditions for the redistribution of solid precipitation (snow) from the windward slopes to form snow catchments. Due to this effect, the slopes of the southern exposures are characterized by better moisturization than the opposite slopes of the northern exposures.

The evapotranspiration values within the downy oak forests approximately coincide with the values of this indicator in the Mediterranean region to average 450 mm per year. The interannual dynamics demonstrates a positive trend towards increased evapotranspiration values. A significant increase in evapotranspiration observes in late summer and early autumn.

Observations of slope runoff conducted at runoff sites suggest that the final value of the water balance is not significantly affected by this factor. The average value of the slope runoff was 0.04 L/km^2; in terms of interannual dynamics, a decrease in this value is recorded for the period under review.

All the analyzed features of the water balance elements form conditions conducive to the displacement of steppe communities by forest species, which is confirmed by the results of field and remote studies.

Author Contributions: Conceptualization, R.G.; methodology, R.G.; validation, R.G.; formal analysis, V.T., T.G. and M.S.; investigation, V.T., T.G.; writing—original draft preparation, V.T., T.G., M.S. and R.G.; writing—review and editing, R.G.; visualization, T.G. and V.T.; supervision, R.G.; project administration, R.G. All authors have read and agreed to the published version of the manuscript.

Funding: This research was funded by Russian Science Foundation, grant number 22-27-00579.

Data Availability Statement: Not accessible.

Acknowledgments: The authors are grateful to Kelip A.A. for technical support of the study.

Conflicts of Interest: The authors declare no conflict of interest. The funders had no role in the design of the study; in the collection, analyses, or interpretation of data; in the writing of the manuscript; or in the decision to publish the results.

References

1. IPCC. Climate Change 2021: The physical science basis. In *Contribution of Working Group I to the Sixth Assessment Report of the Intergovernmental Panel on Climate Change*; Cambridge University Press: Cambridge, UK, 2021.
2. Stefanidis, S. Ability of Different Spatial Resolution Regional Climate Model to Simulate Air Temperature in a Forest Ecosystem of Central Greece. *J. Environ. Prot. Ecol.* **2021**, *22*, 1488–1495.
3. Tolika, K.; Anagnostopoulou, C.; Velikou, K.; Vagenas, C. A comparison of the updated very high resolution model RegCM3_10km with the previous version RegCM3_25km over the complex terrain of Greece: Present and future projections. *Theor. Appl. Climatol.* **2016**, *126*, 715–726. [CrossRef]
4. Stefanidis, S.; Alexandridis, V. Precipitation and potential evapotranspiration temporal variability and their relationship in two forest ecosystems in Greece. *Hydrology* **2021**, *8*, 160. [CrossRef]
5. Baldocchi, D.; Ma, S.; Verfaillie, J. On the inter-and intra-annual variability of ecosystem evapotranspiration and water use efficiency of an oak savanna and annual grassland subjected to booms and busts in rainfall. *Glob. Change Biol.* **2021**, *27*, 359–375. [CrossRef] [PubMed]
6. Giorgi, F.; Lionello, P. Climate change projections for the Mediterranean region. *Glob. Planet. Change* **2008**, *63*, 90–104. [CrossRef]
7. Kochkin, M.A. *Soils, Forests and Climate of the Mountainous Crimea and Ways of Their Rational Use*; Kolos: Moscow, Russia, 1967.
8. Logvinova, K.T.; Barabash, M.B. (Eds.) *Climate and Dangerous Hydrometeorological Phenomena of the Crimea*; Gidrometeoizdat: Leningrad, Russia, 1982.
9. Bokov, V.A. (Ed.) *Landscape and Geophysical Conditions for the Growth of Forests in the Southeastern Part of the Mountainous Crimea*; Tavria-Plus: Simferopol, Ukraine, 2001.
10. Bokov, V.A. (Ed.) *Transformation of the Water Balance in the Crimea in the XX Century—The Beginning of the XXI Century*; Crimean Scientific Center: Simferopol, Ukraine, 2011.
11. Bokov, V.A.; Smirnov, V.O. On the Scope of Methods for the Assessment of Landscapes Humidification Assessment. *Vestn. Mosk. Univ. Ser. 5 Geogr.* **2019**, *1*, 83–92.
12. Bokov, V.A.; Yakovleva, O.V. Causal Relationships in the Formation Processes of Humidification Landscape Complexes. *Geopolit. Ecogeodynamics Reg.* **2020**, *6*, 39–56.
13. Kuznetsov, I.; Bokov, V.; Ustiugov, D.; Pavluk, N. Methodology for Assessing the Water Resource Potential of the Crimea and Priority Tasks of its Study. *Geopolit. Ecogeodynamics Reg.* **2021**, *7*, 155–171.
14. Gorbunov, R.V.; Zuev, A.V.; Snegur, A.V. Precipitation retention by vegetation on the territory of the Karadag landscape-ecological station. Proceedings of V International Scientific and Practical Conference Reserves of the Crimea. Theory, Practice and Prospects of Conservation in the Black Sea Region, Simferopol, Ukraine, 22–23 October 2009; pp. 37–41.
15. Gorbunov, R.; Zuev, A.; Smirnov, V. Water Balance Researches on the Territory of Karadag Landscape-Ecological Stationary. In *100 Years of the T.I. Vyazemsky Karadag Scientific Station*; Gaevskaya, A.V., Morozova, A.L., Eds.; N. Orianda: Simferopol, Russia, 2015; pp. 734–747.
16. Gorbunov, R.; Gorbunova, T.; Kononova, N.; Priymak, A.; Salnikov, A.; Drygval, A.; Lebedev, Y. Spatiotemporal Aspects of Interannual Changes Precipitation in the Crimea. *J. Arid Environ.* **2020**, *183*, 104280. [CrossRef]
17. Bokov, V.A. (Ed.) *Development of Environmentally Balanced Ways to Protect and Restore Water Bodies in the Crimea*; TNU: Simferopol, Ukraine, 2013.
18. Gratsiansky, A.N. Mediterranean Nature. Bokov, V.A., Ed.; Mysl': Moscow, Russia, 1971.
19. Didukh, Y.P.; Litvinskaya, S.A.; Novosad, V.V. Botanico-geographical regionalization of the Crimea-Novorossiysk Province. *Bot. Zhurnal* **1990**, *75*, 494–507.
20. Cordova, C.E. The Mediterraneanization of Crimea. *Méditerranée* **2016**, *126*, 25–36. [CrossRef]
21. Aizenberg, M.M.; Kaganer, M.S. (Eds.) *Resources of Surface Waters of the USSR. T. 6: Ukraine and Moldova. Issue. 4. Crimea*; Gidrometeoizdat: Leningrad, Russia, 1966.
22. Dragan, N.A. *Soils of the Crimea*; SGU: Simferopol, Russia, 1983.
23. Sarkina, I.S.; Mironova, L.P. Annotated List of Basidiomycetes and Ascomycetes of Karadag Nature Reserve. *Sci. Notes Cape Martyan Nat. Reserve* **2015**, *6*, 297–327.
24. Mikhailov, V.N.; Dobrovolsky, A.D.; Dobrolyubov, S.A. *Hydrology*; Vysshaya Shkola: Moscow, Russia, 2007.
25. ERA5 Monthly Averaged Data on Single Levels from 1959 to Present. Available online: https://cds.climate.copernicus.eu/cdsapp#!/dataset/reanalysis-era5-single-levels-monthly-means?tab=overview (accessed on 5 May 2022).
26. Morozova, A.L.; Vronsky, A.A. (Eds.) *Nature of Karadag*; Naukova Dumka: Kiev, Ukraine, 1989.
27. Morozova, A.L. (Ed.) *Reserved Karadag: Popular Science Essays*; N. Orianda: Simferopol, Ukraine, 2011.
28. Zuev, A.V.; Letukhova, V.J.; Zueva, E.A. Climate Change as a Factor of the Vegetation Transformation on the Example of the Karadag Landscape-Ecological Stationary. *Proc. T.I. Vyazemsky Karadag Sci. Stn. Nat. Reserve Russ. Acad. Sci.* **2020**, *1*, 77–98.
29. And, C.D.; Rambal, S. Field study of leaf photosynthetic performance by a Mediterranean deciduous oak tree (Quercus pubescens) during a severe summer drought. *New Phytol.* **1995**, *131*, 159–167. [CrossRef]
30. Baldocchi, D.D.; Ma, S.; Rambal, S.; Misson, L.; Ourcival, J.-M.; Limousin, J.-M.; Pereira, J.; Papale, D. On the differential advantages of evergreenness and deciduousness in mediterranean oak woodlands: A flux perspective. *Ecol. Appl.* **2010**, *20*, 1583–1597. [CrossRef] [PubMed]
31. Klyukin, A.A. *Exogeodynamics of the Crimea*; Tavria: Simferopol, Ukraine, 2007.

32. Bellot, J.; Escarre, A. Stemflow and Throughfall Determination in a Resprouted Mediterranean Holm-Oak Forest. *Ann. Des Sci. For.* **1998**, *55*, 847–865. [CrossRef]
33. Cristóbal, J.; Poyatos, R.; Ninyerola, M.; Llorens, P.; Pons, X. Combining Remote Sensing and GIS Climate Modelling to Estimate Daily Forest Evapotranspiration in a Mediterranean Mountain Area. *Hydrol. Earth Syst. Sci.* **2011**, *15*, 1563–1575. [CrossRef]
34. Llorens, P.; Domingo, F. Rainfall partitioning by vegetation under Mediterranean conditions. *Europe J. Hydrol.* **2007**, *335*, 37–54. [CrossRef]
35. Gorbunov, R.V.; Gorbunova, T.Y. Manifestation of Snow Dissymmetry of Slope Local Landscape Complexes in the South-Eastern Crimea. In Proceedings of the III International Scientific and Practical Conference "Biodiversity and Sustainable Development", Simferopol, Russia, 15–19 September 2014; pp. 91–93.
36. Ved', I.P. (Ed.) *Climatic Atlas of Crimea*; Tavria-Plus: Simferopol, Ukraine, 2000.
37. Ved', I.P. The Role of Ground-Based Hydrometeors in the Water Balance of the Crimean Highlands. *Meteorol. Hydrol.* **1967**, *4*, 68–72.

MDPI
St. Alban-Anlage 66
4052 Basel
Switzerland
www.mdpi.com

Forests Editorial Office
E-mail: forests@mdpi.com
www.mdpi.com/journal/forests

www.ingramcontent.com/pod-product-compliance
Lightning Source LLC
Chambersburg PA
CBHW061500210926
51850CB00008B/1771